British Plant Communities 3

British Plant Communities

VOLUME 3

GRASSLANDS AND MONTANE COMMUNITIES

J.S. Rodwell (editor)

C.D. Pigott, D.A. Ratcliffe

A.J.C. Malloch, H.J.B. Birks

M.C.F. Proctor, D.W. Shimwell

J.P. Huntley, E. Radford

M.J. Wigginton, P. Wilkins

for the

U.K. Joint Nature Conservation Committee

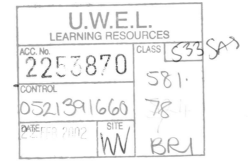

CAMBRIDGE
UNIVERSITY PRESS

VOLUME 3

PUBLISHED BY THE PRESS SYNDICATE OF THE UNIVERSITY OF CAMBRIDGE
The Pitt Building, Trumpington Street, Cambridge CB2 1RP, United Kingdom

CAMBRIDGE UNIVERSITY PRESS
The Edinburgh Building, Cambridge CB2 2RU, UK http://www.cup.cam.ac.uk
40 West 20th Street, New York, NY 10011-4211, USA http://www.cup.org
10 Stamford Road, Oakleigh, Melbourne 3166, Australia

First published 1992
First paperback edition 1998

Printed in the United Kingdom at the University Press, Cambridge

Typeset in Monotype Times New Roman 9/11pt   [SE]

A catalogue record for this book is available from the British Library

Library of Congress Cataloguing in Publication data

British plant communities

Includes bibliographical references (p. [385]–395)
and indexes.
Contents. v. 1. Woodlands and scrub–   –
v. 3. Grasslands and montane communities.
1. Plant communities–Great Britain.   2. Vegetation
classification–Great Britain.   I. Rodwell, J. S.
II. Nature Conservancy Council (Great Britain)
QK306.B857   1992   581.5'247'0941   90-1300

ISBN 0 521 39166 0 hardback
ISBN 0 521 62719 2 paperback

CONTENTS

Calcicolous grasslands

Calcifugous grasslands and montane communities

Contents   vii

FIGURES

GENERAL INTRODUCTION

The background to the work

It is a tribute to the insight of our early ecologists that we can still return with profit to *Types of British Vegetation* which Tansley (1911) edited for the British Vegetation Committee as the first coordinated attempt to recognise and describe different kinds of plant community in this country. The contributors there wrote practically all they knew and a good deal that they guessed, as Tansley himself put it, but they were, on their own admission, far from comprehensive in their coverage. It was to provide this greater breadth, and much more detailed description of the structure and development of plant communities, that Tansley (1939) drew together the wealth of subsequent work in *The British Islands and their Vegetation*, and there must be few ecologists of the generations following who have not been inspired and challenged by the vision of this magisterial book.

Yet, partly because of its greater scope and the uneven understanding of different kinds of vegetation at the time, this is a less systematic work than *Types* in some respects: its narrative thread of explication is authoritative and engaging, but it lacks the light-handed framework of classification which made the earlier volume so very attractive, and within which the plant communities might be related one to another, and to the environmental variables which influence their composition and distribution. Indeed, for the most part, there is a rather self-conscious avoidance of the kind of rigorous taxonomy of vegetation types that had been developing for some time elsewhere in Europe, particularly under the leadership of Braun-Blanquet (1928) and Tüxen (1937). The difference in the scientific temperament of British ecologists that this reflected, their interest in how vegetation works, rather than in exactly what distinguishes plant communities from one another, though refreshing in itself, has been a lasting hindrance to the emergence in this country of any consensus as to how vegetation ought to be described, and whether it ought to be classified at all.

In fact, an impressive demonstration of the value of the traditional phytosociological approach to the description of plant communities in the British Isles was published in German after an international excursion to Ireland in 1949 (Braun-Blanquet & Tüxen 1952), but more immediately productive was a critical test of the techniques among a range of Scottish mountain vegetation by Poore (1955*a*, *b*, *c*). From this, it seemed that the really valuable element in the phytosociological method might be not so much the hierarchical definition of plant associations, as the meticulous sampling of homogeneous stands of vegetation on which this was based, and the possibility of using this to provide a multidimensional framework for the presentation and study of ecological problems. Poore & McVean's (1957) subsequent exercise in the description and mapping of communities defined using this more flexible approach then proved just a prelude to the survey of huge tracts of mountain vegetation by McVean & Ratcliffe (1962), work sponsored and published by the Nature Conservancy (as it then was) as *Plant Communities of the Scottish Highlands*. Here, for the first time, was the application of a systematised sampling technique across the vegetation cover of an extensive and varied landscape in mainland Britain, with assemblages defined in a standard fashion from full floristic data, and interpreted in relation to a complex of climatic, edaphic and biotic factors. The opportunity was taken, too, to relate the classification to other European traditions of vegetation description, particularly that developed in Scandinavia (Nordhagen 1943, Dahl 1956).

McVean & Ratcliffe's study was to prove a continual stimulus to the academic investigation of our mountain vegetation and of abiding value to the development of conservation policy, but their methods were not extended to other parts of the country in any ambitious sponsored surveys in the years immediately following. Despite renewed attempts to commend traditional phytosociology, too (Moore 1962), the attraction of this whole approach was overwhelmed for many by the heated debates that preoccupied British plant ecologists in the 1960s, on the issues of objectivity in the sampling and sorting of data, and the respective values of classification or ordination as analytical techniques. Others, though, found it perfectly possible to integrate multivariate analysis into phytosociological survey, and demonstrated the advantage of computers for the display and interpretation of ecological data, rather than the simple testing of methodologies (Ivimey-Cook & Proctor 1966*b*). New generations of research students also began to draw inspiration from the Scottish and Irish initiatives by applying phytosociology to the solving of particular descriptive and interpretive problems, such as variation among British calcicolous grasslands (Shimwell 1968*a*), heaths (Bridgewater 1970), rich fens (Wheeler 1975) and salt-marshes (Adam 1976), the vegetation of Skye (Birks 1969), Cornish cliffs (Malloch 1971) and Upper Teesdale (Bradshaw & Jones 1976). Meanwhile, too, workers at the Macaulay Institute in Aberdeen had been extending the survey of Scottish vegetation to the lowlands and the Southern Uplands (Birse & Robertson 1976, Birse 1980, 1984).

With an accumulating volume of such data and the appearance of uncoordinated phytosociological perspectives on different kinds of British vegetation, the need for an overall framework of classification became ever more pressing. For some, it was also an increasingly urgent concern that it still proved impossible to integrate a wide variety of ecological research on plants within a generally accepted understanding of their vegetational context in this country. Dr Derek Ratcliffe, as Scientific Assessor of the Nature Conservancy's Reserves Review from the end of 1966, had encountered the problem of the lack of any comprehensive classification of British vegetation types on which to base a systematic selection of habitats for conservation. This same limitation was recognised by Professor Sir Harry Godwin, Professor Donald Pigott and Dr John Phillipson who, as members of the Nature Conservancy, had been asked to read and comment on the Reserves Review. The published version, *A Nature Conservation Review* (Ratcliffe 1977), was able to base the description of only the lowland and upland grasslands and heaths on a phytosociological treatment. In 1971, Dr Ratcliffe, then Deputy Director (Scientific) of the Nature Conservancy, in proposals for development of its research programme, drew attention to 'the need for a national and systematic phytosociological treatment of British vegetation, using standard methods in the field and in analysis/classification of the data'. The intention of setting up a group to examine the issue lapsed through the splitting of the Conservancy which was announced by the Government in 1972. Meanwhile, after discussions with Dr Ratcliffe, Professor Donald Pigott of the University of Lancaster proposed to the Nature Conservancy a programme of research to provide a systematic and comprehensive classification of British plant communities. The new Nature Conservancy Council included it as a priority

item within its proposed commissioned research programme. At its meeting on 24 March 1974, the Council of the British Ecological Society welcomed the proposal. Professor Pigott and Dr Andrew Malloch submitted specific plans for the project and a contract was awarded to Lancaster University, with sub-contractual arrangements with the Universities of Cambridge, Exeter and Manchester, with whom it was intended to share the early stages of the work. A coordinating panel was set up, jointly chaired by Professor Pigott and Dr Ratcliffe, and with research supervisors from the academic staff of the four universities, Drs John Birks, Michael Proctor and David Shimwell joining Dr Malloch. At a later stage, Dr Tim Bines replaced Dr Ratcliffe as nominated officer for the NCC, and Miss Lynne Farrell succeeded him in 1985.

With the appointment of Dr John Rodwell as full-time coordinator of the project, based at Lancaster, the National Vegetation Classification began its work officially in August 1975. Shortly afterwards, four full-time research assistants took up their posts, one based at each of the universities: Mr Martin Wigginton, Miss Jacqueline Paice (later Huntley), Mr Paul Wilkins and Dr Elaine Grindey (later Radford). These remained with the project until the close of the first stage of the work in 1980, sharing with the coordinator the tasks of data collection and analysis in different regions of the country, and beginning to prepare preliminary accounts of the major vegetation types. Drs Michael Lock and Hilary Birks and Miss Katherine Hearn were also able to join the research team for short periods of time. After the departure of the research assistants, the supervisors supplied Dr Rodwell with material for writing the final accounts of the plant communities and their integration within an overall framework. With the completion of this charge in 1989, the handover of the manuscript for publication by the Cambridge University Press began.

The scope and methods of data collection

The contract brief required the production of a classification with standardised descriptions of named and systematically arranged vegetation types and, from the beginning, this was conceived as something much more than an annotated list of interesting and unusual plant communities. It was to be comprehensive in its coverage, taking in the whole of Great Britain but not Northern Ireland, and including vegetation from all natural, semi-natural and major artificial habitats. Around the maritime fringe, interest was to extend to the start of the truly marine zone, and from there to the tops of our remotest mountains, covering virtually all terrestrial plant communities and those of brackish and fresh waters, except where non-vascular plants were the dominants. Only short-term leys were specifically excluded, and, though care was to be taken to sample more pristine and long-established kinds of vegetation, no undue attention was to be given to assemblages of rare plants or to especially rich and varied sites. Thus widespread and dull communities from improved pastures, plantations, run-down mires and neglected heaths were to be extensively sampled, together with the vegetation of paths, verges and recreational swards, walls, man-made waterways and industrial and urban wasteland.

For some vegetation types, we hoped that we might be able to make use, from early on, of existing studies, where these had produced data compatible in style and quality with the requirements of the project. The contract envisaged the abstraction and collation of such material from both published and unpublished sources, and discussions with other workers involved in vegetation survey, so that we could ascertain the precise extent and character of existing coverage and plan our own sampling accordingly. Systematic searches of the literature and research reports revealed many data that we could use in some way and, with scarcely a single exception, the originators of such material allowed us unhindered access to it. Apart from the very few classic phytosociological accounts, the most important sources proved to be postgraduate theses, some of which had already amassed very comprehensive sets of samples of certain kinds of vegetation or from particular areas, and these we were generously permitted to incorporate directly.

Figure 1. Standard NVC sample card.

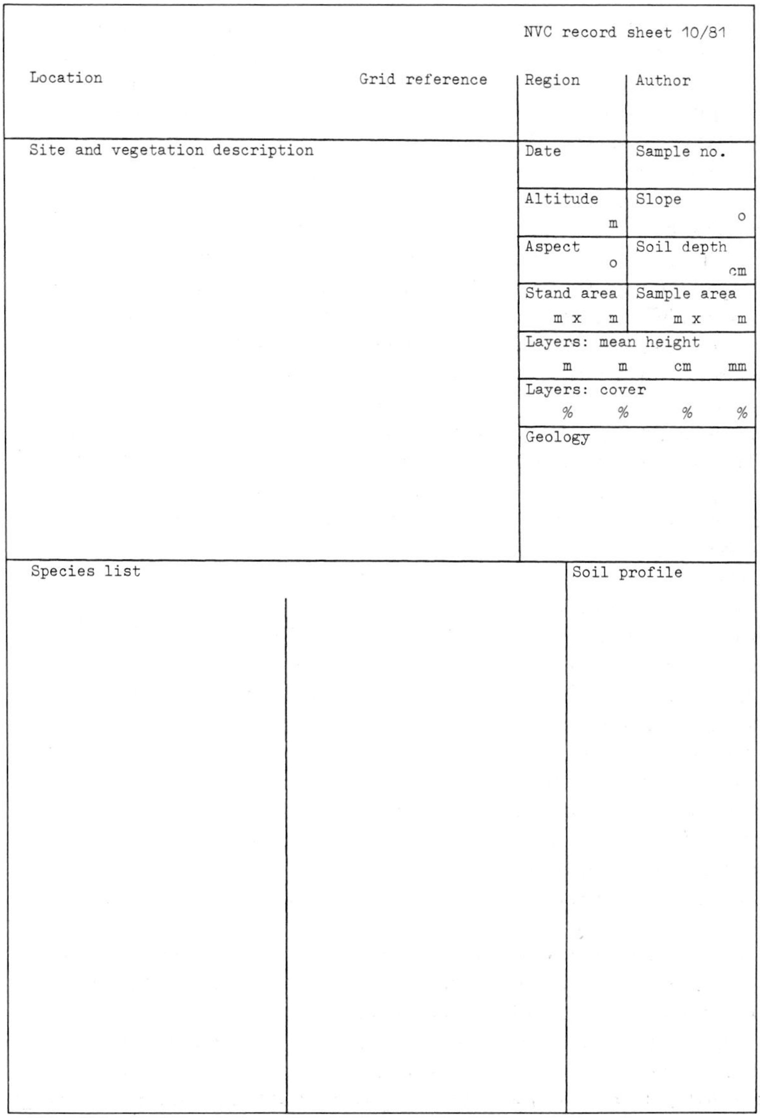
NVC record sheet 10/81

Location
Grid reference
Region
Author
Site and vegetation description
Date
Sample no.
Altitude m
Slope o
Aspect o
Soil depth cm
Stand area m x m
Sample area m x m
Layers: mean height m m cm mm
Layers: cover % % % %
Geology
Species list
Soil profile

Then, from the NCC and some other government agencies, or from individuals who had been engaged in earlier contracts for them, there were some generally smaller bodies of data, occasionally from reports of extensive surveys, more usually from investigations of localised areas. Published papers on particular localities, vegetation types or individual species also provided small numbers of samples. In addition to these sources, the project was able to benefit from and influence ongoing studies by institutions and individuals, and itself to stimulate new work with a similar kind of approach among university researchers, NCC surveyors, local flora recorders and a few suitably qualified amateurs. An initial assessment and annual monitoring of floristic and geographical coverage were designed to ensure that the accumulating data were fairly evenly spread, fully representative of the range of British vegetation, and of a consistently high quality. Full details of the sources of the material, and our acknowledgements of help, are given in the preface and introduction to each volume.

Our own approach to data collection was simple and pragmatic, and a brief period of training at the outset ensured standardisation among the team of five staff who were to carry out the bulk of the sampling for the project in the field seasons of the first four years, 1976–9. The thrust of the approach was phytosociological in its emphasis on the systematic recording of floristic information from stands of vegetation, though these were chosen solely on the basis of their relative homogeneity in composition and structure. Such selection took a little practice, but it was not nearly so difficult as some critics of this approach imply, even in complex vegetation, and not at all mysterious. Thus, crucial guidelines were to avoid obvious vegetation boundaries or unrepresentative floristic or physiognomic features. No prior judgements were necessary about the identity of the vegetation type, nor were stands ever selected because of the presence of species thought characteristic for one reason or another, nor by virtue of any observed uniformity of the environmental context.

From within such homogeneous stands of vegetation, the data were recorded in quadrats, generally square unless the peculiar shape of stands dictated otherwise. A relatively small number of possible sample sizes was used, determined not by any calculation of minimal areas, but by the experienced assessment of their appropriateness to the range of structural scale found among our plant communities. Thus plots of 2 × 2 m were used for most short, herbaceous vegetation and dwarf-shrub heaths, 4 × 4 m for taller or more open herb communities, sub-shrub heaths and low woodland field layers, 10 × 10 m for species-poor or very tall herbaceous vegetation or woodland field layers and dense scrub, and 50 × 50 m for sparse scrub, and woodland canopy and

understorey. Linear vegetation, like that in streams and ditches, on walls or from hedgerow field layers, was sampled in 10 m strips, with 30 m strips for hedgerow shrubs and trees. Quadrats of 1 × 1 m were rejected as being generally inadequate for representative sampling, although some bodies of existing data were used where this, or other sizes different from our own, had been employed. Stands smaller than the relevant sample size were recorded in their entirety, and mosaics were treated as a single vegetation type where they were repeatedly encountered in the same form, or where their scale made it quite impossible to sample their elements separately.

Samples from all different kinds of vegetation were recorded on identical sheets (Figure 1). Priority was always given to the accurate scoring of all vascular plants, bryophytes and macrolichens (*sensu* Dahl 1968), a task which often required assiduous searching in dense and complex vegetation, and the determination of difficult plants in the laboratory or with the help of referees. Critical taxa were treated in as much detail as possible though, with the urgency of sampling, certain groups, like the brambles, hawkweeds, eyebrights and dandelions, often defeated us, and some awkward bryophytes and crusts of lichen squamules had to be referred to just a genus. It is more than likely, too, that some very diminutive mosses and especially hepatics escaped notice in the field and, with much sampling taking place in summer, winter annuals and vernal perennials might have been missed on occasion. In general, nomenclature for vascular plants follows *Flora Europaea* (Tutin *et al.* 1964 *et seq.*) with Corley & Hill (1981) providing the authority for bryophytes and Dahl (1968) for lichens. Any exceptions to this, and details of any difficulties with sampling or identifying particular plants, are given in the introductions to each of the major vegetation types.

A quantitative measure of the abundance of every taxon was recorded using the Domin scale (*sensu* Dahl & Hadač 1941), cover being assessed by eye as a vertical projection on to the ground of all the live, above-ground parts of the plants in the quadrat. On this scale:

Cover of 91–100% is recorded as Domin		10
76–90%		9
51–75%		8
34–50%		7
26–33%		6
11–25%		5
4–10%		4
<4%	with many individuals	3
	with several individuals	2
	with few individuals	1

In heaths, and more especially in woodlands, where the vegetation was obviously layered, the species in the different elements were listed separately as part of the same sample, and any different generations of seedlings or saplings distinguished. A record was made of the total cover and height of the layers, together with the cover of any bare soil, litter, bare rock or open water. Where existing data had been collected using percentage cover or the Braun-Blanquet scale (Braun-Blanquet 1928), it was possible to convert the abundance values to the Domin scale, but we had to reject all samples where DAFOR scoring had been used, because of the inherent confusion within this scale of abundance and frequency.

Each sample was numbered and its location noted using a site name and full grid reference. Altitude was estimated in metres from the Ordnance Survey 1:50 000 series maps, slope estimated by eye or measured using a hand level to the nearest degree, and aspect measured to the nearest degree using a compass. For terrestrial samples, soil depth was measured in centimetres using a probe, and in many cases a soil pit was dug sufficient to allocate the profile to a major soil group (*sensu* Avery 1980). From such profiles, a superficial soil sample was removed for pH determination as soon as possible thereafter using an electric meter on a 1:5 soil:water paste. With aquatic vegetation, water depth was measured in centimetres wherever possible, and some indication of the character of the bottom noted. Details of bedrock and superficial geology were obtained from Geological Survey maps and by field observation.

This basic information was supplemented by notes, with sketches and diagrams where appropriate, on any aspects of the vegetation and the habitat thought likely to help with interpretation of the data. In many cases, for example, the quantitative records for the species were filled out by details of the growth form and patterns of dominance among the plants, and an indication of how they related structurally one to another in finely organised layers, mosaics or phenological sequences within the vegetation. Then, there was often valuable information about the environment to be gained by simple observation of the gross landscape or microrelief, the drainage pattern, signs of erosion or deposition and patterning among rock outcrops, talus slopes or stony soils. Often, too, there were indications of biotic effects including treatments of the vegetation by man, with evidence of grazing or browsing, trampling, dunging, mowing, timber extraction or amenity use. Sometimes, it was possible to detect obvious signs of ongoing change in the vegetation, natural cycles of senescence and regeneration among the plants, or successional shifts consequent upon invasion or particular environmental impacts. In many cases, also, the spatial relationships between the stand and neighbouring vegetation types were highly informative and, where a number of samples was taken from an especially varied or complex site, it often proved useful to draw a map indicating how the various elements in the pattern were interrelated.

The approach to data analysis

At the close of the programme of data collection, we had assembled, through the efforts of the survey team and by the generosity of others, a total of about 35 000 samples of the same basic type, originating from more than 80% of the 10×10 km grid squares of the British mainland and many islands (Figure 2). Thereafter began a coordinated phase of data processing, with each of the four universities taking responsibility for producing preliminary analyses from data sets crudely separated into major vegetation types – mires, calcicolous grasslands, sand-dunes and so on – and liaising with the others where there was a shared interest. We were briefed in the contract to produce accounts of discrete plant communities which could be named and mapped, so our attention was naturally concentrated on techniques of multivariate classification, with the help of computers to sort the very numerous and often complex samples on the basis of their similarity. We were concerned to employ reputable methods of analysis, but the considerable experience of the team in this kind of work led us to resolve at the outset to concentrate on the ecological integrity of the results, rather than on the minutiae of mathematical technique. In fact, each centre was free to

Figure 2. Distribution of samples available for analysis.

some extent to make its own contribution to the development of computer programs for the task, Exeter concentrating on Association and Information Analysis (Ivimey-Cook *et al.* 1975), Cambridge and Manchester on cluster analysis (Huntley *et al.* 1981), Lancaster on Indicator Species Analysis, later Twinspan (Hill *et al.* 1975, Hill 1979), a technique which came to form the core of the VESPAN package, designed, using the experience of the project, to be particularly appropriate for this kind of vegetation survey (Malloch 1988).

Throughout this phase of the work, however, we had some important guiding principles. First, this was to be a new classification, and not an attempt to employ computational analysis to fit groups of samples to some existing scheme, whether phytosociological or otherwise. Second, we were to produce a classification of vegetation types, not of habitats, so only the quantitative floristic records were used to test for similarity between the samples, and not any of the environmental information: this would be reserved, rather, to provide one valuable correlative check on the ecological meaning of the sample groups. Third, no samples were to be rejected at the outset because they appeared nondescript or troublesome, nor removed during the course of analysis or data presentation where they seemed to confuse an otherwise crisply-defined result. Fourth, though, there was to be no slavish adherence to the products of single analyses using arbitrary cut-off points when convenient numbers of end-groups had been produced. In fact, the whole scheme was to be the outcome of many rounds of sorting, with data being pooled and reanalysed repeatedly until optimum stability and sense were achieved within each of the major vegetation types. An important part of the coordination at this stage was to ensure roughly comparable scales of definition among the emerging classifications and to mesh together the work of the separate centres so as to avoid any omissions in the processing or wasteful overlaps.

With the departure from the team of the four research assistants in 1980, the academic supervisors were left to continue the preparation of the preliminary accounts of the vegetation types for the coordinator to bring to completion and integrate into a coherent whole. Throughout the periods of field work and data analysis, we had all been conscious of the charge in the contract that the whole project must gain wide support among ecologists with different attitudes to the descriptive analysis of vegetation. Great efforts were therefore made to establish a regular exchange of information and ideas through the production of progress reports, which gained a wide circulation in Britain and overseas, via contacts with NCC staff and those of other research agencies, and the giving of papers at scientific meetings. This meant that, as we approached the presentation of

the results of the project, we were well informed about the needs of prospective users, and in a good position to offer that balance of concise terminology and broadly-based description that the NCC considered would commend the work, not only to their own personnel, but to others engaged in the assessment and management of vegetation, to plant and animal ecologists in universities and colleges, and to those concerned with land use and planning.

The style of presentation

The presentation of our results thus gives priority to the definition of the vegetation types, rather than to the construction of a hierarchical classification. We have striven to characterise the basic units of the scheme on roughly the same scale as a Braun-Blanquet association, but these have been ordered finally not by any rigid adherence to the higher phytosociological categories of alliance, order and class, but in sections akin to the formations long familiar to British ecologists. In some respects, this is a more untidy arrangement, and even those who find the general approach congenial may be surprised to discover what they have always considered to be, say, a heath, grouped here among the mires, or to search in vain for what they are used to calling 'marsh'. The five volumes of the work gather the major vegetation types into what seem like sensible combinations and provide introductions to the range of communities included: aquatic vegetation, swamps and tall-herb fens; grasslands and montane vegetation; heaths and mires; woodlands and scrub; salt-marsh, sand-dune and sea-cliff communities and weed vegetation. The order of appearance of the volumes, however, reflects more the exigencies of publishing than any ecological viewpoint.

The bulk of the material in the volumes comprises the descriptions of the vegetation types. After much consideration, we decided to call the basic units of the scheme by the rather non-committal term 'community', using 'sub-community' for the first-order sub-groups which could often be distinguished within these, and 'variant' in those very exceptional cases where we have defined a further tier of variation below this. We have also refrained from erecting any novel scheme of complicated nomenclature for the vegetation types, invoking existing names where there is an undisputed phytosociological synonym already in widespread use, but generally using the latin names of one, two or occasionally three of the most frequent species. Among the mesotrophic swards, for example, we have distinguished a *Centaurea nigra-Cynosurus cristatus* grassland, which is fairly obviously identical to what Braun-Blanquet & Tüxen (1952) called *Centaureo-Cynosuretum cristati*, and within which, from our data, we have characterised three sub-communities. For the convenience of shorthand description and mapping, every vegetation type has been given a code letter and number, so that *Centaurea-Cynosurus* grassland for example is MG5, MG referring to its place among the mesotrophic grasslands. The *Galium verum* sub-community of this vegetation type, the second to be distinguished within the description, is thus MG5b.

Vegetation being as variable as it is, it is sometimes expedient to allocate a sample to a community even though the name species are themselves absent. What defines a community as unique are rarely just the plants used to name it, but the particular combination of frequency and abundance values for all the species found in the samples. It is this information which is presented in summary form in the floristic tables for each of the communities in the scheme. Figure 3, for example, shows such a table for MG5 *Centaurea-Cynosurus* grassland. Like all the tables in the volumes, it includes such vascular plants, bryophytes and lichens as occur with a frequency of 5% or more in any one of the sub-communities (or, for vegetation types with no sub-communities, in the community as a whole). Early tests showed that records of species below this level of frequency could be largely considered as noise, but cutting off at any higher level meant that valuable floristic information was lost. The vascular species are not separated from the cryptogams on the table though, for woodlands and scrub, the vegetation is sufficiently complex for it to be sensible to tabulate the species in a way which reflects the layered structure.

Every table has the frequency and abundance values arranged in columns for the species. Here, 'frequency' refers to how often a plant is found on moving from one sample of the vegetation to the next, irrespective of how much of that species is present in each sample. This is summarised in the tables as classes denoted by the Roman numerals I to V: 1–20% frequency (that is, up to one sample in five) = I, 21–40% = II, 41–60% = III, 61–80% = IV and 81–100% = V. We have followed the usual phytosociological convention of referring to species of frequency classes IV and V in a particular community as its constants, and in the text usually refer to those of class III as common or frequent species, of class II as occasional and of class I as scarce. The term 'abundance', on the other hand, is used to describe how much of a plant is present in a sample, irrespective of how frequent or rare it is among the samples, and it is summarised on the tables as bracketed numbers for the Domin ranges, and denoted in the text using terms such as dominant, abundant, plentiful and sparse. Where there are sub-communities, as in this case, the data for these are listed first, with a final column summarising the records for the community as a whole.

The species are arranged in blocks according to their pattern of occurrence among the different sub-communities and within these blocks are generally ordered by decreasing frequency. The first group, *Festuca rubra* to *Trifolium pratense* in this case, is made up of the community constants, that is those species which have an overall frequency IV or V. Generally speaking, such plants tend to maintain their high frequency in each of the sub-communities, though there may be some measure of variation in their representation from one to the next: here, for example, *Plantago lanceolata* is somewhat less common in the last sub-community than the first two, with *Holcus lanatus* and a number of others showing the reverse pattern. More often, there are considerable differences in the abundance of these most frequent species: many of the constants can have very high covers, while others are more consistently sparse, and plants which are not constant can sometimes be numbered among the dominants.

The last group of species on a table, *Ranunculus acris* to *Festuca arundinacea* here, lists the general associates of the community, sometimes referred to as companions. These are plants which occur in the community as a whole with frequencies of III or less, though sometimes they rise to constancy in one or other of the sub-communities, as with *R. acris* in this vegetation. Certain of the companions are consistently common overall like *Rumex acetosa*, some are more occasional throughout as with *Rhinanthus minor*, some are always scarce, for example *Calliergon cuspidatum*. Others, though, are more unevenly represented, like *R. acris*, *Heracleum sphondylium* or *Poa trivialis*, though they do not show any marked affiliation to any particular sub-community. Again, there can be marked variation in the abundance of these associates: *Rumex acetosa*, for example, though quite frequent, is usually of low cover, while *Arrhenatherum elatius* and some of the bryophytes, though more occasional, can be patchily abundant; *Alchemilla xanthochlora* is both uncommon among the samples and sparse within them.

The intervening blocks comprise those species which are distinctly more frequent within one or more of the sub-communities than the others, plants which are referred to as preferential, or differential where their affiliation is more exclusive. For example, the group *Lolium perenne* to *Juncus inflexus* is particularly characteristic of the first sub-community of *Centaurea-Cynosurus* grassland, although some species, like *Leucanthemum vulgare* and, even more so, *Lathyrus pratensis*, are more strongly preferential than others, such as *Lolium*, which continues to be frequent in the second sub-community. Even uncommon plants can be good preferentials, as with *Festuca pratensis* here: it is not often found in *Centaurea-Cynosurus* grassland but, when it does occur, it is generally in this first sub-type.

The species group *Galium verum* to *Festuca ovina* helps to distinguish the second sub-community from the first, though again there is some variation in the strength

Floristic table MG5

	a	b	c	MG5
Festuca rubra	V (1–8)	V (2–8)	V (2–7)	V (1–8)
Cynosurus cristatus	V (1–8)	V (1–7)	V (1–7)	V (1–8)
Lotus corniculatus	V (1–7)	V (1–5)	V (2–4)	V (1–7)
Plantago lanceolata	V (1–7)	V (1–5)	IV (1–4)	V (1–7)
Holcus lanatus	IV (1–6)	IV (1–6)	V (1–5)	IV (1–6)
Dactylis glomerata	IV (1–7)	IV (1–6)	V (1–6)	IV (1–7)
Trifolium repens	IV (1–9)	IV (1–6)	V (1–4)	IV (1–9)
Centaurea nigra	IV (1–5)	IV (1–4)	V (2–4)	IV (1–5)
Agrostis capillaris	IV (1–7)	IV (1–7)	V (3–8)	IV (1–7)
Anthoxanthum odoratum	IV (1–7)	IV (1–8)	V (1–4)	IV (1–8)
Trifolium pratense	IV (1–5)	IV (1–4)	IV (1–3)	IV (1–5)
Lolium perenne	IV (1–8)	III (1–7)	I (2–3)	III (1–8)
Bellis perennis	III (1–7)	II (1–7)	I (4)	II (1–7)
Lathyrus pratensis	III (1–5)	I (1–3)	I (1)	II (1–5)
Leucanthemum vulgare	III (1–3)	I (1–3)	II (1–3)	II (1–3)
Festuca pratensis	II (1–5)	I (2–5)	I (1)	I (1–5)
Knautia arvensis	I (4)			I (4)
Juncus inflexus	I (3–5)			I (3–5)
Galium verum	I (1–6)	V (1–6)		II (1–6)
Trisetum flavescens	II (1–4)	IV (1–6)	II (1–3)	III (1–6)
Achillea millefolium	III (1–6)	V (1–4)	III (1–4)	III (1–6)
Carex flacca	I (1–4)	II (1–4)	I (1)	I (1–4)
Sanguisorba minor	I (4)	II (3–5)		I (3–5)
Koeleria macrantha	I (1)	II (1–6)		I (1–6)
Agrostis stolonifera	I (1–7)	II (1–6)	I (6)	I (1–7)
Festuca ovina		II (1–6)		I (1–6)
Prunella vulgaris	III (1–4)	III (1–4)	IV (1–3)	III (1–4)
Leontodon autumnalis	II (1–5)	II (1–3)	IV (1–4)	III (1–5)
Luzula campestris	II (1–4)	II (1–6)	IV (1–4)	III (1–6)
Danthonia decumbens	I (2–5)	I (1–3)	V (2–5)	I (1–5)
Potentilla erecta	I (1–4)	I (3)	V (1–4)	I (1–4)
Succisa pratensis	I (1–4)	I (1–5)	V (1–4)	I (1–5)
Pimpinella saxifraga	I (1–4)	I (1–4)	III (1–4)	I (1–4)
Stachys betonica	I (1–5)	I (1–4)	III (1–4)	I (1–5)
Carex caryophyllea	I (1–4)	I (1–3)	II (1–2)	I (1–4)
Conopodium majus	I (1–4)	I (1–5)	II (2–3)	I (1–5)
Ranunculus acris	IV (1–4)	II (1–4)	IV (2–4)	III (1–4)
Rumex acetosa	III (1–4)	III (1–4)	III (1–3)	III (1–4)
Hypochoeris radicata	III (1–5)	II (2–4)	III (1–4)	III (1–5)
Ranunculus bulbosus	III (1–7)	II (1–5)	III (1–2)	III (1–7)
Taraxacum officinale agg.	III (1–4)	III (1–4)	III (1–4)	III (1–4)
Brachythecium rutabulum	II (1–6)	III (1–4)	II (2)	III (1–6)
Cerastium fontanum	III (1–3)	II (1–3)	II (1–3)	II (1–3)
Leontodon hispidus	II (1–6)	III (2–4)	III (1–5)	II (1–6)
Rhinanthus minor	II (1–5)	II (1–4)	II (1–3)	II (1–5)
Briza media	II (1–6)	III (1–4)	III (2–3)	II (1–6)
Heracleum sphondylium	II (1–5)	II (1–3)	III (1–3)	II (1–5)
Trifolium dubium	II (1–8)	II (1–5)	I (2)	II (1–8)
Primula veris	II (1–4)	II (2–4)	I (4)	II (1–4)
Arrhenatherum elatius	II (1–6)	II (1–7)	I (3–4)	II (1–7)
Cirsium arvense	II (1–3)	II (1–4)	I (1)	II (1–4)
Eurhynchium praelongum	II (1–5)	II (1–4)	I (1–2)	II (1–5)
Rhytidiadelphus squarrosus	II (1–7)	II (1–5)	III (1–4)	II (1–7)
Poa pratensis	II (1–6)	II (2–5)		II (1–6)
Poa trivialis	II (1–8)	I (1–3)	I (1–2)	II (1–8)
Veronica chamaedrys	II (1–4)	I (1–4)	I (1)	II (1–4)
Alopecurus pratensis	I (1–6)	I (1–4)	I (1)	I (1–6)
Cardamine pratensis	I (1–3)	I (1)	I (3)	I (1–3)
Vicia cracca	I (1–4)	I (1–3)	I (1–2)	I (1–4)
Bromus hordeaceus hordeaceus	I (1–6)	I (2–3)	I (3)	I (1–6)
Phleum pratense pratense	I (1–6)	I (1–5)	I (1)	I (1–6)
Juncus effusus	I (2–3)	I (3)	I (1–2)	I (1–3)
Phleum pratense bertolonii	I (1–3)	I (1–3)	I (1)	I (1–3)
Calliergon cuspidatum	I (1–5)	I (2–4)	I (3)	I (1–5)
Ranunculus repens	II (1–7)	I (2)	II (1–4)	I (1–7)
Pseudoscleropodium purum	I (1–5)	I (3–4)	II (2)	I (1–5)
Ophioglossum vulgatum	I (1–5)	I (1)		I (1–5)
Silaum silaus	I (1–5)	I (1–3)		I (1–5)
Agrimonia eupatoria	I (1–5)	I (1–3)		I (1–5)
Avenula pubescens	I (1–3)	I (2–5)		I (1–5)
Plantago media	I (1–4)	I (1–4)		I (1–4)
Alchemilla glabra	I (2)	I (3)		I (2–3)
Alchemilla filicaulis vestita	I (1–3)	I (3)		I (1–3)
Alchemilla xanthochlora	I (1–3)	I (2)		I (1–3)
Carex panicea	I (1–4)	I (2–4)		I (1–4)
Colchicum autumnale	I (3–4)	I (1–3)		I (1–4)
Crepis capillaris	I (1–5)	I (3)		I (1–5)
Festuca arundinacea	I (1–5)	I (3–5)		I (1–5)

[Handwritten annotations: "Constant Species", "Associates 2 community", "Associates?"]

Figure 3. Floristic table for NVC community MG5 *Centaurea nigra-Cynosurus cristatus* grassland.

of association between these preferentials and the vegetation type, with *Achillea millefolium* being less markedly diagnostic than *Trisetum flavescens* and, particularly, *G. verum*. There are also important negative features, too, because, although some plants typical of the first and third sub-communities, such as *Lolium* and *Prunella vulgaris*, remain quite common here, the disappearance of others, like *Lathyrus pratensis*, *Danthonia decumbens*, *Potentilla erecta* and *Succisa pratensis* is strongly diagnostic. Similarly, with the third subcommunity, there is that same mixture of positive and negative characteristics, and there is, among all the groups of preferentials, that same variation in abundance as is found among the constants and companions. Thus, some plants which can be very marked preferentials are always of rather low cover, as with *Prunella*, whereas others, like *Agrostis stolonifera*, though diagnostic at low frequency, can be locally plentiful.

For the naming of the sub-communities, we have generally used the most strongly preferential species, not necessarily those most frequent in the vegetation type. Sometimes, sub-communities are characterised by no floristic features over and above those of the community as a whole, in which case there will be no block of preferentials on the table. Usually, such vegetation types have been called Typical, although we have tried to avoid this epithet where the sub-community has a very restricted or eccentric distribution.

The tables organise and summarise the floristic variation which we encountered in the vegetation sampled: the text of the community accounts attempts to expound and interpret it in a standardised descriptive format. For each community, there is first a synonymy section which lists those names applied to that particular kind of vegetation where it has figured in some form or another in previous surveys, together with the name of the author and the date of ascription. The list is arranged chronologically, and it includes references to important unpublished studies and to accounts of Irish and Continental associations where these are obviously very similar. It is important to realise that very many synonyms are inexact, our communities corresponding to just part of a previously described vegetation type, in which case the initials *p.p.* (for *pro parte*) follow the name, or being subsumed within an older, more broadly-defined unit. Despite this complexity, however, we hope that this section, together with that on the affinities of the vegetation (see below), will help readers translate our scheme into terms with which they may have been long familiar. A special attempt has been made to indicate correspondence with popular existing schemes and to make sense of venerable but ill-defined terms like 'herb-rich meadow', 'oakwood' or 'general salt-marsh'.

There then follow a list of the constant species of the community, and a list of the rare vascular plants,

bryophytes and lichens which have been encountered in the particular vegetation type, or which are reliably known to occur in it. In this context, 'rare' means, for vascular plants, an A rating in the *Atlas of the British Flora* (Perring & Walters 1962), where scarcity is measured by occurrence in vice-counties, or inclusion on lists compiled by the NCC of plants found in less than 100 10 × 10 km squares. For bryophytes, recorded presence in under 20 vice-counties has been used as a criterion (Corley & Hill 1981), with a necessarily more subjective estimate for lichens.

The first substantial section of text in each community description is an account of the physiognomy, which attempts to communicate the feel of the vegetation in a way which a tabulation of data can never do. Thus, the patterns of frequency and abundance of the different species which characterise the community are here filled out by details of the appearance and structure, variation in dominance and the growth form of the prominent elements of the vegetation, the physiognomic contribution of subordinate plants, and how all these components relate to one another. There is information, too, on important phenological changes that can affect the vegetation through the seasons and an indication of the structural and floristic implications of the progress of the life cycle of the dominants, any patterns of regeneration within the community or obvious signs of competitive interaction between plants. Much of this material is based on observations made during sampling, but it has often been possible to incorporate insights from previous studies, sometimes as brief interpretive notes, in other cases as extended treatments of, say, the biology of particular species such as *Phragmites australis* or *Ammophila arenaria*, the phenology of winter annuals or the demography of turf perennials. We trust that this will help demonstrate the value of this kind of descriptive classification as a framework for integrating all manner of autecological studies (Pigott 1984).

Some indication of the range of floristic and structural variation within each community is given in the discussion of general physiognomy, but where distinct subcommunities have been recognised these are each given a descriptive section of their own. The sub-community name is followed by any synonyms from previous studies, and by a text which concentrates on pointing up the particular features of composition and organisation which distinguish it from the other sub-communities.

Passing reference is often made in these portions of the community accounts to the ways in which the nature of the vegetation reflects the influence of environmental factors upon it, but extended treatment of this is reserved for a section devoted to the habitat. An opening paragraph here attempts to summarise the typical conditions which favour the development and maintenance of the vegetation type, and the major factors which control floristic and structural variation within it. This is followed by as much detail as we have at the present time about the impact of particular climatic, edaphic and biotic variables on the community, or as we suppose to be important to its essential character and distribution. With climate, for example, reference is very frequently made to the influence on the vegetation of the amount and disposition of rainfall through the year, the variation in temperature season by season, differences in cloud cover and sunshine, and how these factors interact in the maintenance of regimes of humidity, drought or frosts. Then, there can be notes of effects attributable to the extent and duration of snow-lie or to the direction and strength of winds, especially where these are icy or salt-laden. In each of these cases, we have tried to draw upon reputable sources of data for interpretation, and to be fully sensitive to the complex operation of topographic climates, where features like aspect and altitude can be of great importance, and of regional patterns, where concepts like continental, oceanic, montane and maritime climates can be of enormous help in understanding vegetation patterns.

Commonly, too, there are interactions between climate and geology that are best perceived in terms of variations in soils. Here again, we have tried to give full weight to the impact of the character of the landscape and its rocks and superficials, their lithology and the ways in which they weather and erode in the processes of pedogenesis. As far as possible, we have employed standardised terminology in the description of soils, trying at least to distinguish the major profile types with which each community is associated, and to draw attention to the influence on its floristics and structure of processes like leaching and podzolisation, gleying and waterlogging, parching, freeze-thaw and solifluction, and inundation by fresh- or salt-waters.

With very many of the communities we have distinguished, it is combinations of climatic and edaphic factors that determine the general character and possible range of the vegetation, but we have often also been able to discern biotic influences, such as the effects of wild herbivores or agents of dispersal, and there are very few instances where the impact of man cannot be seen in the present composition and distribution of the plant communities. Thus, there is frequent reference to the role which treatments such as grazing, mowing and burning have on the floristics and physiognomy of the vegetation, to the influence of manuring and other kinds of eutrophication, of draining and re-seeding for agriculture, of the cropping and planting of trees, of trampling or other disturbance, and of various kinds of recreation.

The amount and quality of the environmental information on which we have been able to draw for interpreting such effects has been very variable. Our own sampling provided just a spare outline of the physical and edaphic conditions at each location, data which we have summarised where appropriate at the foot of the floristic tables; existing sources of samples sometimes offered next to nothing, in other cases very full soil analyses or precise specifications of treatments. In general, we have used what we had, at the risk of great unevenness of understanding, but have tried to bring some shape to the accounts by dealing with the environmental variables in what seems to be their order of importance, irrespective of the amount of detail available, and by pointing up what can already be identified as environmental threats. We have also benefited by being able to draw on the substantial literature on the physiology and reproductive biology of individual species, on the taxonomy and demography of plants, on vegetation history and on farming and forestry techniques. Sometimes, this information provides little more than a provisional substantiation of what must remain for the moment an interpretive hunch. In other cases, it has enabled us to incorporate what amount to small essays on, for example, the past and present role of *Tilia cordata* in our woodlands with variation in climate, the diverse effects of dunging by rabbit, sheep and cattle on calcicolous swards, or the impact of burning on *Calluna-Arctostaphylos* heath on different soils in a boreal climate. Debts of this kind are always acknowledged in the text and, for our part, we hope that the accounts indicate the benefits of being able to locate experimental and historical studies on vegetation within the context of an understanding of plant communities (Pigott 1982).

Mention is often made in the discussion of the habitat of the ways in which stands of communities can show signs of variation in relation to spatial environmental differences, or the beginnings of a response to temporal changes in conditions. Fuller discussion of zonations to other vegetation types follows, with a detailed indication of how shifts in soil, microclimate or treatment affect the composition and structure of each community, and descriptions of the commonest patterns and particularly distinctive ecotones, mosaics and site types in which it and any sub-communities are found. It has also often been possible to give some fuller and more ordered account of the ways in which vegetation types can change through time, with invasion of newly available ground, the progression of communities to maturity, and their regeneration and replacement. Some attempt has been made to identify climax vegetation types and major lines of succession, but we have always been wary of the temptation to extrapolate from spatial patterns to temporal sequences. Once more, we have tried to incorporate the results of existing observational and experimental studies, including some of the classic accounts of patterns and processes among British vegetation, and to point up the great advantages of a reliable scheme of classification as a basis for the monitoring and management of plant communities (Pigott 1977).

Throughout the accounts, we have referred to particular sites and regions wherever we could, many of these visited and sampled by the team, some the location of previous surveys, the results of which we have now been able to redescribe in the terms of the classification we have erected. In this way, we hope that we have begun to make real a scheme which might otherwise remain abstract. We have also tried in the habitat section to provide some indications of how the overall ranges of the vegetation types are determined by environmental conditions. A separate paragraph on distribution summarises what we know of the ranges of the communities and sub-communities, then maps show the location, on the 10×10 km national grid, of the samples that are available to us for each. Much ground, of course, has been thinly covered, and sometimes a dense clustering of samples can reflect intensive sampling rather than locally high frequency of a vegetation type. However, we believe that all the maps we have included are accurate in their general indication of distributions, and we hope that this exercise might encourage the production of a comprehensive atlas of British plant communities.

The last section of each community description considers the floristic affinities of the vegetation types in the scheme, and expands on any particular problems of synonymy with previously described assemblages. Here, too, reference is often given to the equivalent or most closely-related association in continental phytosociological classifications and an attempt made to locate each community in an existing alliance. Where the fuller account of British vegetation that we have been able to provide necessitates a revision of the perspective on European plant communities as a whole, some suggestions are made as to how this might be achieved.

Meanwhile, each reader will bring his or her own needs and commitment to this scheme and perhaps be dismayed by its sheer size and apparent complexity. For those requiring some guidance as to the scope of each volume and the shape of that part of the classification with which it deals, the introductions to the major vegetation types will provide an outline of the variation and how it has been treated. The contents page will then give directions to the particular communities of interest. For readers less sure of the identity of the vegetation types with which they are dealing, a key is provided to each major group of communities which should enable a set of similar samples organised into a constancy table to be taken through a series of questions to a reasonably secure diagnosis. The keys, though, are not infallible short cuts to identification and must be used in conjunction with the floristic tables and community descriptions. An alternative entry to the scheme is provided by the species index which lists the occurrences of all taxa in the communities in which we have recorded them. There is also an index of synonyms which should help readers find the equivalents in our classification of vegetation types already familiar to them.

Finally, we hope that whatever the needs, commitments or even prejudices of those who open these volumes, there will be something here to inform and challenge everyone with an interest in vegetation. We never thought of this work as providing the last word on the classification of British plant communities: indeed, with the limited resources at our disposal, we knew it could offer little more than a first approximation. However, we do feel able to commend the scheme as essentially reliable. We hope that the broad outlines will find wide acceptance and stand the test of time, and that our approach will contribute to setting new standards of vegetation description. At the same time, we have tried to be honest about admitting deficiencies of coverage and recognising much unexplained floristic variation, attempting to make the accounts sufficiently open-textured that new data might be readily incorporated and ecological puzzles clearly seen and pursued. For the classification is meant to be not a static edifice, but a working tool for the description, assessment and study of vegetation. We hope that we have acquitted ourselves of the responsibilities of the contract brief and the expectations of all those who have encouraged us in the task, such that the work might be thought worthy of standing in the tradition of British ecology. Most of all, we trust that our efforts do justice to the vegetation which, for its own sake, deserves understanding and care.

MESOTROPHIC GRASSLANDS

INTRODUCTION TO MESOTROPHIC GRASSLANDS

The sampling of mesotrophic grasslands

Tansley (1911, 1939) understood 'neutral grassland' as comprising semi-natural swards dominated by grasses with associated dicotyledonous herbs but lacking any pronounced calcicole or calcifuge element. Such vegetation was found mostly on lowland clays and loams of acid to neutral reaction and was largely treated as agricultural land. Initially, Tansley (1911) envisaged a single broad association, the *Graminetum neutrale*, and though he later recognised some distinct meadow and pasture communities within this compass (Tansley 1939), it is clear that his 'kernel of neutral grassland', those species especially characteristic of these types of sward, is a somewhat partial and eccentric diagnosis. This is largely because his account was based on just a few stands of rather particular kinds of mesotrophic grassland which had been the subject of classic early studies: Lawes *et al.* (1882) on the Rothamsted grasslands, Fream (1888a) on Hampshire water-meadows and Baker (1937) on the Oxford hay-meads and common pasture.

A renewed interest in these richer grassland types has developed with the urgent desire to conserve them in the face of widespread sward improvement for more productive agriculture. A greater range of communities has been recognised as a result of this concern, though these have often been defined by the occurrence of species of great conservation interest but of relatively little value in understanding more general ecological relationships among the full variety of British mesotrophic grasslands. Some unhelpfully broad categories, like 'herb-rich meadow', have gained common currency, while certain widely distributed sward types, lacking rare or attractive plants, but equally part of our overall grassland economy, have consistently received short shrift (e.g. Duffey *et al.* 1974, Ratcliffe 1977).

The agricultural perspective on mesotrophic grasslands (e.g. Stapledon 1925, Fenton 1931, Davies 1941, 1952, Williams & Davies 1946) has been predominantly an economic one, interested in the abundance of a few widely distributed forage plants and aiming to provide a working basis for sward management and improvement. This kind of approach has tended to ignore associated species in its description of grassland types, even where these might help illuminate the condition or potential of the herbage, and to lump unproductive swards into very broad categories. In general, agriculturalists have been most interested in those mesotrophic grasslands which attract little attention from conservationists, and the latter most concerned for some (though not all) of those swards which farmers want simply to improve or ignore.

In this survey, we wished to be more inclusive than each of these approaches and yet try to meet the needs of both, providing the sort of classification of mesotrophic grasslands long familiar from other parts of Europe (e.g. Koch 1926, Tüxen 1937, LeBrun *et al.* 1949, Oberdorfer 1957, 1983, O'Sullivan 1965, Westhoff & den Held 1969, Ellenberg 1978). And we were especially fortunate, in this part of our work, that Dr Martin Page was beginning a study of southern British swards of these kinds at the same time as we were starting our investigation. Our general approach to vegetation survey and the particular sampling style we had adopted for the project were quite compatible with his needs, so together we were able to ensure a fairly generous coverage of these grasslands through much of the country and to set the exercise within the context of our broader examination of all vegetation types in Britain.

As usual, we located our samples only on the basis of floristic and structural homogeneity of the vegetation, giving no special emphasis to swards that were species-rich or that contained rare plants or those considered indicative of particular histories or treatments. In fact, we were very concerned to set these sorts of grasslands among more widely distributed, impoverished and ill-defined assemblages, so as to produce a single scheme in which the inter-relationships of the rare and commonplace could be investigated together. Much of our sampling was therefore carried out in the highly improved agricultural landscape where only short-term leys were specifically excluded from the survey, among recreational swards and sown grasslands on reclaimed ground and road verges, on wasteland and along paths and hedge-banks, and around the margins of woodlands and mires.

Highly improved swards were often strikingly uniform in composition and structure but, in other situations, homogeneity could be harder to discern. More traditionally treated fields, for example, sometimes enclosed flushed swards around spring lines, or ranker herbage on steeper banks and rocky ground that escaped mowing or assiduous grazing. In other cases, there were gradual or patchy transitions to weedy or tall-herb vegetation or to scrub. With care, it was possible to avoid obvious floristic boundaries or, with intermediate situations, to note the subtle shifts in the make-up and physiognomy of the assemblages. With clearer patterning in the swards, as where distinctive grasslands were disposed over drier ridges and damper furrows, the repeated elements in the mosaics were sampled separately and their inter-relationships noted. In contrast to many field-by-field surveys of mesotrophic grasslands, records from obviously different assemblages within the same enclosure were never included in the same sample.

Almost always, quadrats of 2×2 m were adequate to provide a representative sample of the grasslands, with 4×4 m being occasionally necessary in ranker or rushy swards. Stands of unusual shape, occurring on narrow verges or thin headlands around fields, or along the edges of paths or ditches, were sampled using differently shaped quadrats of the same area and, in those few situations where stands were smaller than the appropriate sample size, the entire stands were treated as the samples.

In the recording of the floristic data, all vascular plants, bryophytes and macrolichens were scored in the usual fashion, although the last two groups, and especially the lichens, were usually poorly represented in the swards. Where the vegetation was mown, sampling was timed before the cut, or sufficiently long afterwards, to ensure ready identification of all the plants, and particular care was taken throughout with grasses and sedges, sometimes very hard to differentiate in close-cropped turf. Wherever possible, distinctions were made between the various taxa included within the *Alchemilla vulgaris* aggregate and between the subspecies of *Phleum pratense* and *Bromus hordeaceus*. With *Taraxacum officinale*, *Euphrasia officinalis* and *Rubus fruticosus*, however, we recorded to the aggregate and we rarely distinguished any rose microspecies. Also, *Lophocolea bidentata sensu lato* may also include infertile *L. cuspidata*.

These floristic data were then supplemented by details of the structure of the vegetation, noting such things as

any pattern of dominance or finer variegation in the sward at the time of sampling or, in ranker grasslands, any development of layering, together with any suggestion of phenological change with the progression of the growing season or in response to treatments. An indication was also often given of the context of the stand in the landscape, detailing zonations to neighbouring vegetation and any suspicion of successional changes which might be in train (Figure 4).

It was also frequently very informative to add to the few basic environmental data any details we could obtain of the habitat of the sample. Thus, notes were frequently made on the relief and drainage conditions, such as any indication of ground-water gleying, flushing or periodic inundation with fresh or brackish waters, and of the influence of aspect or physiography on the climate. Often of greater consequence here, though, was information on the treatment of the vegetation, for these were by and large plagioclimax communities in which succession was deflected by some kind of agricultural or related activity, frequently for the repeated production of a grass crop in enclosed pastures and meadows. In most cases, we had to rely on field observations and casual encounters with farmers to estimate the impact of such treatments, having no time to research site histories, but even those could be very instructive. Information on the intensity and timing of grazing, for example, particularly as to whether it was extended into the growing season, was especially valuable, together with details of the kinds of stock involved, and whether there was any contribution from wild herbivores such as the rabbit (*Oryctolagus cuniculus*), hare (*Lepus capensis* in these largely lowland habitats) or various species of deer. Cattle and sheep were by far the most important herbivores in the grasslands included here, with some grazing by horses. And the different ways in which these defoliated the swards, the animals varying in their bite, selectivity and social behaviour, and being managed in a diversity of free-range, rotational and paddock systems, was often evident in the appearance of the vegetation (Spedding 1971, Fraser 1974). The impact of dung and urine could also frequently be seen, because these stock defaecate in different ways, adding nutrients and fibrous material in dispersed or patchy fashion, with local concentrations producing distinct avoidance mosaics. Then, there were commonly some signs of trampling, particularly where heavier stock had access to moist ground where poaching obviously consolidated the soil and opened up the surface for the spread of weeds.

The timing of mowing – the cutting of herbage by scythe, sickle or machine in a sudden, even and non-selective act of defoliation – was also noted in samples taken from permanent meadows, long-term leys and fields normally treated as pasture but occasionally yielding a hay crop. It was particularly informative here to know how such treatment integrated with any grazing, in or out of the growing season and, where late visits were possible, to observe the condition of the aftermath. Where samples were taken from mown verges, an attempt was made to discover the timing and frequency of the cuts and to see whether the swathes were left or removed.

Evidence was also gathered where possible of any improvement of the ground or sward to encourage more productive cropping of herbage of desired composition, though we deliberately avoided the use of simple categories like 'improved' and 'unimproved': improvement is a complex and diverse process, its various elements affecting the components of swards in different ways, and we are much in need of details of its impact. Here, then, we were concerned to have information on the application of liming materials or artificial fertilisers, on any top-sowing, ploughing or re-seeding that had occurred, particularly where improved varieties of grasses and nutritious dicotyledons had been introduced, and of any herbicide applications or draining operations.

Finally, in trying to locate the samples we collected in their landscape context, we paid some attention to the

Figure 4. Completed sample cards from mesotrophic grasslands.

NVC record sheet 10/81

Location	Grid reference	Region	Author
Holywell Meadow, Warwickshire	SP(42) 195699	MIDS	MP

Site and vegetation description

Sample from somewhat damper area in pronounced ridge-and-furrow meadow with good growth of aftermath. No obvious signs of very recent grazing, although there are horses in the field at present. Sward is species-rich throughout with a striking quantity of Galium verum, but has no single dominant. Neither is there very much variation over the differently-sloping (0-10°) areas of ridges and furrows which run north-south.

Date	Sample no.
5.8.77	372
Altitude: 120 m	Slope: °
Aspect: °	Soil depth: cm
Stand area: m x m	Sample area: 2 m x 2 m
Layers: mean height: m m 45 cm 10 mm	
Layers: cover: % % 100% 8 %	
Geology: Triassic Keuper Marl	

Species list

Galium verum	6	Dactylis glomerata	2
Centaurea nigra	4	Alopecurus pratensis	2
Festuca rubra	5	Carex flacca	3
Agrostis capillaris	4	Carex caryophyllea	3
Trifolium repens	3	Cerastium fontanum	2
Carex hirta	4	Luzula campestris	1
Hypochoeris radicata	2	Prunella vulgaris	2
Lotus corniculatus	3	Avenula pubescens	1
Lolium perenne	2	Chrysanthemum leucanthemum	1
Cynosurus cristatus	3	Succisa pratensis	1
Holcus lanatus	4	Leontodon autumnalis	1
Trifolium pratense	4	Taraxacum officinale	1
Achillea millefolium	2	Stellaria graminea	1
Lathyrus pratensis	2	Cirsium palustre	1
Plantago lanceolata	3	Quercus robur seedling	1
Anthoxanthum odoratum	3	Brachythecium rutabulum	4
Briza media	2	Atrichum undulatum	2
Ranunculus acris	2	Pseudoscleropodium purum	1
Rumex acetosa	2	Rhytidiadelphus squarrosus	2
Poa trivialis	2		

Soil profile: Brown earth, deep, loamy-sandy soil

pH 0-5cm = 5.7

NVC record sheet 10/81

Location	Grid reference	Region	Author
Linbrigg Northumberland	NT(36) 895068	NE	CDP/JR

Site and vegetation description

Tall-herb vegetation in patches over steeply-sloping cliff-base and tumbled debris, among open Fraxinus, Corylus and Prunus padus (see sample 839 for denser woodland vegetation)

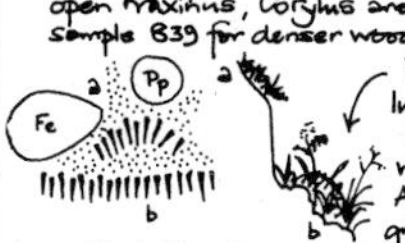

Dominated by large & luxuriant tussocks of Luzula and stools of Dryopteris with scattered clumps of Arrhenatherum & other grasses, some Pteridium and tall dicotyledons with emergent inflorescences. Smaller herbs appear in more open areas and around margins, with patches of mosses over stool-bases and on litter which is quite thick in places. An impoverished version of this vegetation, mainly Luzula & ferns, runs up runnels between rocky knolls, giving way to species-rich Festuca-Agrostis where grazed.

Date	Sample no.
22.6.78	840
Altitude: 210 m	Slope: 45 °
Aspect: 320 °	Soil depth: cm
Stand area: 3 m x 4 m	Sample area: 2 m x 2 m
Layers: mean height: m m 40 cm 10 mm	
Layers: cover: % % 95 % 15 %	
Geology: Andesite lava	

Species list

Arrhenatherum elatius	3	Conopodium majus	1
Dryopteris filix-mas	6	Cystopteris fragilis	1
Luzula sylvatica	6	Vicia sepium	2
Filipendula ulmaria	2	Viola tricolor	2
Angelica sylvestris	2		
Geranium robertianum	3	Calliergon cuspidatum	1
Geranium sylvaticum	2	Plagiothecium denticulatum	2
Holcus lanatus	4	Mnium hornum	2
Pteridium aquilinum	5	Rhizomnium punctatum	1
Mercurialis perennis	2	Brachythecium rutabulum	4
Dactylis glomerata	2	Eurhynchium praelongum	1
Urtica dioica	2	Eurhynchium swartzii	1
Hyacinthoides non-scripta	3	Lophocolea bidentata s.l.	2
Geum rivale	2		
Primula vulgaris	2		
Silene dioica	2		
Stellaria holostea	2		
Valeriana officinalis	1		
Cardamine pratensis	1		

Soil profile: Brown earth, in pockets over irregular cliff-base, highly humic above, mid-brown sandy silt-loam below

pH 0-5cm = 5.0

NVC record sheet 10/81

Location	Grid reference	Region	Author
B5392 near Siddington, Cheshire	SJ(33) 843708	MIDS	ER

Site and vegetation description

Rank verge grassland with Arrhenatherum dominant overall, but patchy local abundance of Urtica and, extending out from bank with Corylus avellana and Rubus idaeus, a striking understorey in some parts of Hyacinthoides. Other stretches of verge have essentially similar vegetation as here but (584) Aegopodium is rather striking in places as a hedge-bottom feature. No recent mowing but shorter swards on drier sandy roadsides nearby have Festuca-Agrostis sward (583,585) with smaller herbs that suggests fairly frequent cutting. Pastures around are improved but have some pools.

Date	Sample no.
30.5.77	591
Altitude: 121 m	Slope: 15 °
Aspect: 190 °	Soil depth: cm
Stand area: 14 m x 2 m	Sample area: 2 m x 2 m
Layers: mean height: m m 120 cm 10 mm	
Layers: cover: % % 100 % 20 %	
Geology: Keuper Marl	

Species list

Urtica dioica	3
Heracleum sphondylium	3
Arrhenatherum elatius	6
Pteridium aquilinum	4
Stellaria media	4
Conopodium majus	3
Hyacinthoides non-scripta	4
Galium aparine	3
Dactylis glomerata	4
Eurhynchium praelongum	4
Plagiothecium denticulatum	4
Mnium hornum	3
Lophocolea bidentata	3

Soil profile: Brown earth, shallow loamy

pattern of enclosure and any other features of historical interest. With ridge-and-furrow, for example, a relic of past arable cultivation seen in many grasslands, we were concerned to note whether this was of the more pronounced and often older, sometimes medieval, type, or more gently cambered, like the 'narrow rig' of Napoleonic and later ploughing. Similarly, we recorded any signs of early irrigation or drainage schemes, as in river-valley flood-pastures and the very local and now usually defunct water-meadows, or on washlands and among the distinctive scenery of coastal reclamations.

Quite apart from the accumulating experience of the research team and the particular contribution of Dr Page, we were able to benefit throughout our sampling programme, and in the analysis and interpretation of the data, from the wisdom of Mr Derek Wells, the lowland grasslands specialist of the NCC Chief Scientist Directorate, and from a variety of NCC surveys, particularly those undertaken by the England Field Unit, then under the direction of Dr Tim Bines. We were also able to make use of published and unpublished data from the work of the Macaulay Institute in the Scottish lowlands (Birse & Robertson 1976, Birse 1980, 1984), from a study of Pennine meadows by Dr Richard Jones (Jones

1984) and from a survey on upper salt-marshes throughout Britain (Adam 1976). Numerous other small reports on particular sites and scientific papers yielded further samples, and where any of these data were fully compatible with our own, we were able to include them in the analyses. Other information, such as data from 1 × 1 m quadrats, which we considered generally too small for adequate recording of grassland vegetation, and samples scored using the DAFOR scale, we rejected from analysis, but sometimes found informative in a more general fashion. In the data set as a whole, we were sometimes conscious of a possible confusion between *Juncus acutiflorus* and *J. articulatus*, rejecting doubtful samples, and of the likely under-recording of *J. conglomeratus*. Also, *Poa pratensis* probably includes some records for *P. subcaerulea*.

Altogether, we had over 2000 samples available for analysis with quite good geographical coverage throughout the British lowlands (Figure 5). Most kinds of mesotrophic grassland seemed well represented in the data set, though we remain concerned about inadequate representation of the full range of variation among certain assemblages and of some transitions to mire vegetation. These are outlined below and noted in the text.

Data analysis and the description of mesotrophic grassland communities

In the usual fashion, only the floristic records for the samples were used to characterise the vegetation types, with such environmental data as were available being reserved for interpretation of the results. The quantitative scores for all vascular plants, bryophytes and macrolichens were employed for analysis, with no special weighting being given to reputed indicators of particular kinds of grassland or environmental conditions.

In its broad outlines and in many areas of detail, the scheme presented here is that developed in Page (1980), with a few major and more minor differences in definition and division among the grasslands, and certain revisions of their boundaries with related calcicole and calcifuge swards and similar vegetation types among the mires and weed and salt-marsh communities. About this latter issue, argument could long continue, because mesotrophic grasslands largely comprise species which show a preference for soils that are neither too acid nor too basic and neither very wet nor very dry. Setting floristic limits around the periphery of the group is thus sometimes a matter of using negative criteria and there are some gradual transitions along which it is very difficult to draw a universally acceptable line. Some of the decisions we have made will no doubt provoke annoyance: for example, it seemed better to us to include certain mesotrophic rush-pasture vegetation in this

Figure 5. Distribution of samples available from mesotrophic grasslands.

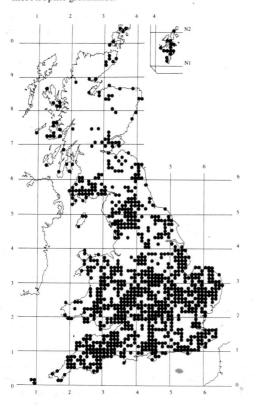

volume, with other communities of a similar type treated among the mires; many slightly improved swards towards the limit of enclosure around the upland fringes, on the other hand, are best understood as modified forms of calcifuge grasslands and are thus located elsewhere. In general, here, it seemed more important to ensure that such intermediates were included somewhere, rather than to agonise overmuch about their exact location in the scheme. Complex relationships and ambiguities of definition are always pointed out in the community accounts.

A further problem with diagnosing relationships among the mesotrophic grasslands themselves arises because they owe a good deal of their specific floristic character to a wide range of agricultural treatments which can be applied in multifarious combinations. In the past, these have often been expressed in distinct local farming traditions, but these days practice is much more universally affected by changes in agricultural fashion and economics. Transitions between different kinds of mesotrophic grassland are frequently related to various combinations of treatments which produce floristic effects that cross-cut distinctions based on edaphic and climatic variation. Furthermore, individual species within communities often respond differently to particular treatments and stands lacking one or more species, but still referable on general grounds to one community rather than another, are widespread. Here, as among other kinds of vegetation, lists of rare plants, which may be especially sensitive to various kinds of treatment, have a limited value in distinguishing communities or sub-communities from one another. This does not detract from the need of such species for protection, but it does affect the framework of understanding within which their conservation is planned and effected.

A third difficulty in defining mesotrophic grasslands from a phytosociological perspective is that certain species that are considered diagnostic of particular associations or high units of classification in Continental Europe do not retain the same fidelity in Britain. Thus, some alliances, such as the Cynosurion cristati and the Calthion palustris, seem very ill-defined when applied to certain British pastures and transitions to fens and this reflects some real and interesting complexities in the relationships between our examples of these kinds of vegetation and their European counterparts and the environmental variables which affect their composition in moving across the Continent.

Bearing these problems in mind, we have distinguished a total of eighteen communities characterised by the general frequency throughout of *Dactylis glomerata, Festuca pratensis, F. rubra, Holcus lanatus, Poa pratensis, P. trivialis, Cerastium fontanum, Plantago lanceolata, Ranunculus acris* and *Trifolium repens*. They are usually found as closed swards on drought-free,

mesotrophic to nutrient-rich mineral soils with a pH of 4.5–6.5 throughout those parts of the British lowlands with a fairly moist and mild climate and a long growing season. All the communities would probably be regarded as falling within the Molinio-Arrhenatheretea which is the class of largely anthropogenic lowland grasslands with some mire and water-margin communities in western Europe and the northern Mediterranean. The list of species above approximates to the floristic definition of this class provided by O'Sullivan (1965) and Shimwell (1968a) for western Europe including the British Isles.

The eighteen communities can be considered under five headings; two types of *Arrhenatherum elatius* grassland, four types of generally well-drained pastures and meadows, six sorts of long-term ley and related swards, three kinds of ill-drained pasture with a poor-fen element and three grass-dominated inundation communities. Some of these vegetation types are very diverse and, in certain cases, we have followed Page (1980) in differentiating variants as well as the usual sub-communities.

Arrhenatherum elatius grasslands

Two communities are very clearly marked off by the consistent occurrence and often great abundance of *A. elatius*, the prominence of coarse Molinio-Arrhenatheretea grasses such as *Holcus lanatus* and *Dactylis glomerata*, and the preferential frequency of tall herbs like *Heracleum sphondylium, Anthriscus sylvestris* and *Urtica dioica*. These communities are thus the major British representatives of the Arrhenatherion elatioris, an alliance of ungrazed, coarse and tussocky swards on free-draining, mesotrophic to eutrophic soils, although many stands in this country are more species-poor and fragmentary than on the Continent and much less commonly mown for hay in meadows.

The bulk of our vegetation of this sort is included here in a single large and rather variable community, the *Arrhenatherum elatius* grassland (MG1, *Arrhenatheretum elatioris* Br.-Bl. 1919). In our scheme, this subsumes some richer *Arrhenatherum* swards previously separated off into a *Centaureo-Arrhenatheretum* (O'Sullivan 1965, Shimwell 1968a, Birse 1980) and a more calcicolous *Pastinaco-Arrhenatheretum* (Passarge 1964, Page 1980) which extends the coverage of this vegetation on to more calcareous brown soils in the warmer southeast of Britain. By and large, however, the community occurs as rank, species-poor grassland familiar to many from our road and motorway verges. Although rarely yielding an agricultural hay crop, it is in this vegetation that the effects of varying mowing time and frequency are most readily seen, because the herbage is generally cut at least once in the growing season so as to maintain visibility and neatness along highways and to prevent scrub encroachment. With disturbance of various kinds, many stands show transitions to rank, weedy or tall-herb vegetation (Figure 6).

A second community, the *Arrhenatherum elatius-Filipendula ulmaria* grassland (MG2, *Filipendulo-Arrhenatheretum elatioris* Shimwell 1968*a*) has been retained to include some highly distinctive vegetation with abundant *Arrhenatherum* and tall herbs like *F. ulmaria*, *Valeriana officinalis* and *Angelica sylvestris* along with *Mercurialis perennis*, *Geum rivale*, *Silene dioica*, *Cruciata laevipes* and *Dryopteris filix-mas*. Strictly confined to damp, calcareous soils in cool and humid situations in northern Britain, usually on Carboniferous Limestone outcrops, this vegetation has clear affinities with the ungrazed field layers of Alno-Ulmion woodlands. It is also one of the British mesotrophic grasslands which most closely approaches the Cicerbition alpini tall-herb vegetation of ungrazed mountain ledges.

Other communities with an abundance of *Arrhenatherum* are best considered with sand-dune and shingle vegetation.

Well-drained permanent pastures and meadows

The four communities distinguished here share frequent records for the full range of Molinio-Arrhenatheretea plants with the additional preferential occurrence of *Cynosurus cristatus*, *Lolium perenne*, *Bellis perennis*, *Leontodon autumnalis* and *Taraxacum officinale* agg. These are closed swards of grasses and herbaceous dicotyledons and include the bulk of the permanent agricultural grasslands used for grazing and hay production in Britain. The most obvious general affinity of the group is with the Cynosurion cristati, the alliance of north-west European pastures and meadows, but this relationship is far from simple in the more oceanic climate of this country.

Figure 6. Mesotrophic pastures and meadows in relation to treatment.

ARRHENATHERION	CYNOSURION		LOLIO-PLANTAGINION
MG1	MG5	MG6	MG7
Arrhenatheretum elatioris grassland	Centaureo-Cynosuretum grassland	Lolio-Cynosuretum grassland	Lolium perenne leys & related grassland
Mown once or twice annually for amenity, ungrazed and unmanured	Mown annually for hay and autumn- and winter-grazed, manured by stock	Grazed through the year, chemically fertilised and often resown	Sown swards, chemically fertilised and grazed through the year or cut for silage or amenity

A clear distinction can be made among these communities between three generally unimproved grasslands, often treated as meadows under more traditional management, and one highly improved permanent pasture. The former are usually more species-rich swards with a substantial contingent of dicotyledons, including some bulky herbs. Overall, these grasslands are characterised by *Leontodon hispidus*, *Centaurea nigra*, *Lotus corniculatus*, *Trisetum flavescens*, *Luzula campestris*, *Ranunculus bulbosus*, *Rhinanthus minor*, *Leucanthemum vulgare*, *Hypochoeris radicata* and *Primula veris*, with *Anthoxanthum odoratum*, *Agrostis capillaris* and *Rumex acetosa* rather more unevenly represented.

This, then, is the group of species which can be considered generally indicative of older, unimproved, well-drained mesotrophic grasslands in Britain and these three communities include the bulk of our rich and colourful meadows. A summer cut for hay, light winter grazing and an avoidance of artificial fertilisers and pesticides have been the traditional treatment here. Often, time-honoured practices have governed the date of shutting up the meadows in early spring and of the turning out of stock on to the aftermath, and some localities still preserve ancient common rights for the doling out of lots of land and their marking in the sward by stakes or mear stones.

The most widely distributed of these communities is the *Centaurea nigra-Cynosurus cristatus* grassland (MG5, *Centaureo-Cynosuretum cristati* Br.-Bl. & Tx 1952) which has a good representation of most of the listed species and is the characteristic vegetation of traditionally treated meadows on circumneutral brown soils throughout the British lowlands. It can be readily incorporated into the Cynosurion alliance, although it includes some swards which closely approach the sort of Mesobromion grassland found on deeper, calcareous brown earths over lime-rich superficials and clays (Figure 7).

The *Centaureo-Cynosuretum* also extends into the fringes of our uplands in Wales and the Pennines though, where increased rainfall is combined with markedly lower temperatures and a shorter growing season, as in the valley heads of northern England, the community is replaced by the *Anthoxanthum odoratum-Geranium sylvaticum* grassland (MG3). Here, *G. sylvaticum*, *Sanguisorba officinalis*, *Conopodium majus*, *Alchemilla glabra* and *A. xanthochlora* enrich the tall-herb element in what is the nearest approach among our mesotrophic swards to the alpine and Scandinavian meadows of the Polygono-Trisetion alliance. The third type of meadow, the *Alopecurus pratensis-Sanguisorba officinalis* grassland (MG4) shares many floristic features with the *Anthoxanthum-Geranium* grassland, but it is characteristic of periodically flooded meadows on alluvial soils. With the occurrence alongside *A. pratensis* of species such as *Filipendula ulmaria*, *Ranunculus repens* and *Agrostis stolonifera*, it has affinities with the mires of the Molinietalia and Page (1980) considered it to be the equivalent of Molinion caeruleae vegetation on calcareous alluvium in Continental Europe.

With ever gathering pace in recent years, many stands of traditionally treated meadow vegetation of these kinds have been subject to an extension of grazing into the late spring and to agricultural improvement by the application of artificial fertilisers, ploughing and re-seeding, all of which tend to reduce the diversity of the swards and favour dominance by grasses. Much of the improved permanent pasture that results from these processes can be grouped within the *Lolium perenne-Cynosurus cristatus* grassland (MG6, *Lolio-Cynosuretum cristati* Br.-Bl. & De Leeuw 1936) where bulkier palatable herbs such as are characteristic of the unimproved swards and the Arrhenatherion, plants with unprotected apical buds and/or which rely on late seed-set to maintain themselves, are reduced. Among the abundant grass cover, the major fodder plant *L. perenne* is of especial importance, with the smaller Molinio-Arrhenatheretea herbs often supplemented by improved varieties of, for example, *Trifolium repens*. Although intensive treatment often sharpens the floristic boundaries of the *Lolio-Cynosuretum*, it has been derived by the exercise of agricultural ingenuity from an extremely wide range of precursors, not just old meadows but calcicolous and calcifugous grasslands, drained blanket

bog, reclaimed salt-marsh and fixed sand-dunes, and species characteristic of these vegetation types can persist at low frequency and confuse the definition. Generally, however, the affinities of the community are with the Cynosurion.

Long-term leys and related grasslands

Six further assemblages have been characterised by the generally low frequency of many of the Molinio-Arrhenatheretea plants and by the constancy, often in abundance, of *Lolium perenne* in the absence of *Cynosurus cristatus*. These are species-poor, grass-dominated swards which can be considered, for the most part, as highly specialised counterparts of the *Lolio-Cynosuretum*, sometimes derived from it by further improvement, but often specially sown for agricultural or recreational use or naturally developed where other swards are trampled. The alliance Lolio-Plantaginion majoris and the order Plantaginetalia have been erected to contain the European counterparts to these vegetation types and related assemblages of a more open and disturbed character. In this scheme, these latter have been included among the weed communities and, pending further study, the grasslands are all described under a single heading (MG 7).

Among the group, the *Lolium-Trifolium repens*, *Lolium-Poa trivialis*, *Lolium-Alopecurus pratensis* and *Lolium-Alopecurus-Festuca pratensis* grasslands are generally encountered as intensively treated, high-productivity swards, sometimes as long-term leys sown in rotation with arable. Here, the demand for heavy crops of nutritious herbage has resulted in the sowing of

Figure 7. Variation among pastures and meadows with climate and treatment.

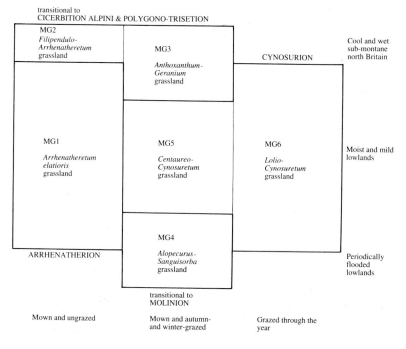

mixtures in which selected long-stalked, erect-leaved and fast-growing cultivars of the grasses predominate, along with improved strains of *T. repens*. With heavy applications of nitrogenous fertilisers, such swards can be mown repeatedly during the growing season to provide fresh grass for zero-grazing in stalls or for silage production.

Two further assemblages, the *Lolium-Poa pratensis* grassland (*Poo-Lolietum pratensis* De Vries & Westhoff *apud* Bakker 1965) and the *Lolium-Plantago lanceolata* community (*Lolio-Plantaginetum* (Link 1921) Beger *emend.* Sissingh 1969) share the general characteristics of this group but are additionally distinct in the high frequency of *Plantago major*, *Bellis perennis*, *Poa annua* and *Trifolium dubium*. The *Poo-Lolietum* is often found as a sown, coarse recreational sward, valued for its resistance to heavy use, and here improved strains of the preferential *P. pratensis* are frequently included in the seed mixtures. The *Lolio-Plantaginetum* is much more a vegetation type of trampled pasture, as around gateways and along footpaths and verges.

Ill-drained permanent pastures

In many mesotrophic soils, particularly those derived from impervious bedrocks such as clays and shales, and from superficials like till and alluvium, the maintenance of moderately high soil moisture is associated with some ground-water gleying. On the more frequently waterlogged and less fertile of these profiles, towards the boundary between agricultural improvement and neglect, mesotrophic grasslands grade into mires and inundation vegetation and among these swards we have characterised three communities.

The general floristic feature of these assemblages is the preponderance of moisture-tolerant or moisture-loving plants. Among the Molinio-Arrhenatheretea species, the most frequent and abundant are *Holcus lanatus* and *Poa trivialis*, while the common occurrence of *Agrostis stolonifera*, *Ranunculus repens*, *Potentilla anserina* and *Rumex crispus* provides a link with the vegetation of periodically flooded ground. The dominants, too, are among those species which occur most commonly on very moist, often waterlogged soils, with *Juncus effusus*, *Deschampsia cespitosa*, *Filipendula ulmaria* and *Caltha palustris* figuring prominently. In contrast with the previously described communities, then, which are best located in the Arrhenatheretalia, these three vegetation types have strong affinities with the grassy, rushy and tall-herb assemblages of the Molinietalia, the other major order of the Molinio-Arrhenatheretea.

The *Holcus lanatus-Deschampsia cespitosa* grassland (MG9) and the *Holcus lanatus-Juncus effusus* rush-pasture (MG10, *Holco-Juncetum effusi* Page 1980) form a pair of rather similar communities in which there is a trend towards dominance by *D. cespitosa* or *J. effusus* (with some *J. inflexus* on more base-rich soils) within a matrix of pasture grasses and dicotyledons. These are very widespread vegetation types of heavy unimproved or neglected agricultural land and their distinctive features have led some to suggest that they should be placed in special higher units such as a Deschampsion alliance and a Calthion sub-alliance, the Holco-Juncion. In fact, the definition of these groups towards the Atlantic coast of Europe is rather indistinct and much in need of reassessment in the light of our increasing knowledge of British vegetation of this type: the *Holco-Juncetum*, for example, comprises just one element in a continuum of rush-dominated meadows and pastures across Britain with complex and shifting relationships to climate, soils and treatment.

A further community, the *Cynosurus cristatus-Caltha palustris* grassland (MG8), has a better representation of Molinio-Arrhenatheretea plants than the above swards and shows clearer affinities with our richer meadows and pastures of better-drained ground. It is, though, a very local community nowadays, surviving as a traditionally treated pasture on seasonally-flooded ground by rivers and streams in a few parts of lowland Britain. Of further interest is the fact that it may represent the naturally occurring vegetation that formed the basis of the highly specialised swards of water-meadows. First created in the sixteenth and seventeenth centuries to supplement spring grazing on our southern Chalklands, these developed into complex systems of irrigated lands along valley bottoms where laborious hand-weeding selected for a grass-dominated herbage that was very productive and palatable. European equivalents of our *Cynosurus-Caltha* grassland have been assigned to the Calthion, though here again the affinities are somewhat ill-defined and further sampling is needed to relate this vegetation to grassier fen-meadows.

Inundation grasslands

The three remaining communities included in this section are distinguished by a generally very poor representation of Molinio-Arrhenatheretea plants, and a continuing rise to prominence of *Agrostis stolonifera* and *Potentilla anserina* with a wide variety of occasionals typical of periodically inundated and disturbed substrates, notably *Rumex crispus* and *Elymus repens*. These assemblages are characteristic of fine-textured mesotrophic soils alongside fluctuating sluggish or standing waters and are best seen as representing the grassier end of the vegetation traditionally included in the Elymo-Rumicion crispi. This is a diverse alliance with a rather chequered history, much revised and sometimes placed with the Lolio-Plantaginion in the Plantaginetalia, in other cases elevated to its own order the Agrostietalia. There is obviously some justification for considering certain vegetation of this kind as falling outside the Molinio-Arrhenatheretea and, in this scheme, the more open and ephemeral assemblages are considered among the weed communities.

Two of the vegetation types included here, the *Festuca rubra-Agrostis stolonifera-Potentilla anserina* grassland (MG11) and the *Festuca arundinacea* grassland (MG12, *Potentillo-Festucetum arundinaceae* Nordhagen 1940) share a high frequency and abundance of *F. rubra* and the quite common occurrence of salt-tolerant plants such as *Atriplex prostrata, Matricaria maritima, Honkenya peploides, Juncus gerardi* and *Glaux maritima*. They extend the range of mesotrophic swards on to reclaimed salt-marshes where there is periodic inundation with brackish or salt waters and in such situations the vegetation has clear affinities with the open halophyte assemblages of the Elymion pycnanthi.

The third community, the *Agrostis stolonifera-Alopecurus geniculatus* grassland (MG13) is more exclusively associated with the fluctuating margins of fresh waters. Here, *A. stolonifera* and *Alopecurus geniculatus* usually dominate with *Holcus lanatus, Poa trivialis, Ranunculus repens, Glyceria* spp. and the tall docks *Rumex crispus* and *R. conglomeratus* in often small stands of vegetation which come close in their composition and structure to Glycerio-Sparganion water-margin vegetation. Other assemblages of this sort can be found among the swamp communities, but we have retained this vegetation type here because of its usually intimate association alongside pasture streams and pools with other mesotrophic swards and its occasional contribution to extensive flood-pastures in washlands.

KEY TO MESOTROPHIC GRASSLANDS

With something as complex and variable as vegetation, no key can pretend to offer an infallible short cut to diagnosis. The following should therefore be seen as but a crude guide to identifying the types of mesotrophic grassland in the scheme and must always be used in conjunction with the data tables and community accounts. It relies on floristic (and, to a minor extent, physiognomic) features of the vegetation and demands a knowledge of the British flora and a few bryophytes. It does not make primary use of any habitat features, though these may provide a valuable confirmation of a diagnosis.

Because the major distinctions between the vegetation types in the classification are based on inter-stand frequency, the key works best when sufficient samples of similar composition are available to construct a constancy table. It is the frequency values in this (and, in some cases, the ranges of abundance) which are then subject to interrogation with the key. Many of the questions are dichotomous and notes are provided at particularly awkward choices or where confusing mosaics and zonations are likely to be found.

Samples should always be taken from homogeneous stands and be of 2×2 m or 4×4 m according to the scale of the vegetation or, where complex patterns occur, of identical size but irregular shape. Very small stands should be sampled in their entirety.

---

1   Coarse grasslands with frequent and often abundant *Arrhenatherum elatius*, *Dactylis glomerata* and *Holcus lanatus*.   2

These species usually occasional at most and rarely abundant in the sward   7

2   *Deschampsia cespitosa* present and usually co-dominant with *Arrhenatherum* and other coarse grasses but usually without large umbellifers

> **MG9** *Holcus lanatus-Deschampsia cespitosa* grassland
> *Arrhenatherum elatius* sub-community

On ill-drained land from which grazing has been withdrawn, this vegetation type may grade to the *Poa trivialis* sub-community of the *Holcus-Deschampsia* grassland but *Arrhenatherum* and *Dactylis* become uncommon there and occasional *Juncus effusus* and *Filipendula ulmaria* appear with the increased frequency of *Poa trivialis*.

D. cespitosa usually absent but *Heracleum sphondylium* frequent and often abundant with one or more of *Anthriscus sylvestris*, *Pastinaca sativa*, *Urtica dioica*, *Filipendula ulmaria* and *Centaurea nigra*   3

3   Species-rich and luxuriant vegetation with frequent *Mercurialis perennis*, *Valeriana officinalis*, *Geum rivale*, *Dryopteris filix-mas* and *Silene dioica*

> **MG2** *Arrhenatherum elatius-Filipendula ulmaria* tall-herb grassland
> *Filipendulo-Arrhenatheretum elatioris* Shimwell 1968*a*   4

Sward can be quite species-rich and luxuriant but never with the above-listed species common   5

4   *Polemonium caeruleum* constant, often with *Oxalis acetosella* and *Stellaria holostea*

> **MG2** *Filipendulo-Arrhenatheretum*
> *Polemonium caeruleum* sub-community

P. caeruleum absent but two or more of *Sanguisorba officinalis*, *Avenula pubescens*, *Senecio jacobaea*, *Poa pratensis* and *Origanum vulgare* usually present

MG2 *Filipendulo-Arrhenatheretum*
Filipendula ulmaria sub-community

5   *Geranium sylvaticum* and *G. pratense* present and sometimes abundant

MG3 *Anthoxanthum odoratum-Geranium sylvaticum* grassland
Arrhenatherum elatius sub-community

Around the less heavily grazed margins of northern meadows, this vegetation type may grade to other sub-communities of the *Anthoxanthum-Geranium* grassland where *Arrhenatherum* is characteristically sparse.

G. sylvaticum absent, though *G. pratense* can be locally prominent

MG1 *Arrhenatherum elatius* grassland
Arrhenatheretum elatioris   Br.-Bl. 1919   6

6   *Pastinaca sativa* constant and sometimes abundant with frequent *Festuca ovina*, *Agrostis capillaris*, *Galium verum* and *Senecio jacobaea*

MG1 *Arrhenatheretum elatioris*
Pastinaca sativa sub-community

Filipendula ulmaria constant and sometimes co-dominant with *Arrhenatherum*

MG1 *Arrhenatheretum elatioris*
Filipendula ulmaria sub-community

Urtica dioica constant and sometimes co-dominant with *Arrhenatherum* without either *Pastinaca* or *F. ulmaria*

MG1 *Arrhenatheretum elatioris*
Urtica dioica sub-community

Centaurea nigra and *Lotus corniculatus* constant with one or more of *Leucanthemum vulgare*, *Veronica chamaedrys*, *Anthoxanthum odoratum* and *Trisetum flavescens*

MG1 *Arrhenatheretum elatioris*
Centaurea nigra sub-community

Generally grass-dominated vegetation, often rank and species-poor, with frequent and abundant *Festuca rubra* in the general absence of the above species

MG1 *Arrhenatheretum elatioris*
Festuca rubra sub-community

The *Centaurea* and *Festuca* sub-communities of the *Arrhenatheretum* often occur in mosaics and seral transitions with the *Centaureo-Cynosuretum*, the *Lolio-Cynosuretum* and the *Anthoxanthum-Geranium* grassland where grazing is patchy or changing in its intensity.

7   *Deschampsia cespitosa* constant and abundant with *Holcus lanatus* and *Poa trivialis*

MG9 *Holcus lanatus-Deschampsia cespitosa* grassland
Poa trivialis sub-community

D. cespitosa generally infrequent and not usually abundant   8

D. cespitosa can be locally prominent in badly-draining stretches of the *Alopecurus-Sanguisorba* grassland and the *Lolio-Cynosuretum*, and ill-defined transitions to the latter are especially common. It can also be difficult to separate this vegetation type from certain kinds of *Juncus subnodulosus-Cirsium palustre* fen-meadow.

8   *Juncus effusus* and/or *J. inflexus* constant and usually dominant in a grassy ground with frequent *Holcus lanatus*, *Agrostis stolonifera* and *Ranunculus repens*

MG10 *Holcus lanatus-Juncus effusus* rush-pasture
Holco-Juncetum effusi   Page 1980   9

J. effusus and *J. inflexus* generally infrequent and not usually abundant   10

J. effusus and, to a lesser extent, *J. inflexus* can be locally prominent in badly-draining stretches of other grasslands, particularly the *Lolio-Cynosuretum* and *Centaureo-Cynosuretum* and it may be difficult then to effect this separation

9   *J. inflexus* constant and abundant usually with some *J. effusus*

MG10 *Holco-Juncetum*
Juncus inflexus sub-community

Iris pseudacorus constant and sometimes dominant with an often reduced cover of *J. effusus*

MG10 *Holco-Juncetum*
Iris pseudacorus sub-community

Juncus effusus abundant without *J. inflexus* or *I. pseudacorus*

MG10 *Holco-Juncetum*
Typical sub-community

Towards the west of Britain, it becomes increasingly difficult to separate Typical *Holco-Juncetum* from the *J. effusus* sub-community of the *Juncus-Galium* rush-pasture, but *Galium palustre*, *Lotus uliginosus*, *Cirsium palustre* and *Agrostis canina* ssp. *canina* become much commoner there.

---

10   *Caltha palustris* constant with one or more of *Carex panicea*, *Filipendula ulmaria*, *Agrostis stolonifera* and *Ranunculus repens*

    MG8 *Cynosurus   cristatus-Caltha   palustris* grassland

C. palustris and *Carex panicea* generally very scarce   11

11   *Festuca arundinacea* constant and usually dominant in a coarse sward with *F. rubra*, *Agrostis stolonifera* and *Potentilla anserina*

    MG12 *Festuca arundinacea* grassland
    Potentillo-Festucetum arundinaceae   Nordhagen
    1940   12

F. rubra, *A. stolonifera* and *P. anserina* can be frequent but *F. arundinacea* generally infrequent and not abundant   13

12   One or more of *Oenanthe lachenalii*, *Juncus gerardi*, *Glaux maritima*, *Carex otrubae* and *Sonchus arvensis* present in usually small quantities

    MG12 *Potentillo-Festucetum*
    Oenanthe lachenalii sub-community

On the upper reaches of salt-marshes, this kind of vegetation can form mosaics with and grade into *Juncus maritimus* vegetation, but *J. maritimus*, generally only occasional here, is constant and abundant there.

Above-listed species at most occasional but *Lolium perenne* and/or *Holcus lanatus* frequent and abundant

    MG12 *Potentillo-Festucetum*
    Lolium perenne-Holcus lanatus sub-community

13   Species-poor, sometimes quite open, swards dominated by mixtures of *F. rubra*, *A. stolonifera* and *P. anserina*

    MG11 *Festuca rubra-Agrostis stolonifera-Potentilla anserina* grassland   14

F. rubra and *A. stolonifera* can be frequent and abundant but not with *P. anserina*   15

14   One or more of *Honkenya peploides*, *Carex arenaria*, *Sagina procumbens* and *Silene dioica* occasional to frequent

    MG11 *Festuca rubra-Agrostis stolonifera-Potentilla anserina* grassland
    Honkenya peploides sub-community

On disturbed strandlines, this vegetation grades to more open halophyte assemblages.

One or more of *Atriplex prostrata*, *Matricaria maritima*, *Polygonum aviculare* and *Oenenthe lachenalii* occasional

    MG11 *Festuca rubra-Agrostis stolonifera-Potentilla anserina* grassland
    Atriplex prostrata sub-community

Towards the upper reaches of salt-marshes, this kind of vegetation can grade to either the *Juncus maritimus* community or the *Juncetum gerardi*, but *J. maritimus* and/or *J. gerardi* become very frequent and often abundant there with such species as *Triglochin maritima*, *Glaux maritima* and *Plantago maritima*.

L. perenne and/or *H. lanatus* frequent and abundant in the general absence of the above sets of species

    MG11 *Festuca rubra-Agrostis stolonifera-Potentilla anserina* grassland
    Lolium perenne sub-community

15   Species-poor swards, sometimes rather open, dominated by mixtures of *Agrostis stolonifera* and *Alopecurus geniculatus*, often with *Glyceria fluitans*, *G. plicata* or *G. declinata*

    MG13 *Agrostis stolonifera-Alopecurus geniculatus* grassland

Combinations of the above species not a prominent element in the vegetation   16

16   Generally species-rich swards dominated by mixtures of bulky perennial dicotyledons such as *Sanguisorba officinalis*, *Geranium sylvaticum*, *Filipendula ulmaria*, *Alchemilla glabra* and *Lathyrus pratensis*   17

Sward can be species-rich but above-listed species are occasional at most   19

Improved stands of meadow vegetation which

have lost one or more of their characteristic dominants may be hard to place here.

---

17   *G. sylvaticum, A. glabra, A. xanthochlora* and *Conopodium majus* frequent with often abundant *Anthoxanthum odoratum* but usually little *Alopecurus pratensis*

> **MG3** *Anthoxanthum odoratum-Geranium sylvaticum* grassland   18

G. sylvaticum, A. glabra, A. xanthochlora and *Conopodium majus* usually absent and *Anthoxanthum* scarce, but *Alopecurus pratensis, Filipendula ulmaria* and *Lathyrus pratensis* frequent

> **MG4** *Alopecurus pratensis-Sanguisorba officinalis* grassland

---

On alluvial soils in valleys in northern England, this can be a difficult separation to make.

---

18   *Bromus hordeaceus* ssp. *hordeaceus, Lolium perenne* and *Phleum pratense* ssp. *pratense* frequent

> **MG3** *Anthoxanthum odoratum-Geranium sylvaticum* grassland
> *Bromus hordeaceus* ssp. *hordeaceus* sub-community

Above species usually no more than infrequent but some of the following present: *Briza media, Lotus corniculatus, Luzula campestris, Hypochoeris radicata, Centaurea nigra, Trifolium pratense*

> **MG3** *Anthoxanthum odoratum-Geranium sylvaticum* grassland
> *Briza media* sub-community

There is a continuous gradation between these vegetation types and from the *Bromus* sub-community to the *Lolio-Cynosuretum* according to the degree and kind of improvement to which northern meadows have been subjected.

---

19   Generally species-rich swards with an abundance of herbaceous dicotyledons including *Lotus corniculatus* and some of *Leontodon hispidus, Ranunculus bulbosus, Leucanthemum vulgare, Primula veris, Rumex acetosa, Trifolium pratense* and with frequent and sometimes abundant *Anthoxanthum odoratum* and *Agrostis capillaris*

> **MG5** *Cynosurus cristatus-Centaurea nigra* grassland

Centaureo-Cynosuretum cristati   Br.-Bl. & Tx 1952   20

Generally species-poor, grass-dominated swards with constant and usually abundant *Lolium perenne* and few of the above species   21

There is a complete gradation between rich, unimproved stands of the *Centaureo-Cynosuretum* and the very species-poor swards of the *Lolium* leys which have been ploughed and re-seeded, fertilised and drained. The above list of dicotyledons is a generally satisfactory means of separating the *Centaureo-Cynosuretum* from richer stands of the *Lolio-Cynosuretum* but, in many cases, the best that can be hoped for is to place a stand at particular points along a line of continuous variation.

---

20   *Galium verum* and *Trisetum flavescens* frequent and sometimes abundant with occasional records for *Sanguisorba minor, Carex flacca* and *Koeleria macrantha*

> **MG5** *Centaureo-Cynosuretum*
> *Galium verum* sub-community

On brown calcareous soils over lime-rich parent materials, it may be difficult to partition samples between this vegetation type and more mesotrophic swards of the Mesobromion. In general, the abundance of *F. rubra, Agrostis capillaris* and *Anthoxanthum* in the *Centaureo-Cynosuretum* will serve as a distinguishing feature, although even this is of less value on northern limestones in a wetter, cooler climate.

Danthonia decumbens, Luzula campestris, Succisa pratensis and *Potentilla erecta* present without *Galium verum*

> **MG5** *Centaureo-Cynosuretum*
> *Danthonia decumbens* sub-community

On brown earths with a tendency to eluviation, as over pervious lime-poor strata in regions of higher rainfall, it may sometimes be difficult to partition samples between this vegetation type and more mesotrophic swards of the Nardo-Galion. In general, the abundance here of *Holcus lanatus, Trifolium repens* and *T. pratense* will serve as a distinguishing feature, although even these species are of less value on more ill-drained soils at lower altitudes.

Lolium perenne and *Lathyrus pratensis* frequent in the general absence of the above combinations of species

> **MG5** *Centaureo-Cynosuretum*
> *Lathyrus pratensis* sub-community

21   *L. perenne* abundant with varying amounts of *Cynosurus cristatus*

> **MG6** *Lolium   perenne-Cynosurus   cristatus* grassland
> *Lolio-Cynosuretum cristati*   (Br.-Bl. & De Leeuw 1936) R.Tx 1937   22

L. perenne abundant in the absence of *Cynosurus*

> **MG7** *Lolium perenne* leys and related grasslands
>   23

It is sometimes difficult to effect this separation where *Cynosurus* is invading longer-established sown swards

---

22   *Trisetum flavescens* and *Phleum pratense* ssp. *bertolonii* frequent without *Anthoxanthum*

> **MG6** *Lolio-Cynosuretum*
> *Trisetum flavescens* sub-community

Anthoxanthum odoratum and *Rumex acetosa* frequent with *Trisetum* and *P. pratense* ssp. *bertolonii* scarce

> **MG6** *Lolio-Cynosuretum*
> *Anthoxanthum odoratum* sub-community

As with the *Centaureo-Cynosuretum*, it is sometimes difficult to partition samples between these two vegetation types and grasslands of the Meso-bromion or Nardo-Galion respectively on soils which are somewhat more base-rich or acidic.

Above combinations of species lacking

> **MG6** *Lolio-Cynosuretum*
> Typical sub-community

23   *Poa trivialis* and *Phleum pratense* ssp. *pratense* frequent and sometimes co-dominant with *L. perenne*

> **MG7** *Lolium perenne* leys and related grasslands
> *Lolium perenne-Poa trivialis* leys

Alopecurus pratensis and *Festuca pratensis* frequent and sometimes co-dominant with *L. perenne*

> **MG7** *Lolium perenne* leys and related grasslands
> *Lolium perenne-Alopecurus pratensis-Festuca pratensis* grassland

Alopecurus pratensis frequent and sometimes co-dominant with *L. perenne* without *F. pratensis*

> **MG7** *Lolium perenne* leys and related grasslands
> *Lolium perenne-Alopecurus pratensis* grassland

L. perenne and *Trifolium repens* co-dominant with infrequent *P. trivialis*, *A. pratensis* and *F. pratensis*

> **MG7** *Lolium perenne* leys and related grasslands
> *Lolium perenne-Trifolium repens* leys

Plantago lanceolata and *Poa pratensis* frequent and the latter sometimes co-dominant with *L. perenne*

> **MG7** *Lolium perenne* leys and related grasslands
> *Poo-Lolietum perennis* De Vries & Westhoff *apud* Bakker 1965

Plantago lanceolata frequent with *Poa pratensis* uncommon

> **MG7** *Lolium perenne* leys and related grasslands
> *Lolio-Plantaginetum* (Link 1921) Beger 1930 *emend.* Sissingh 1969

COMMUNITY DESCRIPTIONS

MG1
Arrhenatherum elatius grassland
Arrhenatheretum elatioris   Br.-Bl. 1919

Synonymy
Arrhenatheretum elatioris sub-atlanticum   R.Tx. (1937) 1955; *Arrhenatherum elatius* stands   Pfitzenmeyer 1962 *p.p.*; *Pastinaco-Arrhenatheretum* (Knapp 1954) Passarge 1964; *Centaureo-Arrhenatheretum* O'Sullivan 1965; Slope grassland Packham *et al.* 1966, 1967; *Arrhenatheretum elatioris inops* Neijenhuijs & Westhoff 1968; Tall herb community Lloyd 1968; Species-rich *Arrhenatherum* grassland Lloyd 1972 *p.p.*; *Arrhenatherum elatius/Festuca rubra/Helictotrichon pubescens* grassland Wells *et al.* 1976.

Constant species
Arrhenatherum elatius, Dactylis glomerata.

Rare species
Silene nutans.

Physiognomy
The *Arrhenatheretum elatioris* is a community in which coarse-leaved tussock grasses, notably *Arrhenatherum elatius* with usually smaller amounts of *Dactylis glomerata* and *Holcus lanatus*, are always conspicuous and generally dominant. Large umbellifers are frequent throughout and sometimes abundant and the sequential flowering of first, *Anthriscus sylvestris* and later, *Heracleum sphondylium* and *Chaerophyllum temulentum*, can give stands a distinctive creamy-white haze throughout most of late spring and summer. Apart from *Cirsium arvense, Centaurea nigra* and *Urtica dioica*, other tall herbs are generally infrequent, though a variety of species may attain dominance locally (see variants below).

Beneath these taller species there is usually a layer of fine-leaved grasses, most frequently *Festuca rubra, Poa pratensis, P. trivialis, Lolium perenne* and *Elymus repens*, and small dicotyledons, notably *Trifolium pratense, T. repens, Achillea millefolium, Taraxacum officinale* agg., *Plantago lanceolata, Lotus corniculatus* and *Rumex acetosa*. At the height of the growing season the vegetation often becomes choked by sprawling legumes such as *Lathyrus pratensis, Vicia sativa* ssp. *nigra* (= *V. angustifolia* L. = *V. sativa* ssp. *angustifolia* (L.) Gaud.), *V. cracca* and *V. sepium* and trailing stems of *Galium aparine* and *Rubus fruticosus* agg. Certain species, for example *Taraxacum officinale* agg. and occasionals characteristic of woodlands like *Ranunculus ficaria* and *Mercurialis perennis*, complete most of their growth and flower before this substantial bulk of vegetation has developed.

There is often a large amount of standing dead material in winter in the *Arrhenatheretum* and some ground litter generally persists throughout the year. Bryophytes are usually confined to thin wefts of pleurocarpous mosses over this decaying vegetation and on sparse patches of bare soil. *Brachythecium rutabulum* is the most frequent species throughout with some *Eurhynchium praelongum, Pseudoscleropodium purum* and *Rhytidiadelphus squarrosus*.

Sub-communities
Five sub-communities are recognised. In the first two in particular, occasional species may attain local abundance or even dominance and a number of variants have been characterised on this basis. Further work is needed to establish the validity of these vegetation types and their hierarchical relationships to the sub-communities.

Festuca rubra **sub-community.** Here the vegetation is generally grass-dominated with abundant *Arrhenatherum, D. glomerata* and *F. rubra*. Stands on recently-established sites, such as newly-created road verges, may be especially species-poor with up to 90% *F. rubra* at first and only occasional tussocks of *Arrhenatherum* and scattered dicotyledons; *Vicia sativa* ssp. *nigra, Cirsium arvense, Taraxacum officinale* agg. and *Plantago lanceolata* tend to be characteristic of such early stages. As *Arrhenatherum* and *D. glomerata* seed in, they overcome the dominance of the *F. rubra* and a wider range of dicotyledons is established: *Achillea millefolium, Ceras-*

tium fontanum, Lathyrus pratensis, Rumex acetosa, R. obtusifolius and the large umbellifers. Species-richness increases with age (provided the vegetation is cut: see below) but, even with its full complement of species, this sub-community is one of the poorest of the British Arrhenathereta.

***Centaurea scabiosa* variant.** The presence of *C. scabiosa* tends to be associated with the replacement of *F. rubra* by *F. ovina* as the major grass of the understorey and an increase in records for *Hypericum perforatum, Galium verum* and *Agrimonia eupatoria*.

***Geranium pratense* variant.** *G. pratense, Vicia sepium, Lathyrus pratensis, Centaurea nigra* and *Cruciata laevipes* sometimes occur together in abundance in a particularly lush and colourful vegetation.

***Bromus sterilis* variant.** *B. sterilis* may be abundant in stands of this sub-community and it is often accompanied by *Alliaria petiolata* and a sparse ground cover of *Hedera helix* and *Glechoma hederacea*.

***Myrrhis odorata* variant.** *M. odorata* sometimes augments or replaces the large umbellifers of this sub-community to form a tall lush canopy with a somewhat sparse understorey.

***Epilobium angustifolium* variant.** A dense tall canopy of *E. angustifolium* may replace the tussock grasses as the dominant.

***Urtica dioica* sub-community:** Inner Wayside Bates 1937. In general floristics, this sub-community resembles the latter but here *U. dioica* is a constant and usually conspicuous component of the upper layer of the vegetation. The large umbellifers, too, are especially frequent and may account for up to 90% of the cover. *Galium aparine* is often present as a dense tangle.

Under the denser shade of this canopy, there is a generally more open cover of smaller grasses and dicotyledons than is usual in the *F. rubra* sub-community but a more extensive bryophyte layer. *F. rubra* itself and *Poa pratensis* and *P. trivialis* are all markedly less frequent. The leaf litter of *U. dioica* decays rapidly and, after the growing season, open patches in the vegetation expose conspicuous pads of pleurocarpous mosses, especially *Brachythecium rutabulum* on the less rapidly decaying fibrous stems.

***Papaver rhoeas* variant.** In the more open vegetation of this variant, patches of bare soil are colonised by annual ruderals such as *Papaver rhoeas, P. dubium, Capsella bursa-pastoris* and *Sonchus asper. Lolium perenne* attains constancy and it may be abundant.

***Artemisia vulgaris* variant.** Here, *A. vulgaris* is constant and it may form a bushy canopy. *Cirsium arvense, Elymus repens* and *Lolium perenne* are also more frequent.

***Epilobium hirsutum* variant.** Patches of *E. hirsutum* make a distinctive contribution to the canopy in this variant though the species does not attain dominance.

***Filipendula ulmaria* sub-community.** *Arrhenatherum, U. dioica* and *H. sphondylium* all remain frequent in this sub-community but here *F. ulmaria* is a constant and sometimes abundant component of the canopy. *Holcus lanatus, Dactylis glomerata* and *Poa trivialis* are the major grasses beneath although *Alopecurus pratensis, Deschampsia cespitosa* and *Festuca arundinacea* are occasionally abundant. *Ranunculus repens* is slightly preferential for this sub-community and there are occasional records for *Symphytum officinale* and *Pulicaria dysenterica. Rubus fruticosus* agg. is also frequent and, where stands are adjacent to hedgerows (a common occurrence), woodland species such as *Mercurialis perennis, Silene dioica* and *Stellaria holostea* may occur. In general, the vegetation is slightly richer than in previous sub-communities.

***Pastinaca sativa* sub-community:** *Arrhenatheretum elatioris* Sub-Association Group B Br.-Bl. 1919; Coulter's Dean Wasteland 1920 Tansley & Adamson 1925; Coulter's Dean Wasteland 1936B Hope-Simpson 1940; *Arrhenatheretum elatioris agrimonietosum eupatoriae* LeBrun 1949; *Pastinaco-Arrhenatheretum* (Knapp 1954) Passarge 1964; *Arrhenatheretum elatioris picridetosum* Neijenhuijs & Westhoff 1968; *Arrhenatherum elatius/Festuca rubra/Helictotrichon pubescens* grassland Wells *et al.* 1976 *p.p.* In this more distinct sub-community, the upper layer is usually dominated by *Arrhenatherum, D. glomerata, H. lanatus* and *Pastinaca sativa* which here tends to replace *H. sphondylium* as the major umbellifer and which may lend stands a very striking appearance in the flowering season. Beneath, *F. ovina* generally replaces *F. rubra* and *Agrostis capillaris* appears as a frequent and sometimes abundant component. *Plantago lanceolata, Achillea millefolium, Galium verum* and *Senecio jacobaea* are the most frequent associates and there are occasional records for species characteristic of the coarse Mesobromion swards, e.g. *Sanguisorba minor, Knautia arvensis, Agrimonia eupatoria, Bromus erectus, Clinopodium vulgare, Centaurea scabiosa* and *Origanum vulgare*. Patches dominated by *U. dioica* with *Cirsium arvense* and *Galium aparine* may approach the *Urtica* sub-community in composition. Again, there is often a substantial litter accumulation below and bryophytes are sparse.

***Centaurea nigra* sub-community:** Broadbalk Wilderness 1913 Brenchley & Adam 1915; Rothamsted Plots 7 & 8 Brenchley rev. Warington 1958; *Centaureo-Arrhenatheretum* O'Sullivan 1965; Tall herb community Lloyd 1968; Species-rich *Arrhenatherum* grassland B$_c$ Lloyd 1972; *Arrhenatherum elatius/Festuca rubra/Helictotrichon pubescens* grassland Wells *et al.* 1976 *p.p.* This sub-community is richer and more varied than other British Arrhenathereta. *Arrhenatherum* itself is not so consistently dominant here although, with *D. glomerata* and *H. lanatus*, it remains constant in the upper layer of the vegetation. *Trisetum flavescens* and *Avenula pubescens* occur less frequently but are preferential for this sub-community and are occasionally abundant. Among the taller dicotyledons, *H. sphondylium* is joined by *Centaurea nigra* and, less frequently, by *Leucanthemum vulgare* and (shared with the *Pastinaca* sub-community) *Senecio jacobaea*, *Knautia arvensis* and *Agrimonia eupatoria*. Beneath is a rich and extensive cover of grasses (the more usual *F. rubra*, *Poa trivialis* and *P. pratensis* with *Anthoxanthum odoratum* and *Agrostis capillaris*) and small dicotyledons (notably *Plantago lanceolata*, *Lotus corniculatus*, *Achillea millefolium*, *Galium verum*, *Veronica chamaedrys*, *Lathyrus pratensis* and *Luzula campestris*). *Primula veris*, *Malva moschata* and *Hypericum perforatum* are distinctive at low frequency. There is generally a bryophyte layer and *Pseudoscleropodium purum* and *Rhytidiadelphus squarrosus* are preferentially frequent here.

***Pimpinella saxifraga* variant:** *Centaureo-Arrhenatheretum* O'Sullivan 1965 *sensu* Shimwell 1968a; Habitat Study 42 Elkington 1969. This highly distinctive vegetation is best considered as a variant of the *Centaurea* sub-community. The taller species *Arrhenatherum*, *D. glomerata*, *H. lanatus*, *C. nigra* and *H. sphondylium* remain constant and *Avenula pubescens*, *Knautia arvensis* and *Agrimonia eupatoria* increase in frequency. *Pimpinella saxifraga*, generally infrequent in the sub-community as a whole, is a particularly distinctive preferential. Many of the smaller grasses and dicotyledons are, however, reduced in frequency, notably *Anthoxanthum odoratum*, *Agrostis capillaris*, *Trifolium repens* and *Luzula campestris* while *Briza media*, *Pimpinella major* and *Brachypodium sylvaticum* show an increased occurrence. *Silene nutans* is a national rarity which occurs in this variant. Bryophytes are more varied here with records for *Calliergon cuspidatum*, *Hylocomium splendens* and *Lophocolea bidentata s.l.* In general, the vegetation of this variant is more open and heterogeneous and is often disposed irregularly over steep and rocky slopes.

Habitat

The *Arrhenatheretum* is, above all, an ungrazed grassland. It is characteristic of circumneutral soils throughout the British lowlands and occurs on road verges, railway embankments and churchyards and in neglected agricultural and industrial habitats such as badly-managed pastures and meadows, building sites, disused quarries and rubbish dumps.

The key factor in the development of the *Arrhenatheretum* on otherwise suitable sites is the absence or irregularity of grazing. However, without mowing, ungrazed stands of the community are rapidly invaded by shrubs and, in such cases, the *Arrhenatheretum* is a temporary

stage in the succession to scrub and woodland. The community is maintained, in the absence of grazing, by regular but infrequent cutting. Road verges, which represent one of the main reserves of permanent stands, are generally cut annually in early summer by mechanical mowers or, still occasionally, by the traditional scythe and sickle. On bends of roads and along major routes there may be two (or even more) cuts each year. Provided cut material is removed, the species-richness and diversity of the vegetation is maintained under such a regime. Early cutting or the use of herbicides or growth retardants may, however, drastically reduce the dicotyledonous component of the community and very frequent cutting or the resumption of grazing can convert the vegetation into something resembling a Cynosurion sward.

The *Arrhenatheretum* can become established and be maintained on a variety of circumneutral soils although it thrives best on well-structured and freely-draining loams. The most important edaphic variables influencing the floristic variation within the community are probably pH, the amount of the nutrients N, P and K and drainage. The *Festuca rubra*, *Urtica* and *Filipendula* sub-communities are generally associated with brown earths of neutral pH and, in artificial habitats, these may be shallow and somewhat stony or compacted and clayey. The *Centaurea scabiosa* and *Geranium* variants of the *Festuca rubra* sub-community and, more especially, the *Pastinaca* sub-community are more characteristic of brown calcareous earths of higher pH developed over calcareous bedrocks or superficials. The *Centaurea nigra* sub-community seems to be associated with more mesotrophic soils. These may be calcareous and of high pH but the sub-community as a whole occurs over a wide range of bedrock types. There may be some developmental relationship between the typical variant of this sub-community and the long-unploughed pastures and meadows of the *Centaureo-Cynosuretum*. In the Park Grass Plot experiments at Rothamsted, application of mixed mineral (P, K, Na, Mg) manure to what is essentially a *Centaureo-Cynosuretum* sward (Plot 3) has produced an *Arrhenatheretum* very similar to the *Centaurea* sub-community (Plot 7: see Brenchley rev. Warington 1958; also Pigott 1982). Fertilising of unimproved meadows with subsequent lack of grazing may thus account for some natural stands of this sub-community. Interestingly, the typical variant of the *Centaurea* sub-community is also frequent in churchyards where there is a distinctive release of nutrients into often deep profiles. The most striking of all the British Arrhenathereta, the *Pimpinella* variant of the *Centaurea* sub-community, is very closely related to moist calcareous soils on steep rocky slopes of Carboniferous and Magnesian limestones.

Verges and banks are frequently enriched with

mineral N, P and K in a somewhat raw fashion by run-
off of ground water from fields treated with artificial
fertilisers and the *Urtica* sub-community is especially
frequent in areas of intensive arable agriculture. It also
occurs as a narrow belt beneath hedges, fences and walls
where perching or roosting birds provide a more natural
input of these minerals in their droppings (Figure 8).

The soil moisture regime in the habitats of the *Arrhenatheretum* is most obviously influenced by the slope of
adjacent land and the provision of banks and ditches as
on verges and field margins. Soils are generally freely
draining but moister soils typically carry the *Filipendula*
sub-community (especially in roadside ditches or

Figure 8. Typical pattern of grasslands in and around
a run-down lowland pasture.

MG1a   *Arrhenatheretum, Festuca* sub-community on
verge bank
MG1b   *Arrhenatheretum, Urtica* sub-community on
disturbed verge
MG1c   *Arrhenatheretum, Filipendula* sub-community
in verge ditch
MG6a   *Lolio-Cynosuretum*, Typical sub-community
on frequently mown verge edge
MG6b   *Lolio-Cynosuretum, Anthoxanthum* sub-
community with avoidance mosaic
MG7e   *Lolio-Plantaginetum* towards gateway
MG7f   *Poo-Lolietum* in gateway
MG10a   *Holco-Juncetum*, Typical sub-community in
ill-drained field hollow
W24b   *Rubus-Holcus* underscrub, *Arrhenatherum-
Heracleum* sub-community invading around field
margin

choked streams) or the *Epilobium hirsutum* variant of
the *Urtica* sub-community (particularly on fen margins
and on the banks of droves or rhynes across
flood-meadows).

Sporadic interference or informal treatment is generally responsible for the local prominence of the variants
of the *Festuca* and *Urtica* sub-communities. Physical
disturbance, sometimes with mineral or organic enrichment, is common in the habitats of the *Arrhenatheretum*
and may lead to the development of a ruderal element as
in the *Papaver* and *Artemisia* variants of the *Urtica* sub-
community. The former is frequent around road works
and the gateways of arable fields and the latter occurs
commonly on building sites and on the central reserva-
tions of motorways. The *Myrrhis* variant of the *Festuca
rubra* sub-community is frequently associated with old
habitations and this may reflect the once popular use of
Myrrhis as a pot-herb (Tutin 1980).

The often substantial amounts of litter and, in winter,
of standing dead material render the *Arrhenatheretum*
highly susceptible to fire. If burning occurs outside or
early in the growing season, when the bulk of the
resources of the vegetation are under ground, even the
short-term effects may be slight and repeated burning of
this kind would probably not drastically alter the
community (Lloyd 1968). Frequent burning, especially
during the growing season, may permit colonisation by
Epilobium angustifolium and stands of the variant of this
species may be temporarily prominent on fire sites.
Railway embankments (though less so now than in the
days of steam locomotion) and amenity verges often
carry this vegetation.

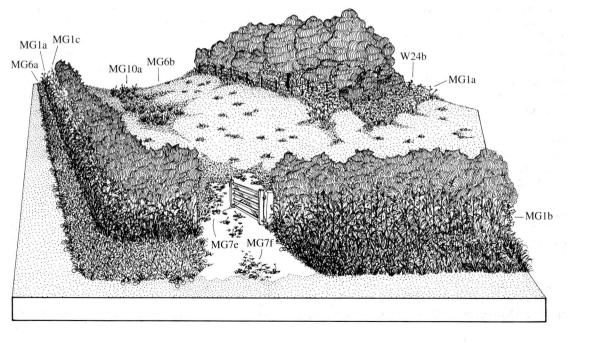

The occurrence of and variation within the *Arrhenatheretum* often reflect patterns of past treatment which have continuing effects. The sites of many permanent stands are artificial to some degree and the particular conjunction of verge, bank, ditch and boundary hedge, fence or wall can exert an influence through the modification of soil drainage or by the close juxtaposition of grassland, swamp and hedgerow vegetation. The *Bromus* variant of the *Festuca rubra* sub-community, for example, is very characteristic of the inner fringe of verges which directly front hedgerows and many stands of the *Arrhenatheretum* have a minor component of woodland field-layer species which reflect long association with adjacent hedges or wood margins.

Zonation and succession

Some of the most frequent zonations involving *Arrhenatheretum* stands are the often sharp juxtapositions of the various sub-communities and variants with one another or with other vegetation types along road margins. Such patterns are very varied and sometimes reflect long, local traditions of boundary construction and maintenance.

Where stands are not maintained by cutting, zonations are generally a reflection of stages in succession to mixed deciduous woodland. This succession is most often mediated by grazing. *Arrhenatherum* is a ready invader of soils and calcareous talus exposed by landslip, construction work or the abandonment of ploughed land. It can also spread into and increase in certain other grassland types where grazing is absent or withdrawn or where there is injudicious pastoral treatment. Resumption of grazing can once more reduce its cover. Hope-Simpson (1940*b*) noted that, within four years of the removal of rabbits from a primitive *Festuca ovina-Avenula pratensis* sward at War Down, Hampshire, *Arrhenatherum* previously no more than occasional, had become very abundant. Within seven years of the resumption of grazing by sheep and rabbits on an *Arrhenatheretum* at nearby Coulter's Dean Wasteland, the vegetation had been converted to a *Festuca ovina-Avenula pratensis* grassland with no *Arrhenatherum* (Hope-Simpson 1940*b*). Thomas (1960, 1963) showed that, after the virtual elimination of rabbits by myxomatosis, the coarser grasses characteristic of the *Arrhenatheretum* were among those species which spread in grasslands and heaths on chalk. These had previously been present in small amounts and had provided a valuable winter-green bite for the rabbits. They were reduced in cover when rabbit numbers rose again. *Arrhenatherum* can also spread into seeded verges and some of the floristic variation within the *Festuca rubra* sub-community (see above) is attributable to this advance. Increase of *Arrhenatherum* in *Centaureo-Cynosuretum* grasslands seems to follow careless grazing, perhaps by the use of such grasslands as horse-pad-docks, and the *Centaurea nigra* sub-community of the *Arrhenatheretum* may originate partly in this way.

Ploughing and subsequent abandonment of land after unsuccessful arable cultivation seems to have been a fairly widespread factor in the development of some of the more extensive stands of the community, especially on deeper soils over chalk. The *Arrhenatheretum* at Coulter's Dean Wasteland had developed on land ploughed and abandoned some 30 years before (Tansley & Adamson 1925). Brenchley & Adam (1915) had reported a similar occurrence of the community on the abandoned and ungrazed Broadbalk Wilderness at Rothamsted. Several subsequent studies (Lloyd & Pigott 1967 in the Chilterns, Grubb *et al.* 1969 at Lullington Heath, Sussex and Wells *et al.* 1976 at Porton Down on the Hampshire/Wiltshire border) have confirmed the view that the *Arrhenatheretum* is one of the communities dominated by coarse grasses which can, in the absence of grazing, develop naturally on the chalk without the prior development of a *Festuca ovina-Avenula pratensis* sward.

The Arrhenathereta described in these studies are of the kind classified here as the *Pastinaca* and *Centaurea nigra* sub-communities but the lists provided are generally insufficiently precise to allocate stands unequivocally to one or the other. The available data suggest that the *Pastinaca* sub-community develops on the more calcareous and oligotrophic soils where these have been exposed by ploughing or where there has been a relaxation of grazing of calcicolous grassland. This sub-community is most common on the chalk of the south and east where *P. sativa* has its centre of distribution in Britain and where there has been a history of unsuccessful arable cultivation of marginal chalkland and pastoral neglect (see, for example, Smith 1980). The *Centaurea* sub-community is more widespread and seems to develop on more mesotrophic soils which are generally less calcareous. It is perhaps the more natural sub-community on deeper soils over chalk, such as those derived from superficial deposits, but it may also develop by the relaxation of grazing of mesotrophic grasslands throughout the lowlands, especially where this has been combined with application of fertilisers.

Similar differences in the trophic state of soils may be an important factor in controlling whether it is *Arrhenatheretum* (of any kind) or grassland dominated by other coarse species such as *Avenula pubescens*, *Festuca rubra*, *Bromus erectus* and *Brachypodium pinnatum* which develops in any given situation. Wells *et al.* (1976) showed that, at Porton, the *Arrhenatheretum* occurred on land ploughed relatively recently where, they suggested, there was still a comparatively small accumulation of organic matter and mineral nutrients. Other coarse grassland types had developed on soils ploughed less recently which were consequently richer.

If any kind of *Arrhenatheretum* remains ungrazed and

uncut it eventually develops a pronounced tussock physiognomy and litter accumulation depresses species-richness. Shrubs invade and the frequently close hedge-rows and field boundaries provide a seed source. Fore-most among the invaders is *Crataegus monogyna* with *Prunus spinosa*, *Sambucus nigra* (especially where there has been disturbance and soil eutrophication), *Rubus fruticosus* agg., *Rosa* spp. and, on the more calcareous soils, *Cornus sanguinea*, *Viburnum lantana* and *Ligustrum vulgare*. *Clematis vitabla* may form a tangle among the developing scrub. Eventually, succession progresses to some form of mixed deciduous woodland but even very coarse and scrubby *Arrhenatheretum* can be converted to a low sward by grazing. At Aston Rowant NNR, Oxfordshire, four years of sheep-grazing in winter and spring drastically reduced the height of a scrubby *Arrhenatheretum* and converted *A. elatius* itself from large tussocks to small scattered shoots in a ground of *Festuca ovina* (Wells 1969: compare especially Figures 1 and 2).

Distribution

The *Arrhenatheretum* is virtually ubiquitous throughout the lowlands of Britain, although the bulk of Scotland has been under-sampled. Many of the stands are frag-mentary and the distributions of the sub-communities generally reflect differences in verge and grassland man-agement or neglect which may vary greatly from one locality to the next. The *Festuca rubra* sub-community is the most widely distributed; the *Urtica* sub-community is especially prominent in areas of intensive arable agriculture and the *Filipendula* sub-community occurs wherever local drainage requirements have necessitated the provision of roadside and field ditches. The *Pastinaca* and *Centaurea nigra* sub-communities are more restricted by soil conditions, the former to more calcar-eous soils in the south and east, the typical variant of the latter to more mesotrophic soils throughout the country. The *Pimpinella* variant of the *Centaurea* sub-community has been recorded only from the Carbon-iferous and Magnesian limestones of the Mendips, Derbyshire, Yorkshire and Durham. The occurrence of some of the other variants is also influenced by geology and soil: the *Centaurea scabiosa* and *Geranium pratense* variants are particularly well developed over limestones with good examples of the former on Salisbury Plain

verges and of the latter in the Yorkshire Dales.

With an increased intensity of land management, the *Arrhenatheretum* has become restricted in some areas: the richest stands of the *Centaurea* sub-community, for example, are now confined to churchyards. However, road verges continue to provide a valuable resource and the gradual maturation of motorway verges is increasing the potential extent of the community.

Affinities

There is no existing comprehensive description of the *Arrhenatheretum elatioris* in the British literature, although the community is obviously very similar to the *Arrhenatheretum* of western Europe, where this vege-tation is also widespread (e.g. Tüxen 1955). In general, however, the core of the Continental *Arrhenatheretum* corresponds to the vegetation described here as the *Centaurea nigra* sub-community. It is for this reason that we have not separated off this vegetation as a distinct community (cf. O'Sullivan 1965, Shimwell 1968a). We have also retained the floristically less dis-tinctive *Pastinaca sativa* type of *Arrhenatheretum* within the ambit of a single community, although its counter-part in Europe has sometimes been distinguished as a separate association (e.g. Knapp 1954, Passarge 1964; cf. LeBrun *et al.* 1949). In a Continental perspective, the bulk of the British Arrhenathereta would be regarded as impoverished forms of a rich grassland type which is frequently managed in parts of western Europe as hay-meadow. In Britain, the continuing management of the widely-distributed roadside verge habitat maintains these poorer forms in abundance.

The *Arrhenatheretum* has clear floristic affinities with other major vegetation types. It grades, through an increase in ruderal elements, to the communities of the Secalinetea and Artemisietea and, through species such as *Filipendula ulmaria* and *Epilobium hirsutum*, to Fili-pendulion mires. The presence of Mesobromion or Cynosurion species within the *Pastinaca* and *Centaurea* sub-communities represents floristic transitions to calci-colous grasslands and mesotrophic pastures. An herba-ceous woodland element is frequently present in Arrhe-nathereta and there is a clear affinity with the damp mixed deciduous woodlands of the Alno-Ulmion through the *Filipendula* sub-community and the closely-related *Filipendulo-Arrhenatheretum*.

Floristic table MG1

	a	b	c	d	e	1
Arrhenatherum elatius	V (2–9)	V (1–9)	V (1–8)	V (2–9)	V (2–8)	V (1–9)
Dactylis glomerata	IV (2–8)	IV (1–8)	IV (1–7)	V (3–5)	V (1–7)	IV (1–8)
Holcus lanatus	II (2–8)	II (2–6)	IV (2–7)	IV (3–7)	IV (1–8)	III (1–8)
Heracleum sphondylium	III (1–6)	IV (2–7)	III (1–5)	II (3–4)	V (1–5)	III (1–7)
Anthriscus sylvestris	II (1–7)	II (2–7)	I (2–8)	I (3)	I (1–5)	II (1–8)
Agrostis stolonifera	II (2–6)	I (1–9)	I (2–5)	I (3–4)	I (3–8)	I (1–9)
Lamium album	I (3–4)	I (2–4)	I (1–2)			I (1–4)
Papaver rhoeas	I (2)	I (3–4)		I (4)		I (2–4)
Papaver dubium	I (1)	I (2)				I (1–2)
Capsella bursa-pastoris	I (3–4)	I (2–3)				I (2–4)
Sonchus asper	I (3)	I (2–4)				I (2–4)
Aegopodium podagraria	I (5)	I (4)				I (4–5)
Urtica dioica	I (1–3)	V (1–7)	III (1–6)	II (2–5)	I (2)	III (1–7)
Galium aparine	I (1–5)	II (1–5)	III (2–5)	I (4)		II (1–5)
Epilobium hirsutum		I (2–6)	I (1–3)			I (1–6)
Artemisia vulgaris		I (3–7)				I (3–7)
Filipendula ulmaria		I (1–2)	V (1–6)		I (6)	I (1–6)
Pastinaca sativa	I (2)			V (2–5)		II (2–5)
Achillea millefolium	II (1–5)	II (2–6)	I (3)	V (2–5)	III (1–5)	III (1–6)
Plantago lanceolata	III (1–5)	II (2–4)	II (2–4)	IV (2–4)	IV (1–4)	III (1–5)
Galium verum	I (2–7)	I (3–4)	I (3)	III (2–4)	III (1–5)	II (1–7)
Agrostis capillaris	I (2–8)	I (2–4)	I (3)	III (3–7)	II (1–7)	I (1–8)
Festuca ovina	I (2–5)	I (3–5)		III (3–6)	I (1–5)	I (1–6)
Senecio jacobaea	I (2–3)			III (1–4)	II (1–4)	I (1–4)
Knautia arvensis	I (3–4)	I (3–5)		II (2–3)	II (1–4)	I (1–5)
Sanguisorba minor	I (1–4)	I (3)		II (1–4)	I (5)	I (1–5)
Prunella vulgaris	I (5)		I (1)	II (1–2)	I (2–4)	I (1–5)
Agrimonia eupatoria	I (1–3)			II (4)	II (1–5)	I (1–5)
Clinopodium vulgare	I (2–3)			II (3)	I (1)	I (1–3)
Centaurea scabiosa	I (1–7)			II (1–4)		I (1–7)
Pimpinella saxifraga				II (1–3)	II (1–3)	I (1–3)

Species	1	2	3	4	5	6
Origanum vulgare				II (2–3)		I (2–3)
Campanula rotundifolia				I (1–3)	I (1–4)	I (1–4)
Scabiosa columbaria				I (1–3)		I (1–3)
Bromus erectus				I (1–4)		I (1–4)
Helianthemum nummularium				I (1–3)		I (1–3)
Linum catharticum				I (1–2)		I (1–2)
Teucrium scorodonia				I (1–3)		I (1–3)
Thymus praecox				I (1–2)		I (1–2)
Festuca rubra	III (1–9)	I (4–6)	I (3–5)	I (4–7)	IV (1–7)	II (1–9)
Lotus corniculatus	I (2–5)	I (3)	I (3)	II (3–4)	IV (1–7)	II (1–7)
Centaurea nigra	II (1–6)	I (1–7)	II (1–5)	III (3–5)	V (1–7)	III (1–7)
Trisetum flavescens	I (3–4)	I (2–3)	I (2–3)	I (2–5)	III (1–5)	I (1–5)
Veronica chamaedrys	I (1–4)	I (2–3)	I (1–2)	II (2–3)	III (1–4)	I (1–4)
Leucanthemum vulgare	I (3–6)	I (4)	I (2)		II (1–7)	I (1–7)
Trifolium pratense	I (3–7)	I (6)	I (2–4)	II (2–4)	II (1–4)	II (1–7)
Anthoxanthum odoratum				II (1–4)	III (1–7)	I (1–7)
Avenula pubescens		I (4)			II (1–5)	I (1–5)
Hypericum perforatum	I (2–8)				II (1–4)	I (1–8)
Primula veris		I (2)			I (1–3)	I (1–3)
Silene nutans					I (1–4)	I (1–4)
Cirsium arvense	III (1–5)	III (1–6)	III (1–5)	III (3–4)	I (1–5)	III (1–6)
Poa pratensis	II (1–6)	I (2–7)	I (2–5)	II (3–8)	II (1–4)	II (1–8)
Poa trivialis	II (2–8)	I (2–7)	III (3–6)		II (1–4)	II (1–8)
Rumex acetosa	I (1–3)	I (1–4)	III (1–4)		II (1–3)	II (1–4)
Trifolium repens	I (2–4)	I (2–3)	II (2–6)	I (3)	II (1–5)	II (1–6)
Lathyrus pratensis	II (1–7)	I (1–5)	III (2–4)	II (2–3)	II (1–5)	II (1–7)
Elymus repens	II (2–8)	II (1–8)	II (2–5)	I (1–2)	III (1–5)	II (1–8)
Lolium perenne	II (1–8)	II (1–6)	II (1–7)		I (2–6)	II (1–8)
Rubus fruticosus agg.	II (2–6)	II (1–6)	III (1–4)	II (2–6)	I (2–6)	II (1–8)
Taraxacum officinale agg.	II (1–5)	I (1–4)	II (1–3)	I (2–8)	II (2–4)	II (1–5)
Vicia sativa nigra	III (2–4)	I (1–3)	II (4–5)	I (2–3)	II (1–3)	II (1–5)
Brachythecium rutabulum	II (1–7)	I (2–4)	I (1–2)	II (1–4)	II (1–5)	II (1–7)
Eurhynchium praelongum	I (2–7)	I (2–6)	II (1–9)	I (4–5)	I (1–2)	I (1–9)
Pseudoscleropodium purum	I (3)	I (1)	I (1)	I (3)	II (1–4)	I (1–4)
Rhytidiadelphus squarrosus	I (2–3)	I (3)	I (1)	I (3)	I (1)	I (1–3)
Alopecurus pratensis	I (4–8)	I (1–8)	I (4–7)	I (3)	I (2–3)	I (1–8)

Floristic table MG1 (*cont.*)

	a	b	c	d	e	1
Cerastium fontanum	I (1–4)	I (2)	I (1–3)	II (1–3)	II (1–3)	I (1–4)
Ranunculus acris	I (1–4)	I (3–5)	II (1–3)	I (3)	II (1–4)	I (1–5)
Vicia cracca	I (2–4)	I (3)	I (3–4)	I (2)	I (2)	I (2–4)
Galium mollugo	I (3–4)		I (4–5)	I (3)	I (3)	I (3–5)
Bromus sterilis	I (1–8)	I (2–7)	I (1)	I (3)		I (1–8)
Brachypodium sylvaticum	I (2–5)	I (5)	I (1–4)	I (1–2)	I (1–7)	I (1–7)
Convolvulus arvensis	I (2–6)	I (2–6)	I (1–3)	II (3–5)	I (3)	I (1–6)
Cynosurus cristatus	I (2–8)	I (3)	I (2–3)	I (2)	I (4–5)	I (2–8)
Glechoma hederacea	I (1–4)	I (1–6)	I (4)	II (1–5)	I (3–4)	I (1–6)
Pteridium aquilinum	I (3–6)	I (2–5)	I (4–5)	I (3)		I (2–6)
Ranunculus repens	I (1–6)	I (2–5)	II (1–5)	II (2–5)		I (1–6)
Lotus uliginosus	I (2)	I (1–3)	I (3)	I (3)	I (1–3)	I (1–3)
Leontodon hispidus	I (3)	I (2)	I (1)		I (1–5)	I (1–5)
Potentilla reptans	I (1–7)	I (2)	I (2–4)		I (3–6)	I (1–7)
Festuca pratensis	I (1–3)	I (1–3)	I (3)		I (1–5)	I (1–5)
Stellaria graminea	I (2–3)	I (1)	I (2–3)		I (2–4)	I (2–4)
Ranunculus ficaria	I (1–4)	I (1–3)	I (4)		I (1–4)	I (1–4)
Vicia sepium	I (3–4)	I (1–2)	I (1–3)		I (1)	I (1–4)
Epilobium angustifolium	I (2–6)	I (4)	I (1)	I (1–2)		I (1–6)
Phleum pratense pratense	I (3–4)	I (1–4)	I (1)	I (1–2)		I (1–4)
Cirsium vulgare	I (3–5)	I (1–2)	I (4)	I (2)		I (1–5)
Rumex crispus	I (4)	I (1–4)	I (2–3)	I (1–2)		I (1–4)
Hypochoeris radicata	I (2)			I (1–2)	I (1–4)	I (1–4)
Rhinanthus minor	I (2)		I (4)	I (3)		I (2–4)
Phleum pratense bertolonii	I (2–3)			I (6)	I (3)	I (2–6)
Carduus nutans	I (4)			I (1)	I (1–4)	I (1–4)
Luzula campestris	I (1)		I (4)		II (1–4)	I (1–4)
Holcus mollis	I (3–4)		I (2–7)		I (1–6)	I (1–7)
Ranunculus bulbosus	I (3)				I (1–3)	I (1–3)
Daucus carota carota	I (3–4)			I (1–3)	I (1–3)	I (1–4)
Bromus hordeaceus hordeaceus	I (1–4)		I (2–3)	I (1)		I (1–4)
Geranium dissectum	I (2–5)		I (3)	I (1)		I (1–5)
Hedera helix	I (3–5)	I (2–4)	I (3)			I (2–5)

	a	b	c	d	e	1
Rumex obtusifolius	I (1–4)	I (1–4)	I (1–4)			I (1–4)
Symphytum officinale		I (3–5)	I (3–5)			I (3–5)
Briza media		I (1)		I (1–3)	I (1–4)	I (1–4)
Conopodium majus	I (3)				I (1–4)	I (1–4)
Pimpinella major	I (4)				I (1–7)	I (1–7)
Number of samples	85	118	27	46	40	316
Number of species/sample	12 (4–19)	12 (3–18)	15 (4–21)	16 (5–20)	21 (11–30)	14 (3–30)

a *Festuca rubra* sub-community
b *Urtica dioica* sub-community
c *Filipendula ulmaria* sub-community
d *Pastinaca sativa* sub-community
e *Centaurea nigra* sub-community
1 *Arrhenatheretum elatioris* (total)

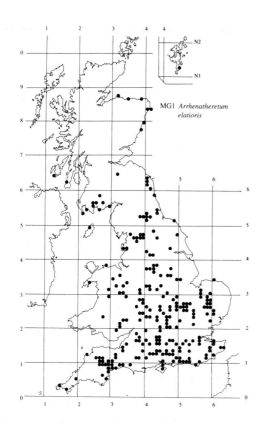

MG1 *Arrhenatheretum*
       elatioris

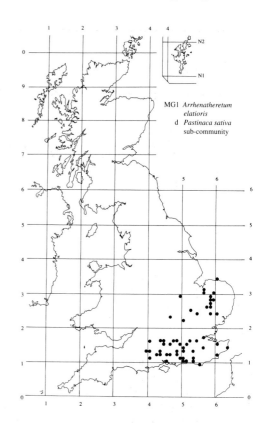

MG1 *Arrhenatheretum*
       elatioris
   d *Pastinaca sativa*
       sub-community

Pimpinella major	II (1–7)	I (3–5)	II (1–7)
Ranunculus acris	I (1)	II (1–4)	II (1–4)
Lathyrus pratensis	II (1–2)	I (2)	I (1–2)
Prunella vulgaris	I (1)	II (1)	I (1)
Pseudoscleropodium purum	I (1–3)	I (1–3)	I (1–3)
Eurhynchium praelongum	I (4)	I (1–2)	I (1–4)
Holcus mollis	I (7)	I (1)	I (1–7)
Centaurea nigra	I (2–3)	I (2)	I (2–3)
Alchemilla xanthochlora	I (1–2)	I (2)	I (1–2)
Stachys sylvatica	I (5)	I (3–5)	I (3–5)
Viola riviniana	I (1–3)	I (1–2)	I (1–3)
Briza media	I (1)	I (1)	I (1)
Galium verum	I (1–2)	I (1)	I (1–2)
Festuca ovina	I (1)	I (2–5)	I (1–5)
Eurhynchium striatum	I (2–3)	I (3)	I (2–3)
Teucrium scorodonia	I (1)	I (1–4)	I (1–4)
Mnium hornum	I (1–2)	I (1–2)	I (1–2)
Dicranum scoparium	I (1)	I (1–4)	I (1–4)
Sesleria albicans	I (1)	I (8)	I (1–8)
Rhytidiadelphus loreus	I (1)	I (3)	I (1–3)
Geranium sylvaticum	I (1)	I (2–5)	I (1–5)
Number of samples	12	27	39
Number of species/sample	33 (17–38)	33 (23–48)	33 (17–48)

a   *Filipendula ulmaria* sub-community
b   *Polemonium caeruleum* sub-community
2   *Filipendulo-Arrhenatheretum elatioris* (total)

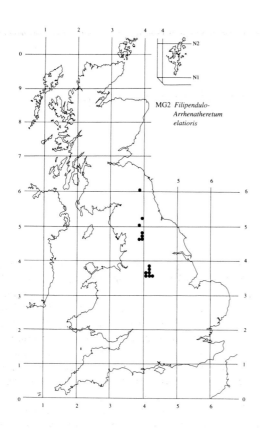

MG2 *Filipendulo-
Arrhenatheretum
elatioris*

MG3
Anthoxanthum odoratum-Geranium sylvaticum grassland

Synonymy

Trisetetum flavescentis Beger 1922; Hay meadows Pigott 1956a, Bradshaw 1962, Bradshaw & Clark 1965, Ratcliffe 1978; Dales hay meadows Duffey *et al.* 1974; Northern hay meadows Ratcliffe 1977; *Helictotricho-Trisetetum* Jones 1984 *p.p.*; *Dactylo-Geranietum* Jones 1984.

Constant species

Agrostis capillaris, Alchemilla glabra, Anthoxanthum odoratum, Cerastium fontanum, Conopodium majus, Dactylis glomerata, Festuca rubra, Geranium sylvaticum, Holcus lanatus, Plantago lanceolata, Poa trivialis, Ranunculus acris, Rumex acetosa, Sanguisorba officinalis, Trifolium repens.

Rare species

Alchemilla acutiloba, A. glomerulans, A. monticola, A. subcrenata, A. wichurae.

Physiognomy

The *Anthoxanthum odoratum-Geranium sylvaticum* community has a dense growth of grasses and herbaceous dicotyledons up to 60–80 cm high. Among the former, *Anthoxanthum odoratum, Dactylis glomerata, Festuca rubra, Holcus lanatus, Agrostis capillaris* and *Poa trivialis* are constant with *Cynosurus cristatus, Trisetum flavescens, Poa pratensis, Alopecurus pratensis* and *Festuca pratensis* occurring throughout though less frequently. Some of these grasses may attain local abundance (especially the constant species) and there is also some phenological variation with fine-leaved species such as *F. rubra, P. trivialis* and *A. odoratum* prominent early in the season but later being overtopped by the more tussocky coarse-leaved species such as *H. lanatus* and *D. glomerata*. However, it is a feature of the community that no single grass species is consistently dominant. Indeed, grasses as a whole commonly comprise a relatively small proportion of the herbage and the most obvious physiognomic feature of the vegetation is generally the variety and abundance of dicotyledons. In summer, the vegetation presents a colourful spectacle and has earned itself the local name of 'herbie meadow'.

Among the constant dicotyledons, *Geranium sylvaticum* is usually the most prominent and exceptionally may account for 90% of the cover. *Sanguisorba officinalis, Conopodium majus, Alchemilla glabra* and *A. xanthochlora* are also occasionally abundant and by the end of June these five species generally comprise the bulk of the herbage. Beneath, *Plantago lanceolata, Rumex acetosa, Ranunculus acris, Cerastium fontanum* and *Trifolium repens* are constant components with *Taraxacum officinale* agg., *Lathyrus pratensis, Bellis perennis* and *Ranunculus bulbosus* frequent. Tall herbs are generally uncommon, although *Heracleum sphondylium* is occasionally conspicuous. Among the low-frequency associates are *Cardamine pratensis, Anemone nemorosa, Cirsium helenioides, Trollius europaeus, Leucanthemum vulgare, Euphrasia montana, Polygonum viviparum, Trifolium medium, Dactylorhiza fuchsii, Coeloglossum viride, Orchis mascula, Gymnadenia conopsea, Leucorchis albida* and *Listera ovata*. Three ancient introductions, *Rumex longifolius, Polygonum bistorta* and *Peucedanum ostruthium* occur locally, sometimes in dense patches.

There is sometimes a hazy mosaic in stands of the community with areas tending towards calcifugous grassland or mires. In the former, *Anthoxanthum odoratum* may be locally abundant with *Viola lutea* and even *Deschampsia flexuosa* and *Vaccinium myrtillus*. In the latter, *Crepis paludosa* and *Carex panicea* are often conspicuous with occasional records for *Filipendula ulmaria, Geum rivale, Cochlearia alpina, Lychnis floscuculi, Potentilla palustris, Valeriana dioica, Pedicularis palustris, Dactylorhiza incarnata* and *D. majalis* ssp. *purpurella*.

We do not have systematically collected bryophyte data for all the available samples but mosses are frequent throughout. The most common species are

Brachythecium rutabulum, Eurhynchium praelongum and *Rhytidiadelphus squarrosus* and each of these may be locally abundant.

Sub-communities

***Bromus hordeaceus* ssp. *hordeaceus* sub-community:** *Helictotricho-Trisetetum, Lolium perenne* Subassociation Jones 1984. In general, this sub-community is characterised by the replacement of species indicative of lack of improvement by those typical of re-seeded grasslands, e.g. *Lolium perenne* and *Phleum pratense* ssp. *pratense*, and attendant weeds, among which *Bromus hordeaceus* ssp. *hordeaceus* is the most conspicuous. Many of the occasionals of the *Briza* sub-community are much rarer or absent here and the vegetation is poorer and less diverse. Although the dicotyledons may remain conspicuous, grasses are generally more abundant and *Dactylis glomerata* and *Poa trivialis* sometimes share dominance. *P. trivialis* is often particularly noticeable early in the season when young plants appear on areas of bare soil. In more extreme cases, *Geranium sylvaticum* and *Sanguisorba officinalis* show greatly reduced vitality and may persist only as isolated patches. *Alchemilla xanthochlora* and (especially) *A. glabra* may remain more abundant but, in some stands, these too are much less conspicuous than usual.

***Briza media* sub-community:** Hay meadows Bradshaw 1962, Bradshaw & Clark 1965; *Helictotricho-Trisetetum, Lathyrus pratensis* Subassociation Jones 1984. This is the richest and most diverse of the sub-communities. It is characterised partly by the markedly preferential frequency of species common in unimproved grasslands, e.g. *Briza media, Lotus corniculatus, Luzula campestris, Rhinanthus minor, Centaurea nigra* and *Leontodon hispidus*, and the virtual absence of species associated with improvement and/or re-seeding. There is also a wide variety of occasionals, some of which reflect the occurrence of undisturbed transitions to mires and other grassland types which are more common in this sub-community.

Others, however, belong to the meadow vegetation proper and include some species of local or very restricted national distribution. Most prominent among these are various of the microspecies within the *Alchemilla vulgaris* aggregate (Walters 1949, 1952) which have been the subject of ecological study by Bradshaw (1962). Three are Northern Montane species (Hultén 1950, Matthews 1955) confined in Britain to the Teesdale/Weardale area of Durham and almost always found in this sub-community of the *Anthoxanthum-Geranium* grassland. *Alchemilla monticola* and *A. acutiloba* are both fairly widespread in this area: the former occurs in varying degrees of abundance in meadow stands and, towards the fringe of its distribution, on or near to roads; the latter is more frequent along verges, often in dense patches. *A. subcrenata* is a further, much rarer, Northern Montane species occurring here. Two other members of the aggregate, the Arctic-Subarctic *A. wichurae* and the Arctic-Alpine *A. glomerulans*, are more widely distributed in northern Britain and, though sometimes conspicuous in this sub-community, also occur in calcicolous grasslands and ledge vegetation. *A. filicaulis* ssp. *filicaulis* is another Arctic-Alpine taxon but it is much rarer here than in calcicolous grassland. The much more widespred *A. filicaulis* ssp. *vestita* is also frequent and occasionally abundant in this sub-community but it occurs, too, in the meadows and pastures of the *Centaureo-Cynosuretum*. In exceptional cases, samples have been recorded with four or five of these species in addition to *A. glabra* and *A. xanthochlora*.

***Arrhenatherum elatius* sub-community:** *Dactylo-Geranietum* Jones 1984. In the coarser sward of this sub-community, tussock grasses such as *Arrhenatherum elatius, Dactylis glomerata* and *Avenula pubescens* are more abundant, though they rarely dominate. In general, the vegetation resembles that of the poorer *Bromus* sub-community but *Geranium pratense, Cruciata laevipes* and *Holcus mollis* are preferential here and there are sometimes dense patches of *Alchemilla acutiloba* and, more rarely, *A. monticola*.

Habitat

The *Anthoxanthum-Geranium* community is an upland grassland confined to areas where traditional hay-meadow treatment has been applied in a harsh sub-montane climate. It is most characteristic of brown soils on level to moderately sloping sites and is now almost entirely restricted to a few valley heads, between 200 and 400 m, in northern England. Many stands are still used as hay-meadows but the community also occurs on river banks and road verges, in churchyards and in woodland clearings.

The climate of the sites is cold, wet, windy and cloudy (Manley 1936, 1942) and the major effects on the vegetation are felt through temperature and precipitation. Winters are bitter and stormy with up to 50 days observed snow or sleet (Manley 1940) and a long spring snow-lie. The growing season starts later and is shorter than in any other agroclimatic area of England and Wales (Smith 1976). Growth normally begins in late April to early May but late frosts are frequent and, in the brief cloudy summer, temperatures may be below the critical mean for plant growth for more than half the time (Pigott 1956a). The autumn is windy and very wet.

Annual precipitation ranges from 900 to 1800 mm (*Climatological Atlas* 1952) with 180–200 wet days yr^{-1} (Ratcliffe 1968), but the effect of this high rainfall is modified by the physical and chemical properties of the soils. Although these may occur over permeable sandstones and limestones, they are frequently derived from superficial deposits: alluvium, head, glacio-fluvial material or, most commonly, till, often laid down in moraines or as drumlins. Such material usually has a very substantial fine sand fraction which, especially when compacted, has little pore space (Pigott 1978*b*). In this region, such soils show an early autumn return to field capacity (Smith 1976) and poor to impeded drainage throughout the year with gleying below. Where the soil parent material is more free-draining, the predominant direction of soil water movement is downwards and there is a superficial removal of any free calcium carbonate and a fall in pH and even the development of an illuvial horizon (Pigott 1956*a*). Generations of the traditional application of lime and farmyard manure (see below) have tended to offset the loss of minerals by such leaching (and by the annual removal of the hay crop).

The chemical composition, especially the calcium content, of the soil parent material further influences the effect of the soil water on the vegetation. On material derived from siliceous rocks, especially where this is permeable, soils tend to a typical brown earth profile and it is in such situations that the vegetation has species characteristic of calcifugous grasslands. On more calcareous material, the typical profile is of the brown calcareous earth type and, where this is impermeable below, high calcium content and base-status may be maintained by flushing with ground water from adjacent limestones. It is around the springs which develop in such circumstances that the *Anthoxanthum-Geranium* community attains its greatest richness and diversity with occasional records for a wide variety of species of poor fens and base-rich mires. In general, this kind of heterogeneity is most characteristic of situations where mixed till in a complex morainic topography abuts on to limestone exposures.

Most of the differences between the three sub-communities are, however, attributable directly to variations in the treatment of the vegetation. The *Anthoxanthum-Geranium* community is essentially a hay-meadow and comprises part of the 'in-by' land of Pennine and Lakes hill farms. These valley fields are grazed in winter, mainly by sheep, except in very unfavourable weather when the stock are kept indoors. In late April to early May, the meadows are shut up for hay and the stock, apart from animals in poor condition, transferred to the 'out-by' summer grazing on the open moorland. Mowing takes place generally in late July to early August

though, in unfavourable seasons, it may be delayed as late as September. The aftermath is then grazed once more until the weather deteriorates.

Traditionally, the meadows have been given a light dressing of farmyard manure after being shut up and it is this, together with liming, which has helped maintain the richness and diversity of the *Briza* sub-community. In some areas, the old practice of stacking the hay in a mound within the meadows (in a different place each year) has probably contributed to the local abundance of particular species seeding in from the cut material (Bailey 1810, Bradshaw 1962). Traditional methods of seeding in from barn-sweepings or with seed collections from rich meadows or by the transplanting of turfs (Bradshaw 1962) have probably also helped maintain the frequency and abundance of a wide range of species.

Such long-continued practices have been abandoned at an increasing rate in recent years. It is now commonplace to use chemical fertilisers, especially nitrogen, to increase hay yield and this, more than any other factor, is responsible for the widespread conversion of the *Briza* sub-community to the much poorer *Bromus* sub-community. Many of the stands described as recently as the 1960s have shown such a floristic impoverishment and, in some areas, only substantial subsidies maintain the traditional methods of treatment (Ratcliffe 1978).

Ploughing and subsequent abandonment and/or re-seeding have also contributed to the loss of the richer *Briza* sub-community in some areas. In the Lake District, many meadows were ploughed up for potato cultivation during the Second World War and afterwards reverted to the *Bromus* sub-community. More systematic and repeated ploughing and re-seeding, especially when combined with application of artificial fertilisers, has the more drastic effect of converting the *Anthoxanthum-Geranium* community to other more productive types of grassland. However, both *Geranium sylvaticum* and *Sanguisorba officinalis* are remarkably persistent after ploughing alone and can regenerate from buried rhizome fragments. *Alchemilla glabra* and *A. xanthochlora* also seem able to reappear rapidly from dormant seed. Even some of the rare *Alchemilla* spp. may spread on to road verges and along disturbed paths; indeed, *A. acutiloba* is particularly associated with such habitats in Durham (Bradshaw 1962).

Stands of the community along road verges are a valuable supplement to the enclosed grazing of upland farms and may be used for irregular winter pasturing. Even in summer, sheep may escape from the out-by land over the cattle grids and graze the verges lightly. Many verges are not cut and this enables more of the dicotyledons to set seed, thus helping to maintain the richness of adjacent meadows. However, where such stands remain ungrazed, they may be converted to the coarser

Arrhenatherum sub-community which can persist even with annual cutting.

Zonation and succession

Stands of the *Anthoxanthum-Geranium* community occur most frequently in fields bounded by walls or fences and each subject to a more or less uniform treatment regime. Spatial zonations between the sub-communities are therefore infrequent, although neglected corners and margins of fields with either the *Briza* or the *Bromus* sub-community may show a narrow transition to the *Arrhenatherum* sub-community. The blue-violet flowers of *Geranium pratense* often mark out this sub-community in summer. On verges which have a less systematic treatment such transitions are more common and irregular.

The local appearance of mire species in stands of the *Anthoxanthum-Geranium* community may form part of a transition to flush vegetation where drainage of soil water is strongly impeded. The types of mire involved in such sequences depend on the calcium status and pH of the ground water and include the *Pinguiculo-Caricetum* and the *Molinia-Crepis* fen. Along stream sides where there is seasonal inundation a tall-herb Filipendulion mire may terminate the sequence.

The striking floristic similarity between the *Anthoxanthum-Geranium* community and the field layer of open stands of the *Fraxinus-Sorbus-Mercurialis* and *Alnus-Fraxinus-Lysimachia* woodlands strongly suggests that this meadow vegetation has developed by canopy clearance and is maintained by annual mowing which excludes tall dominants (Pigott 1956*a*, Bradshaw 1962). Zonations between the community and such woodland are rare but occasionally the two vegetation types occur contiguously on identical sites separated only by a boundary wall. In Scandinavia, very similar meadow vegetation is mown in the clearings of such open woodland (Nordhagen 1928, Sjörs 1954; see 'Affinities' below).

Traditional treatment has maintained the community in all its richness but the vegetation of the *Bromus* sub-community represents one stage in a process of increasing the productivity and decreasing the floristic diversity of hay-meadows in response to various combinations of fertiliser application, ploughing and re-seeding. This seems eventually to convert the vegetation to the *Holcus-Trifolium* sub-community of *Festuca-Agrostis-Galium* grassland or the *Anthoxanthum* sub-community of the *Lolio-Cynosuretum*.

Distribution

This is a northern sub-montane community now almost entirely restricted to a few upland valleys in northern England, notably Teesdale and Weardale in Durham, Swaledale and Wensleydale in North Yorkshire and parts of the Lake District. We have no Scottish samples but fragments of the community survive along river banks in Tayside. It was undoubtedly more widespread in the past and, in areas of agricultural improvement, verge stands provide a valuable reserve.

Affinities

The *Anthoxanthum-Geranium* community has long been recognised for its distinctive richness but its exact floristic affinities are unclear. On the one hand, it shares many species with the 'Park Meadow' communities described from Norway (Nordhagen 1928, 1936, 1943), Sweden (Sjörs 1954) and Greenland (Böcher 1954). These have traditionally been placed in the Cicerbition alpini which comprises vegetation approached most closely in Britain by the tall-herb communities of mountain ledges. On the other hand, there are clear affinities with other meadow types in the Arrhenatheretalia allocated to the Polygono-Trisetion, which replaces the Arrhenatherion at higher altitudes in Europe (Beger 1922, LeBrun *et al.* 1949). A further difficulty is that complexities of treatment styles have served to blur the boundary between this alliance and the pastures of the Cynosurion in Britain.

Floristic table MG3

	a	b	3
Plantago lanceolata	V (2–5)	V (2–4)	V (2–5)
Rumex acetosa	V (1–5)	V (1–3)	V (1–5)
Ranunculus acris	V (1–5)	V (1–3)	V (1–5)
Geranium sylvaticum	V (2–9)	IV (1–4)	V (1–9)
Anthoxanthum odoratum	IV (1–5)	V (3–6)	V (1–6)
Conopodium majus	IV (2–7)	V (2–6)	V (2–7)
Cerastium fontanum	IV (1–3)	V (1–3)	V (1–3)
Dactylis glomerata	V (1–7)	IV (1–5)	IV (1–7)
Alchemilla glabra	V (1–4)	IV (1–4)	IV (1–4)

Trifolium repens	V (1–4)	IV (1–3)	IV (1–4)
Poa trivialis	V (1–7)	IV (1–3)	IV (1–7)
Festuca rubra	IV (2–7)	IV (3–8)	IV (2–8)
Agrostis capillaris	IV (1–7)	IV (1–5)	IV (1–7)
Holcus lanatus	IV (2–6)	III (1–5)	IV (1–6)
Sanguisorba officinalis	IV (1–6)	III (1–3)	IV (1–6)
Bromus hordeaceus hordeaceus	III (1–5)		II (1–5)
Lolium perenne	III (1–5)	I (2–6)	II (1–6)
Phleum pratense pratense	II (2–3)	I (1–2)	I (1–3)
Ranunculus repens	II (1–4)		I (1–4)
Arrhenatherum elatius	I (2–3)		I (2–3)
Vicia sativa nigra	I (2)		I (2)
Veronica serpyllifolia	I (1)		I (1)
Myosotis discolor	I (1)		I (1)
Stellaria media	I (1)		I (1)
Rhinanthus minor	II (1–6)	IV (2–3)	III (1–6)
Bellis perennis	III (1–4)	IV (1–4)	III (1–4)
Cynosurus cristatus	III (1–6)	IV (2–5)	III (1–6)
Ranunculus bulbosus	III (1–4)	IV (1–3)	III (1–4)
Leontodon hispidus	II (1–5)	IV (1–4)	III (1–5)
Luzula campestris	I (1–4)	IV (1–3)	III (1–4)
Trifolium pratense	II (1–3)	IV (1–4)	II (1–4)
Hypochoeris radicata	I (1–2)	III (1–3)	II (1–3)
Lotus corniculatus	I (1–4)	III (1–3)	II (1–4)
Centaurea nigra	I (1–3)	III (1–3)	II (1–3)
Briza media		III (2–3)	I (2–3)
Alchemilla filicaulis vestita		II (1–4)	I (1–4)
Alchemilla monticola		II (1–4)	I (1–4)
Succisa pratensis		II (1–3)	I (1–3)
Orchis mascula		II (1)	I (1)
Alchemilla acutiloba		I (3–6)	I (3–6)
Alchemilla subcrenata		I (2–3)	I (2–3)
Alchemilla filicaulis filicaulis		I (3)	I (3)
Alchemilla glomerulans		I (1–5)	I (1–5)
Alchemilla wichurae		I (1–4)	I (1–4)
Caltha palustris		I (1)	I (1)
Primula veris		I (1–2)	I (1–2)
Thymus praecox		I (2)	I (2)
Viola riviniana		I (1)	I (1)
Viola lutea		I (1)	I (1)
Saxifraga granulata		I (1–2)	I (1–2)
Senecio jacobaea		I (1)	I (1)
Crepis paludosa		I (1)	I (1)
Alchemilla xanthochlora	II (1–5)	III (1–3)	III (1–5)
Taraxacum officinale agg.	III (1–3)	III (1–2)	III (1–3)
Lathyrus pratensis	III (1–3)	III (1–2)	III (1–3)
Poa pratensis	II (2–6)	II (3–6)	II (2–6)
Heracleum sphondylium	II (1–5)	III (1–5)	II (1–5)
Alopecurus pratensis	II (1–7)	II (1–3)	II (1–7)

Floristic table MG3 (*cont.*)

	a	b	3
Cardamine pratensis	II (1–3)	II (1–2)	II (1–3)
Anemone nemorosa	I (2–4)	II (1–3)	II (1–4)
Trisetum flavescens	II (1–3)	II (1)	II (1–3)
Leucanthemum vulgare	II (1–3)	I (1)	I (1–3)
Festuca pratensis	II (1–4)	II (1–3)	II (1–3)
Achillea millefolium	I (1–3)	II (1–2)	I (1–2)
Potentilla erecta	I (3–4)	II (1–2)	I (1–4)
Carex panicea	I (3)	II (1–3)	I (1–3)
Carex caryophyllea	I (2)	II (1)	I (1–2)
Deschampsia cespitosa	I (1)	II (1–3)	I (1–3)
Filipendula ulmaria	I (3–4)	II (1–2)	I (1–4)
Geum rivale	I (1)	II (1–3)	I (1–3)
Euphrasia officinalis agg.	I (1–2)	II (1–3)	I (1–3)
Trollius europaeus	I (4)	II (1–2)	I (1–4)
Avenula pubescens	I (2–4)	II (3)	I (2–4)
Leontodon autumnalis	I (1–3)	I (2–3)	I (1–3)
Prunella vulgaris	I (1–6)	I (2–3)	I (1–6)
Campanula rotundifolia	I (2)	I (1)	I (1–2)
Cirsium helenioides	I (1)	I (1–2)	I (1–2)
Lathyrus montanus	I (3)	I (1)	I (1–3)
Equisetum arvense	I (3–4)	I (1)	I (1–4)
Ajuga reptans	I (1)	I (1)	I (1)
Trifolium medium	I (2)	I (1–2)	I (1–2)
Anthriscus sylvestris	I (4)	I (1–2)	I (1–4)
Vicia sepium	I (4)	I (1)	I (1–4)
Brachythecium rutabulum	III (1–5)		
Eurhynchium praelongum	III (1–5)		
Rhytidiadelphus squarrosus	III (1–7)		
Calliergon cuspidatum	II (1–3)		
Mnium hornum	II (1–3)		
Plagiomnium cuspidatum	I (2–4)		
Number of samples	34	40	74
Number of species/sample	23 (12–33)	35 (19–43)	26 (12–43)

a   *Bromus hordeaceus hordeaceus* sub-community
b   *Briza media* sub-community
3   *Anthoxanthum odoratum-Geranium sylvaticum* grassland (total)

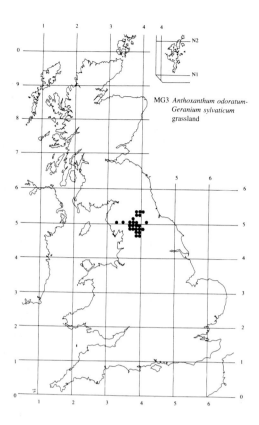

MG3 *Anthoxanthum odoratum-Geranium sylvaticum* grassland

MG4
Alopecurus pratensis-Sanguisorba officinalis grassland

Synonymy
Hay Meads Baker 1937; *Fritillario-Alopecuretum pratensis* Westhoff & den Held 1969 *p.p.*; Flood-Meadows Duffey *et al.* 1974; Alluvial meadows Ratcliffe 1977; *Fritillario-Sanguisorbetum officinalis* Page 1980.

Constant species
Alopecurus pratensis, Cerastium fontanum, Cynosurus cristatis, Festuca rubra, Filipendula ulmaria, Holcus lanatus, Lathyrus pratensis, Leontodon autumnalis, Lolium perenne, Plantago lanceolata, Ranunculus acris, Rumex acetosa, Sanguisorba officinalis, Taraxacum officinale agg., *Trifolium pratense, T. repens.*

Rare species
Fritillaria meleagris, Taraxacum fulgidum, T. haematicum, T. melanthoides, T. sublaeticolor, T. subundulatum, T. tamesense.

Physiognomy
The *Alopecurus pratensis-Sanguisorba officinalis* community has a species-rich and somewhat varied sward of grasses and herbaceous dicotyledons. Among the former, there is generally no single dominant and, by June, when most of the grasses are flowering, *Festuca rubra, Alopecurus pratensis, Cynosurus cristatus* and *Lolium perenne* may all be abundant with, less frequently and usually in smaller amounts, *Holcus lanatus, Anthoxanthum odoratum, Dactylis glomerata* and *Trisetum flavescens.* Many other grasses occur occasionally and some of these may attain local abundance (e.g. *Bromus hordeaceus* ssp. *hordeaceus* and *Agrostis stolonifera*) or be conspicuous by virtue of a dense tussock habit (e.g. *Arrhenatherum elatius, Deschampsia cespitosa* and *Festuca arundinacea*). *Carex acutiformis* is occasional and it may be abundant but other sedges (e.g. *C. panicea* and *C. hirta*) are less frequent and never prominent. *Juncus articulatus* and *J. inflexus* occur patchily at low frequency.

Herbaceous dicotyledons are always an important component of the herbage. When growth commences in April or May, rosette species such as *Leontodon autumnalis, L. hispidus, Plantago lanceolata* and *Bellis perennis* are often prominent in the short sward. *Taraxacum officinale* agg. is a constant and sometimes abundant member of the community and is often especially conspicuous at this time when flowering. Older stands have a rich and varied dandelion flora and a number of species seem to be confined to the community (Richards 1972). It is at this time, too, that the most renowned occasional of the vegetation, *Fritillaria meleagris*, makes a spectacular display with its flowers, occurring patchily in the sward or sometimes as extensive sheets.

By July, the grasses have generally been overtopped by *Sanguisorba officinalis* and *Filipendula ulmaria*, both of which, though especially the former, may be abundant. By this time, the vegetation forms a dense herbage up to 70 cm or more tall. Below, *Trifolium pratense, T. repens, Ranunculus acris, R. repens, Rumex acetosa* and *Rhinanthus minor* may all be plentiful with smaller amounts of *Cerastium fontanum, Lotus corniculatus, Primula veris* and *Luzula campestris. Lathyrus pratensis* is a constant and sometimes abundant sprawler and the inflorescences of tall hemicryptophytes such as *Centaurea nigra, Leontodon autumnalis, L. hispidus, Silaum silaus, Leucanthemum vulgare* and *Succisa pratensis* may make a colourful show.

There are almost always some bryophytes and both *Brachythecium rutabulum* and *Calliergon cuspidatum* can occur as extensive patches over soil and litter. *Eurhynchium praelongum* and *Plagiomnium elatum* are less frequent.

Habitat
The *Alopecurus-Sanguisorba* community is a lowland grassland especially characteristic of areas where traditional hay-meadow treatment has been applied to seasonally-flooded land with alluvial soils. With almost universal improvement of grasslands and river drainage,

it is now of restricted occurrence and some of the richest stands remain where common rights have kept treatment unchanged for many generations.

As with the *Anthoxanthum-Geranium* community, hay-meadow treatment has traditionally comprised the taking of an annual hay crop, light winter grazing and light application of organic manures. Typically, stands are shut up for hay in spring, sometimes as early as February, and mown in July. At some sites, the ancient practice of distributing hay lots or doles persists, the hay from the parcels of land within the common meadow being mown and carted individually. Wooden stakes or mear stones sometimes remain as markers of the permanent lot boundaries within the open fields. The aftermath is generally grazed by cattle and, in some places, the turning out of stock takes place on or around Lammas Day, 1 August or, according to the older calendar, 12 August. North Meadow at Cricklade in Wiltshire and Pixey and Yarnton Meads in Oxford seem to have been managed in this way for centuries (Baker 1937, Hoskins & Stamp 1963, Ratcliffe 1977).

Under this regime, the meadows received no fertiliser apart from the manure of the grazing animals, but of great importance in the maintenance of fertility here has been winter flooding with its input of salts and deposition of alluvial silt and decaying organic matter. The flood-water which inundates some of the richer stands originates from watersheds with calcareous bedrocks and the calcium content and pH of the soils here are high (Baker 1937). Although pH tends to increase with depth, there is none of the leaching and development of superficial acidity that is characteristic of some stands of the *Anthoxanthum-Geranium* community.

Deep alluvial profiles have accumulated under the meadows with the repeated deposition of silt. Winter flooding leaves the land waterlogged, sometimes for many months, and the soils are frequently gleyed below. However, the profiles are generally free-draining and, as the water-table subsides, they gradually dry out above except in hollows and alongside ditches. Deep-rooted species probably always have access to soil water and, except in the driest summers, water availability is probably not limiting to plant growth.

Zonation and succession
Stands of the community frequently show considerable variation in the abundance of particular species. Some of this is related to the distinctive bushy (e.g. *S. officinalis* and *F. ulmaria*) or tussock (e.g. larger grasses) habit of certain components but local variations in treatment, such as differences in mowing-time, continued over long periods, may also have an effect (Ratcliffe 1977). Zonations to other communities are most frequently related to differences in soil moisture status. Damp hollows in meadows sometimes show gradations to the *Holcus*

lanatus-Deschampsia cespitosa community or to the *Holco-Juncetum*, especially the *Juncus inflexus* sub-community on more base-rich soils. These in turn may give way to *Carex acutiformis* swamp. Sharper zonations of this kind can sometimes be seen bordering the older drainage ditches which frequently traverse the meadows or in such places there may be an abrupt switch to the *Agrostis stolonifera-Alopecurus geniculatus* inundation grassland. Modern ditches around the meadows often have linear stands of *Salix* spp. or *Alnus* on their banks.

Changes in treatment practice can alter the composition of the *Alopecurus-Sanguisorba* community and may initiate successions to other grassland types. An extension of grazing into the spring reduces the abundance of some of the most distinctive species of the community such as *Sanguisorba officinalis*, *Silaum silaus* and *Fritillaria meleagris*, all of which are highly palatable. This is the major community in Britain for *F. meleagris* and, if grazing is continued into its flowering period in late April to early May, it cannot set seed. At Marston Meadows in Staffordshire, a change from mowing to late spring grazing over the last 20 years has extensively reduced its cover (Ratcliffe 1977). It can, however, remain dormant in the soil for a number of years and seems able to recolonise fields from margins or ditch edges. If stands are ungrazed over the winter months, coarse grasses such as *Arrhenatherum elatius* (in drier places) and *Deschampsia cespitosa* (in wetter areas) may expand and eventually form stands of the *Arrhenatheretum* or the *Holcus lanatus-Deschampsia cespitosa* community.

Drainage, ploughing, re-seeding or the addition of artificial fertilisers can all have a more drastic effect on the vegetation. Combined with an increase in grazing, such changes probably convert the community to the *Lolium perenne-Alopecurus pratensis-Festuca pratensis* flood-pasture or to the *Lolio-Cynosuretum*.

Distribution
The *Alopecurus-Sanguisorba* community is the lowland counterpart of the *Anthoxanthum-Geranium* community and it was probably once a common vegetation type of traditionally treated alluvial meadows. It now has a widespread but local distribution and the richer stands are very sparsely scattered in the Midlands and southern England. Continuing agricultural improvement, the neglect of common meadow rights and the extraction of river gravels present beneath alluvium at some sites could further reduce its extent.

Affinities
The richer stands of the community have long been the subject of admiration and study and descriptive accounts (e.g. Baker 1937, Tansley 1939, Ratcliffe 1977)

have clearly recognised some of the distinctive floristic features of the vegetation. Even where species of more restricted distribution are lacking, the community remains well defined in relation to other meadow types. It differs from the *Anthoxanthum-Geranium* community in the absence of northern species such as *Geranium sylvaticum, Conopodium majus, Alchemilla glabra* and *A. xanthochlora* and from the *Centaureo-Cynosuretum* in the presence of tall dicotyledons such as *Sanguisorba officinalis* and *Filipendula ulmaria*. Also, *Alopecurus pratensis* and *Lolium perenne* tend here to replace *Anthoxanthum odoratum, Agrostis capillaris, Poa trivialis* and *Dactylis glomerata* which are prominent in these other communities.

However, the affinities of the *Alopecurus-Sanguisorba*

community are somewhat mixed and it appears to straddle the Cynosurion pastures, the coarse ungrazed swards of the Arrhenatherion and the more grassy poor fens of the Molinietalia. This accurately reflects the rather particular combination of treatment factors which maintain meadows on more calcareous alluvium in Britain. Unimproved communities described from similar habitats in other parts of western Europe tend to be totally ungrazed (e.g. the *Molinietum caeruleae* of Braun-Blanquet 1948 and the *Arrhenatheretum elatioris colchicetosum* of LeBrun *et al.* 1949) or unmown (e.g. the *Fritillario-Alopecuretum* of Westhoff & den Held 1969) and these show much clearer affinities at the alliance level.

Floristic table MG4

Festuca rubra	V (2–6)	*Leucanthemum vulgare*	II (2–4)
Cynosurus cristatus	V (1–6)	*Deschampsia cespitosa*	II (2–4)
Sanguisorba officinalis	V (2–7)	*Succisa pratensis*	II (2–3)
Plantago lanceolata	V (1–5)	*Calliergon cuspidatum*	II (1–6)
Ranunculus acris	V (2–5)	*Cardamine pratensis*	II (1–2)
Rumex acetosa	V (1–4)	*Bromus hordeaceus hordeaceus*	II (1–5)
Filipendula ulmaria	V (1–6)	*Carex acutiformis*	II (2–5)
Taraxacum officinale agg.	V (1–5)	*Festuca arundinacea*	II (2–9)
Trifolium pratense	V (1–5)	*Juncus articulatus*	II (1–3)
Alopecurus pratensis	IV (1–6)	*Leontodon hispidus*	II (1–4)
Cerastium fontanum	IV (1–4)	*Luzula campestris*	II (2–3)
Holcus lanatus	IV (2–5)	*Primula veris*	II (2–4)
Lathyrus pratensis	IV (2–5)	*Prunella vulgaris*	II (1–2)
Leontodon autumnalis	IV (1–5)	*Trifolium dubium*	II (1–3)
Trifolium repens	IV (2–5)	*Eurhynchium praelongum*	II (2–3)
Lolium perenne	IV (2–7)	*Poa trivialis*	I (1–7)
		Carex panicea	I (1)
Rhinanthus minor	III (1–5)	*Phleum pratense pratense*	I (2–4)
Anthoxanthum odoratum	III (2–6)	*Carex hirta*	I (1–3)
Bellis perennis	III (1–4)	*Ranunculus bulbosus*	I (1–4)
Silaum silaus	III (1–4)	*Veronica serpyllifolia*	I (1–2)
Centaurea nigra	III (1–4)	*Juncus inflexus*	I (3)
Dactylis glomerata	III (1–5)	*Cirsium palustre*	I (2)
Lotus corniculatus	III (1–3)	*Bromus erectus*	I (4–8)
Ranunculus repens	III (1–5)	*Plagiomnium elatum*	I (2–3)
Fritillaria meleagris	III (2–3)	*Poa pratensis*	I (3)
Trisetum flavescens	III (2–4)	*Achillea millefolium*	I (3–5)
Brachythecium rutabulum	III (2–6)	*Thalictrum flavum*	I (3–7)
Festuca pratensis	II (2–5)	*Serratula tinctoria*	I (3)
Vicia cracca	II (2–3)	*Stachys betonica*	I (3)
Agrostis stolonifera	II (2–6)	*Potentilla anglica*	I (2–3)
Agrostis capillaris	II (3–6)	*Hypochoeris radicata*	I (2)
Arrhenatherum elatius	II (2–4)	*Briza media*	I (1–2)

Heracleum sphondylium	I (3–4)	*Equisetum arvense*	I (1)
Galium verum	I (3–5)	*Bromus hordeaceus thominii*	I (2)
Cirsium arvense	I (1)		
Caltha palustris	I (2–3)	Number of samples	22
Vicia sepium	I (1)	Number of species/sample	28 (17–38)

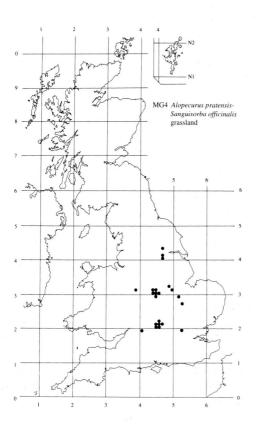

MG4 *Alopecurus pratensis-Sanguisorba officinalis* grassland

MG5

Cynosurus cristatus-Centaurea nigra grassland
Centaureo-Cynosuretum cristati Br.-Bl. & Tx 1952

Synonymy

Dairy Pastures Stapledon 1925; Third-Grade Rye-grass Pastures Williams & Davis 1946; Rothamsted Plot 3 Brenchley rev. Warington 1958; Ridge and Furrow Old Meadow Duffey *et al.* 1974; *Lolio-Cynosuretum luzuletosum* Birse & Robertson 1976 *p.p.*; Calcareous clay pastures Ratcliffe 1977; Calcareous loam pastures Ratcliffe 1977 *p.p.*; *Trifolio-Agrosto-Festucetum* Evans *et al.* 1977 *p.p.*; *Festuco-Centaureo-Brizetum* Jones unpub. *p.p.*

Constant species

Agrostis capillaris, Anthoxanthum odoratum, Centaurea nigra, Cynosurus cristatus, Dactylis glomerata, Festuca rubra, Holcus lanatus, Lotus corniculatus, Plantago lanceolata, Trifolium pratense, T. repens.

Rare species

Cirsium tuberosum.

Physiognomy

The *Centaureo-Cynosuretum* is a dicotyledon-rich grassland of somewhat variable appearance: it may have a tight, low-growing sward or comprise a fairly lush growth up to 60 cm tall. The most frequent grasses are the fine-leaved *Festuca rubra, Cynosurus cristatus* and *Agrostis capillaris.* None of these is consistently the most prominent species in the community, although each may be abundant and occasionally the three are co-dominant. *Anthoxanthum odoratum* and the coarser *Dactylis glomerata* and *Holcus lanatus* are rather less frequent and usually not so abundant. *Lolium perenne* and *Trisetum flavescens* occur throughout but are preferential for particular sub-communities and almost always have low cover. *Briza media* is a distinctive occasional easily overlooked in early sampling. *Arrhenatherum elatius* and *Festuca arundinacea* are uncommon but their robust tussock habit may make them conspicuous. Carices may be abundant in some stands with occasional records for *Carex caryophyllea, C. flacca, C. panicea* and *C. hirta.*

Dicotyledons always comprise a substantial proportion of the herbage and exceptionally may account for 95% of the cover. Among these, legumes and rosette hemicryptophytes are particularly prominent. *Lotus corniculatus, Plantago lanceolata* and *Trifolium repens* are the most frequent and generally the most abundant species, with *T. pratense* and *Centaurea nigra* rather less so. Other species frequent throughout are *Ranunculus acris, R. bulbosus, Rumex acetosa, Hypochoeris radicata, Taraxacum officinale* agg., *Achillea millefolium, Prunella vulgaris* and *Leontodon autumnalis.* In particular stands *Leontodon hispidus, Primula veris, Leucanthemum vulgare* (e.g. Birks 1973) and *Rhinanthus minor* may be prominent. Severe infestation with *R. minor* may greatly reduce the vigour of the grasses and give rise to a sward in which rosette species are dominant.

In the *Lathyrus pratensis* and *Galium verum* sub-communities, and especially where there is a long history of freedom from improvement and disturbance, the sward is enriched by occasional records for woodland field-layer species such as *Hyacinthoides non-scripta, Ranunculus auricomus* and *Anemone nemorosa* and a variety of meadow and pasture species which are of somewhat restricted distribution: *Silaum silaus, Colchicum autumnale, Ophioglossum vulgatum, Genista tinctoria, Alchemilla filicaulis* ssp. *vestita, Cirsium tuberosum, Orchis morio, Coeloglossum viride, Listera ovata* and *Platanthera chlorantha.*

Bryophytes are generally present although their total cover is very variable. The most frequent species are *Brachythecium rutabulum, Eurhynchium praelongum* and *Rhytidiadelphus squarrosus. Pseudoscleropodium purum* and *Calliergon cuspidatum* are less common but any of these species may be abundant in particular stands.

Sub-communities

***Lathyrus pratensis* sub-community:** *Centaureo-Cynosuretum typicum* Br.-Bl. & Tx 1952 *p.p.*; *Centaureo-Cyno-*

suretum typicum and *juncetosum* O'Sullivan 1965; Calcareous clay pastures Ratcliffe 1977. In this sub-community, legumes are particularly prominent. Apart from the abundant community constants *L. corniculatus* and *Trifolium* spp., *Lathyrus pratensis* is preferential here though it is rarely abundant. Grasses may have a substantial cover but their growth is often poor and, although *Lolium perenne* is preferential, it is generally sparse and of low vitality.

Some stands of the *Lathyrus* sub-community show a heterogeneity based on the presence of *Filipendula ulmaria* or *Juncus* spp. and such vegetation could be recognised as variants. *F. ulmaria* never attains dominance here but, with *Vicia cracca* and sometimes *Silaum silaus*, it can form quite conspicuous patches (*Filipendula ulmaria* variant Page 1980). *Juncus effusus, J. inflexus* and/or, less frequently, *J. articulatus* also occur occasionally, sometimes with *Potentilla erecta, Lotus uliginosus* and *Carex ovalis* (Sub-Association *juncetosum effusi* O'Sullivan 1965).

Galium verum **sub-community:** Rothamsted Plot 3 Brenchley rev. Warington 1958; *Centaureo-Cynosuretum galietosum* O'Sullivan 1965, including *Centaureo-Cynosuretum*, Sub-Association of *Thymus drucei* Br.-Bl. & Tx 1952; Calcareous loam pastures Ratcliffe 1977 *p.p.* The grass component of this sub-community is somewhat more varied than in the *Lathyrus* sub-community. *Festuca ovina* partly replaces *F. rubra*, *Trisetum flavescens* and *Avenula pubescens* are preferential, generally at low cover, and *Koeleria macrantha* is occasional. In general, however, grasses remain less prominent than dicotyledons.

Galium verum is the most distinctive preferential for this sub-community. In summer, it gives the sward a prominent yellow aspect and its bushy growth may make the vegetation somewhat uneven. Other preferentials are species characteristic of Mesobromion grasslands such as *Agrimonia eupatoria, Carex flacca, Plantago media* and *Sanguisorba minor*. In some cases, the vegetation closely resembles the *Festuca ovina-Avenula pratensis* grassland in its floristics and physiognomy with a close sward with much *F. ovina* and *K. macrantha* and a variety of smaller calcicolous dicotyledons (Variants of *Poterium sanguisorba* and *Koeleria cristata* Page 1980).

Danthonia decumbens **sub-community:** *Centaureo-Cynosuretum typicum, Sieglingia decumbens* Variant Br.-Bl. & Tx 1952; *Trifolio-Agrosto-Festucetum* Evans *et al.* 1977 *p.p.*; *Lolio-Cynosuretum luzuletosum* Birse & Robertson 1976 p.p.; *Centaureo-Cynosuretum sieglingietosum* Page 1980. Here, grasses are rather more prominent than in the other two sub-communities with *F. rubra* and/or *A. capillaris* usually abundant. *A. odora-*

tum, C. cristatus, H. lanatus and *D. glomerata* remain constant and *Danthonia decumbens* is strongly preferential, though generally at low cover. The dicotyledons typical of the community remain very frequent but *Lathyrus pratensis* and *Galium verum* are both rare here, being replaced by a block of species more characteristic of calcifugous grasslands: *Potentilla erecta, Succisa pratensis, Leontodon autumnalis, Luzula campestris, Stachys betonica* and, less frequently, *Lathyrus montanus*. Where stands occur near heath, seedlings of *Calluna vulgaris* are occasional in the sward and the sub-community often has clumps of *Ulex europaeus* and stands of *Pteridium aquilinum* within it. *Juncus articulatus* and *J. acutiflorus* may also occur in patches. Bryophytes remain frequent and *Pseudoscleropodium purum* and *Rhytidiadelphus squarrosus* are often conspicuous with some *Hypnum cupressiforme s.l.*

Habitat

The *Centaureo-Cynosuretum* is the typical grassland of grazed hay-meadows treated in the traditional fashion on circumneutral brown soils throughout the lowlands of Britain. It is becoming increasingly rare as a result of agricultural improvement but is still widespread in farm fields with more fragmentary stands in churchyards, on road verges, railway embankments and lawns and in disused quarries.

This community is the major grassland type included within the popular, but often loosely applied, category of 'Old Meadow'. The term has generally been used to describe species-rich stands of lowland meadow vegetation for which there is sometimes external evidence of (or often a presumption of) a long history of lack of disturbance or improvement. The presence of ridge-and-furrow under such stands is frequently accepted as absolute confirmation of an absence of ploughing since medieval or even earlier times. Although many examples of the *Centaureo-Cynosuretum* have probably been treated in the traditional manner for many generations, the epithet 'Old Meadow' should not be applied uncritically to the community. In the first place, other rich grassland types (most notably the *Alopecurus-Sanguisorba* community) have long been managed for hay in the lowlands. Second, the *Centaureo-Cynosuretum* is most frequently characterised by common, rather than rare, species indicative of lack of improvement. Although the full richness and variety of the vegetation is undoubtedly maintained by unbroken traditional treatment, the absence of rare species is not a simple indicator of agricultural improvement.

Furthermore, although the community often occurs in fields with ridge-and-furrow, the presence of this distinctive topography is no guarantee of the antiquity of the sward. Much ridge-and-furrow is a relic of the ploughing of medieval open fields or even earlier enclosures, but some is relatively recent, like the 'narrow rig' of eighteenth- and nineteenth-century Parliamentary enclosures (see, for example, Orwin & Orwin 1938, Baker & Butlin 1973, Taylor 1975). Moreover, many grasslands with even very old ridge-and-furrow have been improved without disturbing the ancient plough-ing patterns. Long-existing swards, including stands of the *Centaureo-Cynosuretum*, may, of course, occur on land that has no such signs of ploughing whatever.

What is characteristic of the community is a history of treatment which has traditionally comprised grazing, the taking of a hay crop and the light application of natural organic manures. The climate in the English lowlands is generally sufficiently mild to permit grazing throughout the winter, and stock are usually left on the grassland until the end of April when the fields are shut up for hay and lightly dressed, traditionally with farm-yard manure. The hay is mown in June and the stock turned out again to graze the aftermath. Where there has been an unbroken regime of this kind, stands of the community may be of great age and show a pleasing richness and diversity. In some areas, as in the south-west Midlands, such stands often occur on land which bears the wide ridge-and-furrow typical of medieval ploughing and this gives sites an additional archaeologi-cal interest. However, the community can be maintained wherever an approximation to the traditional scheme has been applied on otherwise suitable sites and fragm-entary stands are common in relatively recently created artificial habitats. These, too, may be species-rich.

Most of the floristic variation between the sub-com-munities is related to edaphic differences. The commun-ity as a whole is characteristic of deep brown soils of generally loamy to clayey texture but these vary consi-derably in their pH and calcium content, in trophic state and in their soil moisture regime. The *Lathyrus* sub-community is typically found on fairly mesotrophic and circumneutral brown earths of usually heavy texture, derived from clays and shales (such as the Lower Lias and Keuper Marl) or superficial deposits of low calcium content. On deep soils which are more calcareous but less mesotrophic, the *Galium* sub-community is charac-teristic. Such brown calcareous earths are often of lighter texture and develop from alluvium, loess and head, frequently, though not exclusively, over calcar-eous bedrocks such as Chalk and Carboniferous Lime-stone. The *Danthonia* sub-community extends the range of the community on to the upland margins where oligotrophic and calcium-deficient brown earths, some-times of rather sandy texture and usually derived from siliceous material, have been bulked up and improved by repeated manuring. Such soils are superficially acid but show no signs of podzolisation.

Substantial proportions of finer particles in these soils may impede drainage and lead to gleying, especially in hollows, beside streams or in areas of higher rainfall. Such profiles will be colder in spring and the start of new growth may be held back but they are also frequently marked out by the patchy occurrence of *Filipendula ulmaria* (especially where there is no grazing) or *Juncus* spp. (particularly where stock have access).

Zonation and succession

Zonations in the *Centaureo-Cynosuretum* are usually related to edaphic patterns and, even in enclosed fields subject to a uniform treatment, there may be soil differ-ences which reflect heterogeneities in the parent mater-ial. Within the *Galium* sub-community, an increase in Mesobromion species (such as *Koeleria macrantha* and *Sanguisorba minor*) is often related to a rise in calcium content and pH as the soil thins towards limestone exposures. Such patterns are commonly part of a com-plete transition from the *Centaureo-Cynosuretum* to calcicolous grassland on the Carboniferous Limestone of the Yorkshire Dales where meadows on till or head abut on to, or surround, rocky outcrops (cf. Ivimey-Cook & Proctor 1966*b* on comparable zonations in The Burren). In a similar fashion, there may be an increase of species typical of calcifugous grasslands and heaths in the *Danthonia* sub-community where its characteristic brown earths grade to rankers or podzolised soils over siliceous material.

The patchy occurrence of *Juncus* spp. or *F. ulmaria* within stands of the *Centaureo-Cynosuretum* represents truncated zonations to the *Holco-Juncetum* or Filipen-dulion mires which are mediated by the soil moisture status. Such patterns are a common feature of meadows on undulating topography and frequently accentuate ridge-and-furrow with fragmentary strips of *F. ulmaria* or *Juncus* spp. on the less well-drained soils of the furrows. Alongside streams there may be a more com-plete gradation to poor-fen vegetation.

Occasionally, differences in treatment may be evident as zonations within enclosed meadows as, for example, where narrow belts of rich *Centaureo-Cynosuretum* remain on steeper banks within fields or around margins which have escaped improvement. Generally, however, differences in treatment style have a gradual succession-al effect upon the community which is evident in a range of intermediates between the *Centaureo-Cynosur-etum* and more improved and productive grassland types. Two changes of practice are of particular importance.

The first is an increase of grazing pressure. Growth starts quite early in the regions where the community occurs and stands may provide a valuable supplemen-tary bite towards the end of April. If grazing is continued into the summer, and especially if it is heavy, there is an eventual decrease in the richness of the sward. Early-flowering species such as *Orchis morio* are then unable to set seed and there is a gradual expansion of rosette hemicryptophytes. The occurrence of the poisonous *Colchicum autumnale* in stands of the community has sometimes led to their being set aside for hay but some farmers have destroyed this species so as to allow an expansion of grazing. In fact, *C. autumnale* is more toxic during the early part of the season than in summer (Butcher 1954). Frequent mowing of the community, as occurs in some churchyards and on some verges and lawns, appears to have a similar effect to an increase in grazing. Although such treatment may permit the survival of an impoverished form of the *Centaureo-Cynosuretum*, the trend is for the sward to be converted eventually to the *Lolio-Cynosuretum*.

This change may be further encouraged by the second, now very widespread, alteration in treatment. This is the replacement of the traditional farmyard manure by artificial mineral fertilisers. These enhance the growth of the grasses to the detriment of the dicotyledons. *Lotus corniculatus* and *Leontodon hispidus* are often the first species to be lost from the vegetation but a much greater impoverishment accompanies the rise in productivity in the long term.

More drastic treatments of the community involve ploughing or the use of total weedkillers such as paraquat and re-seeding to produce *Lolio-Cynosuretum* or various kinds of ley.

Such artificial successions to pasture depend ultimately on the maintenance of grazing. When the *Centaureo-Cynosuretum* or its derivatives are ungrazed, there is an expansion of coarser grasses and an eventual invasion of shrubs. The *Lathyrus* and *Galium* sub-communities seem to progress to various types of *Arrhenatheretum* or, on more calcareous soils, to one of the coarser Mesobromion swards. The exact nature of the succession may also be influenced by the fertiliser regime on the original meadow. Ungrazed stands of the *Danthonia* sub-community may be directly invaded by heath shrubs or *Ulex europeaus*. The results of careless grazing are sometimes evident as mosaics of *Centaureo-Cynosuretum* with patches of coarse grassland or heath.

Distribution

Stands of the community occur throughout the British lowlands but the centre of distribution is on the claylands of the Midlands. Even here, however, agricultural improvement has drastically reduced the extent of the community. The sub-communities are distributed largely in relation to local and regional variations in soil type. The *Lathyrus* sub-community is the most widespread type with the *Galium* sub-community showing a more restricted occurrence, largely over calcareous bedrocks. The *Danthonia* sub-community extends the altitudinal range of the community on to the upland margins of the Welsh borderlands and northern England.

In the harsh montane climate of the northern Pennines the community is replaced by the *Anthoxanthum-Geranium* community. As with this other increasingly restricted meadow type, verge stands provide a valuable reserve.

Affinities

The *Centaureo-Cynosuretum* has been widely recognised in recent descriptive accounts with the increasing interest in the conservation of 'Old Meadows', although attention has generally been concentrated on richer stands and rarer species have been used for its characterisation. Phytosociologically, the community is distinct in its particular balance of more common and widespread species of agricultural grasslands and it clearly belongs among the pastures and meadows of the Cynosurion.

Although the great variety of agricultural treatments may blur the floristic distinctions between the community and the closely-related *Lolio-Cynosuretum*, the higher frequency (and often abundance) of the following species are especially characteristic of the *Centaureo-Cynosuretum*: *Lotus corniculatus, Centaurea nigra, Rhinanthus minor, Briza media, Carex flacca, Lathyrus pratensis, Leontodon hispidus, Chrysanthemum leucanthemum* and *Primula veris*. Most of these species are also characteristic of the *Anthoxanthum-Geranium* community but there the northern species *Geranium sylvaticum* and *Alchemilla glabra* effect an adequate separation and represent a transition to the alpine meadows of the Trisetion.

There is a considerable floristic overlap between the *Centaureo-Cynosuretum* and the coarse swards of the Arrhenatherion on the one hand and the calcicolous grasslands of the Mesobromion on the other. Distinctions in the former case are rarely difficult to make because *A. elatius* and tall herbs are usually totally excluded or held in check by grazing. There is, however, a close similarity between some stands of the *Galium* sub-community of the *Centaureo-Cynosuretum* and some Mesobromion swards such as the *Holcus-Trifolium* sub-community of the *Festuca ovina-Avenula pratensis* grassland. Similarly, it is sometimes difficult to make a distinction between the *Danthonia* sub-community and certain types of calcifugous grasslands. Problems of separation here reflect real lines of continuous floristic variation in relation to soil characteristics.

The occasional presence of species such as *Filipendula ulmaria* and *Juncus* spp. in the *Centaureo-Cynosuretum* represents floristic transitions to mires of the Filipendulion and Holco-Juncion. The lack of dominance by such species and the absence of the full range of associates usually serve to distinguish between the vegetation types involved.

The *Centaureo-Cynosuretum* was first described from

Eire by Braun-Blanquet & Tüxen (1952) and later, in greater detail, by O'Sullivan (1965). Phytosociological accounts have been provided from British localities by Birks (1973) and Shimwell (1968a, b). These descriptions all clearly correspond with the community as defined here although the status of the various subdivisions erected in the Irish accounts differs somewhat in the above. No equivalent vegetation type has been described from the Continent.

Floristic table MG5

	a	b	c	5
Festuca rubra	V (1–8)	V (2–8)	V (2–7)	V (1–8)
Cynosurus cristatus	V (1–8)	V (1–7)	V (1–7)	V (1–8)
Lotus corniculatus	V (1–7)	V (1–5)	V (2–4)	V (1–7)
Plantago lanceolata	V (1–7)	V (1–5)	IV (1–4)	V (1–7)
Holcus lanatus	IV (1–6)	IV (1–6)	V (1–5)	IV (1–6)
Dactylis glomerata	IV (1–7)	IV (1–6)	V (1–6)	IV (1–7)
Trifolium repens	IV (1–9)	IV (1–6)	V (1–4)	IV (1–9)
Centaurea nigra	IV (1–5)	IV (1–4)	V (2–4)	IV (1–5)
Agrostis capillaris	IV (1–7)	IV (1–7)	V (3–8)	IV (1–8)
Anthoxanthum odoratum	IV (1–7)	IV (1–8)	V (1–4)	IV (1–8)
Trifolium pratense	IV (1–5)	IV (1–4)	IV (1–3)	IV (1–5)
Lolium perenne	IV (1–8)	III (1–7)	I (2–3)	III (1–8)
Bellis perennis	III (1–7)	II (1–7)	I (4)	II (1–7)
Lathyrus pratensis	III (1–5)	I (1–3)	I (1)	II (1–5)
Leucanthemum vulgare	III (1–3)	I (1–3)	II (1–3)	II (1–3)
Festuca pratensis	II (1–5)	I (2–5)	I (1)	I (1–5)
Knautia arvensis	I (4)			I (4)
Juncus inflexus	I (3–5)			I (3–5)
Galium verum	I (1–6)	V (1–6)		II (1–6)
Trisetum flavescens	II (1–4)	IV (1–6)	II (1–3)	III (1–6)
Achillea millefolium	III (1–6)	V (1–4)	III (1–4)	III (1–6)
Carex flacca	I (1–4)	II (1–4)	I (1)	I (1–4)
Sanguisorba minor	I (4)	II (3–5)		I (3–5)
Koeleria macrantha	I (1)	II (1–6)		I (1–6)
Agrostis stolonifera	I (1–7)	II (1–6)	I (6)	I (1–7)
Festuca ovina		II (1–6)		I (1–6)
Prunella vulgaris	III (1–4)	III (1–4)	IV (1–3)	III (1–4)
Leontodon autumnalis	II (1–5)	II (1–3)	IV (1–4)	III (1–5)
Luzula campestris	II (1–4)	II (1–6)	IV (1–4)	III (1–6)
Danthonia decumbens	I (2–5)	I (1–3)	V (2–5)	I (1–5)
Potentilla erecta	I (1–4)	I (3)	V (1–4)	I (1–4)
Succisa pratensis	I (1–4)	I (1–5)	V (1–4)	I (1–5)
Pimpinella saxifraga	I (1–4)	I (1–4)	III (1–4)	I (1–4)
Stachys betonica	I (1–5)	I (1–4)	III (1–4)	I (1–5)
Carex caryophyllea	I (1–4)	I (1–3)	II (1–2)	I (1–4)
Conopodium majus	I (1–4)	I (1–5)	II (2–3)	I (1–5)
Ranunculus acris	IV (1–4)	II (1–4)	IV (2–4)	III (1–4)
Rumex acetosa	III (1–4)	III (1–4)	III (1–3)	III (1–4)
Hypochoeris radicata	III (1–5)	II (2–4)	III (1–4)	III (1–5)
Ranunculus bulbosus	III (1–7)	II (1–5)	III (1–2)	III (1–7)
Taraxacum officinale agg.	III (1–4)	III (1–4)	III (1–3)	III (1–4)

	a	b	c	5
Brachythecium rutabulum	II (1–6)	III (1–4)	II (2)	III (1–6)
Cerastium fontanum	III (1–3)	II (1–3)	II (1–3)	II (1–3)
Leontodon hispidus	II (1–6)	III (2–4)	III (1–5)	II (1–6)
Rhinanthus minor	II (1–5)	II (1–4)	II (1–3)	II (1–5)
Briza media	II (1–6)	III (1–4)	III (2–3)	II (1–6)
Heracleum sphondylium	II (1–5)	II (1–3)	III (1–3)	II (1–5)
Trifolium dubium	II (1–8)	II (1–5)	I (2)	II (1–8)
Primula veris	II (1–4)	II (2–4)	I (2)	II (1–4)
Arrhenatherum elatius	II (1–6)	II (1–7)	I (3–4)	II (1–7)
Cirsium arvense	II (1–3)	II (1–4)	I (1)	II (1–4)
Eurhynchium praelongum	II (1–5)	II (1–4)	I (1–2)	II (1–5)
Rhytidiadelphus squarrosus	II (1–7)	II (1–5)	III (1–4)	II (1–7)
Poa pratensis	II (1–6)	II (2–5)		II (1–6)
Poa trivialis	II (1–8)	I (1–3)	I (1–2)	II (1–8)
Veronica chamaedrys	II (1–4)	I (1–4)	I (1)	II (1–4)
Alopecurus pratensis	I (1–6)	I (1–4)	I (1)	I (1–6)
Cardamine pratensis	I (1–3)	I (1)	I (3)	I (1–3)
Vicia cracca	I (1–4)	I (1–3)	I (1–2)	I (1–4)
Bromus hordeaceus hordeaceus	I (1–6)	I (2–3)	I (3)	I (1–6)
Phleum pratense pratense	I (1–6)	I (1–5)	I (1)	I (1–6)
Juncus effusus	I (2–3)	I (3)	I (1–2)	I (1–3)
Phleum pratense bertolonii	I (1–3)	I (1–3)	I (1)	I (1–3)
Calliergon cuspidatum	I (1–5)	I (2–4)	I (3)	I (1–5)
Ranunculus repens	II (1–7)	I (2)	II (1–4)	I (1–7)
Pseudoscleropodium purum	I (1–5)	I (3–4)	II (2)	I (1–5)
Ophioglossum vulgatum	I (1–5)	I (1)		I (1–5)
Silaum silaus	I (1–5)	I (1–3)		I (1–5)
Agrimonia eupatoria	I (1–5)	I (1–3)		I (1–5)
Avenula pubescens	I (1–3)	I (2–5)		I (1–5)
Plantago media	I (1–4)	I (1–4)		I (1–4)
Alchemilla glabra	I (2)	I (3)		I (2–3)
Alchemilla filicaulis vestita	I (1–3)	I (3)		I (1–3)
Alchemilla xanthochlora	I (1–3)	I (2)		I (1–3)
Carex panicea	I (1–4)	I (2–4)		I (1–4)
Colchicum autumnale	I (3–4)	I (1–3)		I (1–4)
Crepis capillaris	I (1–5)	I (3)		I (1–5)
Festuca arundinacea	I (1–5)	I (3–5)		I (1–5)
Potentilla reptans	I (1–6)	I (1–4)		I (1–6)
Senecio jacobaea	I (1–3)	II (1–4)		I (1–4)
Filipendula ulmaria	I (1–5)		I (1)	I (1–5)
Juncus articulatus	I (1–5)		II (2–3)	I (1–5)
Number of samples	137	42	15	194
Number of species/sample	22 (13–32)	26 (12–38)	22 (18–27)	23 (12–38)

a   *Lathyrus pratensis* sub-community
b   *Galium verum* sub-community
c   *Danthonia decumbens* sub-community
5   *Centaureo-Cynosuretum cristati* (total)

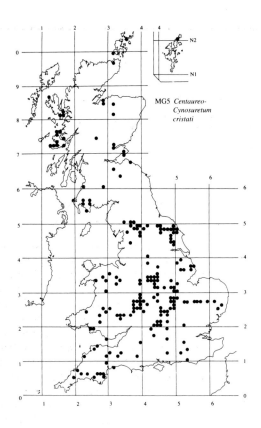

MG5 *Centaureo-*
Cynosuretum
cristati

MG6
Lolium perenne-Cynosurus cristatus grassland
Lolio-Cynosuretum cristati (Br.-Bl. & De Leeuw 1936) R.Tx. 1937

Synonymy
Fatting Pastures and Dairy Pastures Stapledon 1925; First-Grade Ryegrass Pasture and Second-Grade Ryegrass Pasture Davies 1941 & 1952, Beddows 1967; Reverted Pasture Duffey et al. 1974; Ordinary dry meadows Ratcliffe 1977; Trifolio-Agrosto-Festucetum Evans et al. 1977 p.p.

Constant species
Cerastium fontanum, Cynosurus cristatus, Festuca rubra, Holcus lanatus, Lolium perenne, Trifolium repens.

Physiognomy
The Lolio-Cynosuretum generally has a short, tight sward which is grass-dominated. Lolium perenne is usually the most abundant grass with varying amounts of Cynosurus cristatus. In younger, re-sown grasslands, C. cristatus may be rare and, with good management, it generally remains a subsidiary species. With less intensive grazing, however, it may attain dominance and it is then especially conspicuous in winter by the abundance of its dead wiry flowering stems. Festuca rubra and Agrostis capillaris are frequent throughout and, in long-established pasture, they may be abundant. Holcus lanatus and Dactylis glomerata are also frequent but of somewhat patchy distribution. They may become more prominent as coarse tussocks if pasture is under-grazed and H. lanatus is often abundant and vigorous around cattle dung which the animals avoid. Poa pratensis and P. trivialis are the only other grasses frequent throughout.

The dicotyledonous component of the sward is rather unvaried in its composition but individual species may be prominent in particular stands. The most frequent and abundant species overall is Trifolium repens which often attains co-dominance with L. perenne in well-managed swards. Cerastium fontanum, Plantago lanceolata, Ranunculus acris, Achillea millefolium and Bellis perennis are also frequent but generally at low cover. Diversity may be increased by the persistence of meadow species where the community has been derived from such grasslands as the Centaureo-Cynosuretum or by the occurrence of a variety of ephemerals alongside footpaths, around gateways or on patches of bare soil exposed by poaching. Bromus hordeaceus ssp. hordeaceus, Medicago lupulina, Trifolium dubium, Poa annua and Hordeum murinum may be locally abundant in such situations. Tall herbs are generally rare but Urtica dioica, Heracleum sphondylium and Anthriscus sylvestris may be prominent where there is soil eutrophication and disturbance around gateways or along field margins.

Of greater importance for the management of the Lolio-Cynosuretum is the frequent occurrence of Senecio jacobaea and Cirsium arvense as abundant and tenacious weeds. S. jacobaea is a biennial which seeds prolifically and which can rapidly colonise over-grazed pastures and areas of bare soil, especially in times of drought or where there is infestation by rabbits. It is not unpalatable to all stock (sheep can eat it) but in quantity it is lethal. Regeneration can occur from root-buds so that defoliation with herbicides may be unsuccessful as a means of eradication. Cirsium arvense is a perennial which spreads readily by the production of nodal shoots from extensive adventitious roots. This more than compensates for its usually dioecious sexuality and its frequent occurrence in single-sex populations. Again, herbicide treatment or, in this case, cutting or even ploughing, may be inadequate for control (Salisbury 1964).

Bryophytes are generally present in the sward at low cover with scattered plants of Brachythecium rutabulum, Eurhynchium praelongum and Rhytidiadelphus squarrosus. B. rutabulum occasionally shows great abundance on moist patches of bare soil.

Sub-communities

Typical sub-community: Lolio-Cynosuretum typicum (Br.-Bl. & De Leeuw 1936) R.Tx. 1937. Here, the general floristics are the same as described above, except that Agrostis stolonifera, Ranunculus repens and Cirsium vulgare are slightly preferential. However, this is a somewhat variable sub-community including, as well as a range of older re-seeded grasslands, a variety of types which could be considered as variants.

Alopecurus pratensis variant. Regularly-mown *Lolio-Cynosuretum* often shows an abundance of *A. pratensis* which may attain co-dominance with taller forms of *L. perenne*. *Ranunculus acris* is slightly preferential.

Alopecurus geniculatus variant. Over the frequently puddled topography of seasonally-inundated areas of pasture alongside ponds and streams, *A. geniculatus* is a constant and sometimes abundant component of an uneven sward.

Deschampsia cespitosa variant. *D. cespitosa* often occurs patchily in undulating ill-drained pastures. It may be grazed, which prevents its coarse tussocks overwhelming the sward. *Potentilla reptans* and *Ranunculus repens* show a slight preference for this variant and there is usually a prominent bryophyte component with sometimes abundant *B. rutabulum*, *Calliergon cuspidatum* and *Plagiomnium cuspidatum*.

Iris pseudacorus variant: *Lolio-Cynosuretum iridetosum* O'Sullivan 1965; *Lolio-Cynosuretum juncetosum, Iris pseudacorus* variant O'Sullivan 1968*b p.p.*; *Lolio-Cynosuretum*, Sub-Association of *Lotus uliginosus* (Br.-Bl. & De Leeuw 1936) R.Tx. 1937 *p.p.* In certain parts of the country, *I. pseudacorus* is a prominent member of river-valley stands of *Lolio-Cynosuretum* which are flooded in winter, sometimes for considerable periods. The shoots may attain a height of 150 cm though they are readily grazed back to the rhizome stumps by cattle and sheep.

Anthoxanthum odoratum sub-community: *Lolio-Cynosuretum*, Sub-Association of *Luzula campestris* (Br.-Bl. & De Leeuw 1936) R.Tx 1937; *Trifolio-Agrosto-Festucetum* Evans *et al.* 1977 *p.p.* In the rather richer vegetation of this sub-community, *L. perenne* and *C. cristatus* generally share dominance with *F. rubra* and *A. capillaris*. *H. lanatus*, *D. glomerata* and *P. pratensis* remain frequent but the distinctive preferential grass is *A. odoratum* which is present in small amounts. Most of the dicotyledons of the community are frequent here, although *T. repens* is not as generally abundant as in the Typical sub-community, and *Rumex acetosa*, *Hypochoeris radicata* and *Luzula campestris* are preferential. There are occasional records for hay-meadow species such as *Centaurea nigra*, *Leucanthemum vulgare* and *Leontodon hispidus* but *Lotus corniculatus* and *Lathyrus pratensis* are usually absent. Bryophytes are frequent with *Rhytidiadelphus squarrosus* preferential and sometimes abundant.

Trisetum flavescens sub-community: *Lolio-Cynosuretum plantaginetosum mediae* Heinemann *p.p.* in LeBrun *et al.* 1949. The grass dominants here are the same as in the

Anthoxanthum sub-community but *T. flavescens* and *Phleum pratense* ssp. *bertolonii* are constant preferentials. There are also occasional records for a variety of more calcicolous species such as *Sanguisorba minor*, *Pimpinella saxifraga*, *Galium verum* and *Saxifraga granulata* as well as the Mesobromion grasses *Festuca ovina* and *Avenula pubescens*.

Habitat

The *Lolio-Cynosuretum* is the major permanent pasture type on moist but freely-draining circumneutral brown soils in lowland Britain. Enclosed stands in farm fields are virtually ubiquitous and form the bulk of dairying and fatting pasture in many parts of the country, as well as providing occasional crops of hay or silage. The community is also widespread as a recreational sward and on village greens, road verges and lawns.

All the major grasses of the community are palatable but *L. perenne*, the usual dominant, has long been considered our most valuable herbage grass (e.g. Marshall 1789, Hubbard 1984). The proportion of *L. perenne* in pasture swards was first used by Davies (1941) to distinguish between first- (30% or more *L. perenne*), second- (15–30%), third- (about 12%) and fourth-grade (trace or absent) types. The more detailed floristic profiles of first- and second-grade pastures given by Beddows (1967, after Jenkin 1923 and Davies 1952) clearly correspond to the *Lolio-Cynosuretum*. Even in its wild forms, *L. perenne* is nutritious and often very productive and rapidly-growing but the selection of first, nineteenth-century varieties and, later, Aberystwyth S cultivars has made available a wide range of forms, some early, others late, some stemmy and relatively short-lived, others compact, very leafy and more persistent (Beddows 1953, 1967, Hubbard 1984). These, and a variety of cultivars of the most important herbage dicotyledon, *T. repens*, have been seeded in to permanent pasture, and into long-leys which have become permanent, to produce a range of grasslands which now approximate to the *Lolio-Cynosuretum*.

Good pastoral treatment is essential for the maintenance of productive pasture stands. Many of the grasses in the community are winter-green and fields can be grazed, in rotation, throughout the year. The usual stock are cattle with some use also for sheep. In the growing season, the number of stock is ideally maintained just to control the growth of the herbage, then reduced if there is any drop in availability of bite and increased once again when there has been a recovery of the sward (Davies 1952, Moore 1966, Beddows 1967). Good pasture stands of the *Lolio-Cynosuretum* have a dense herbage in which there is a balance of the different components and little bare ground. Heavy grazing can encourage the spread of rosette hemicryptophytes at the expense of the more nutritious grasses. Further, although some consolidation of the sward is conducive to the growth of many species (including *L. perenne*: see Beddows 1967), heavy trampling breaks up the surface and may cause severe poaching, especially on wetter ground in winter. This may allow the spread of ephemeral weeds on to the exposed patches of bare soil in the following spring. Where grazing is too light, the low-yielding *C. cristatus* may spread, together with the coarser grasses *H. lanatus* and *D. glomerata*. Tussocky patches of grass with lank rosette species are a common feature of the avoidance-mosaics which develop around dung-pats in cattle-grazed *Lolio-Cynosuretum*. Once there is a taller sward, certain species (e.g. *T. repens*: see Donald 1961) may be shaded out, *L. perenne* may show a reduced ability to tiller (Beddows 1967) and the vegetation may be more susceptible to lose chlorophyll with 'burn' in harsh winter winds (Davies & Fagan 1938, Beddows & Jones 1958).

The *Lolio-Cynosuretum* is not normally treated as permanent hay-meadow but it is a widespread practice for fields to be shut up in spring at irregular intervals for summer-mowing. In such circumstances, there are no floristic differences between mown and unmown stands. However, where there is more regular mowing, *Alopecurus pratensis* tends to be prominent and this species is often sown in, together with taller cultivars of *L. perenne*, where the *Lolio-Cynosuretum* is to be used in this way (see *Alopecurus pratensis* variant above). The occasional presence of species characteristic of traditional hay-meadows is often indicative of the development of the *Lolio-Cynosuretum* from such swards by re-seeding, fertiliser application and an increase in grazing.

Regular grazing, with or without occasional mowing, can maintain sown or derived stands of the community in artificial habitats such as road verges. On coarse recreational swards, village greens and lawns, where the *Lolio-Cynosuretum* occurs widely, regular mowing may approximate to close grazing in its effect on the physiognomy and floristics of the vegetation, though here there is not the input of natural manures that stock provide to counteract the continual removal of clippings. In such situations as these, trampling is often an additional problem and heavy use may cause an increase in such resistant species as *Bellis perennis* and *Plantago major* as well as permitting the spread of ephemerals on scuffed patches of bare soil.

The *Lolio-Cynosuretum* occurs on a variety of brown soils over a wide range of bedrock types and superficial deposits but it thrives best on deep loams which are moist but free-draining, mesotrophic and of circumneutral pH. These features correspond to the edaphic optimum for *L. perenne* (Beddows 1967). The maintenance of fertility is essential in pasture stands and the dung and urine from stock are generally supplemented with natural or, more usually, artificial fertilisers. Where the community occurs on brown earth soils, as is usual in the Typical and *Anthoxanthum* sub-communities, liming may be essential to counteract surface leaching, especially on lighter-textured soils or in areas of higher rainfall. *L. perenne* will not persist in upgraded upland pastures unless adequate supplies of lime and phosphate are provided (Thomas 1936). The brown calcareous soils characteristic of the *Trisetum* sub-community do not normally show such a lime-deficiency but they may be quite oligotrophic without the application of fertilisers. Along roads, soils may be eutrophicated and limed by the application of rock-salt in winter and the throwing up of limestone ballast and dust by traffic. Upland roads sometimes have a strip of *Lolio-Cynosuretum* related to such enrichment along the verge edge and this vegetation is avidly eaten by roaming sheep and ponies. Moderate applications of salt or inundation by sea-water do not appear to affect the growth of *L. perenne* adversely (Chippendale 1954, Beddows 1967).

Although the community will persist on heavy soils or under occasional inundation by fresh-water, effective drainage is essential for maintaining pasture stands of the *Lolio-Cynosuretum* in good heart. Where soils are ill-draining and gleyed, the *Deschampsia* variant of the Typical sub-community is characteristic and this vegetation often occurs in hollows in undulating pastures on till, between the ridges of ridge-and-furrow and in patchy lines between inadequately spaced drains. The *Iris* variant is typical of soils which are inundated, sometimes for long periods, in winter and which remain saturated for much of the summer. More sporadic inundation, often coupled with the effects of trampling by stock around drinking places, leads to the development of the *Alopecurus geniculatus* variant.

Zonation and succession
The application of careful uniform treatment within enclosed pasture stands of the *Lolio-Cynosuretum* tends to minimise the occurrence of zonations. However, where past improvement has been uneven, transitions may remain. Steeper banks inaccessible to ploughing may still carry patches of *Centaureo-Cynosuretum* as remnants of previous meadow vegetation. Hollows resistant to drainage may have the *Holcus-Deschampsia* community or *Holco-Juncetum*; the *Deschampsia* and *Iris* variants of the Typical sub-community can be seen as truncated zonations to such vegetation which occur where drainage is less successful. Where rivers and pools are not embanked, the *Alopecurus geniculatus* variant represents part of a transition to the vegetation of regularly-inundated water margins. Where pastures have been won from rocky upland topography, there may be zonations to calcifugous grasslands (with the *Anthoxanthum* sub-community) or calcicolous grass-

lands (with the *Trisetum* sub-community) as the soil thins to some type of lithomorphic profile around outcrops.

The *Lolio-Cynosuretum* can be derived from a very wide range of vegetation types (Figure 9) including, in extreme cases, communities of stabilised sand-dunes and blanket mire. Certain kinds of agricultural treatment or neglect may allow some of the more closely related types of original unimproved vegetation to replace the *Lolio-Cynosuretum*. In older pastures, a return to more traditional organic manuring and the withdrawal of summer grazing may permit the re-establishment of the meadow vegetation of the *Centaureo-Cynosuretum*. This is perhaps more likely in the *Anthoxanthum* and *Trisetum* sub-communities where certain meadow species persist or where there are fragments of meadow-sward on banks or adjacent verges. Similarly, the choking of drains may permit the development of extensive *Holco-Juncetum*.

Generally, however, successions involving the *Lolio-Cynosuretum* appear to be mediated by grazing. Under-

grazing allows coarser species to increase their cover and this is sometimes a prelude to the invasion and spread of *Arrhenatherum elatius* and the development of an *Arrhenatheretum*. Abandoned pasture often shows a patchy mosaic of the two communities with scattered saplings of *Crataegus monogyna*. Where grazing is too severe, the sward may be opened up for invasion by weeds and rabbit infestation may speed the run-down to some weed-dominated vegetation. Trampling may be an attendant problem with over-grazing by cattle but is especially important in recreational and amenity stands of the *Lolio-Cynosuretum* where heavy use along footpaths, around gateways and in goal-mouths may produce a succession to Lolio-Plantaginion vegetation.

Distribution

The *Lolio-Cynosuretum* is a virtually ubiquitous community of the British lowlands, occurring wherever there has been intensive improvement for pasturing. It is particularly abundant and important in the major dairying areas of the wetter west (the South-West Peninsula, south Wales, the Welsh Borders, parts of the Midlands and the Scottish lowlands) and in the traditional fatting regions (Dorset, Romney Marsh, Leicestershire and Northamptonshire, the Lindsey District of Lincoln-

Figure 9. Convergence and loss of diversity among grasslands with continuing agricultural improvement.

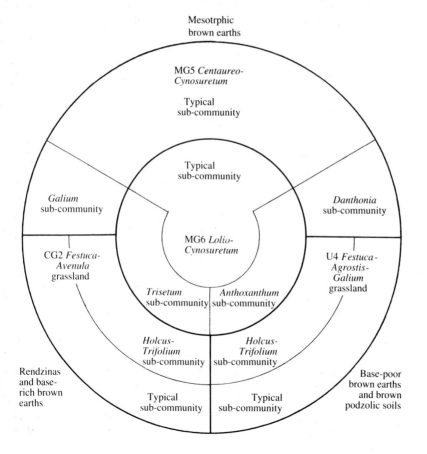

pratensis and *L. perenne*. *A. pratensis* tends to be especially conspicuous early in the growing season, forming almost pure clumps of vigorous, dark green herbage. *F. pratensis* expands later and may assume dominance by mid-summer. *L. perenne* tends to be more prominent where stands are grazed and, where *F. pratensis* and *L. perenne* occur in close proximity, the intergeneric hybrid × *Festulolium loliaceum* may occur as vigorous clumps. Other frequent grasses are *P. trivialis* and *D. glomerata* and, in some older swards, there is abundant *H. lanatus* and *Agrostis stolonifera*. *Alopecurus geniculatus* and *Glyceria fluitans* may be prominent on permanently wet and puddled ground. The dicotyledonous component of the sward is generally not well developed, although it is somewhat more varied than in the *Lolium-Trifolium* and *Lolium-Poa* leys. *T. repens*, *C. fontanum* and *Ranunculus acris* are frequent at low cover and *T. pratense*, *Plantago lanceolata*, *Rumex acetosa* and *Lathyrus pratensis* occur occasionally. *Cirsium arvense* and *Ranunculus repens* are sometimes prominent.

This distinctive grassland is typically found on seasonally-flooded stretches of lowland river valleys in central and southern England and south Wales. It is particularly frequent along the Hampshire Avon, the Sussex Ouse and around the mid-reaches of the Severn in Gloucestershire, Warwickshire and Hereford and Worcester. Here it often forms extensive uniform stands on alluvial soils and drained fen peats. The vegetation is often under water for a considerable part of the winter and soils may remain wet for most of the year, although they are often light-textured and can be readily drained. Most stands are used for grazing by cattle (the ground being often too wet for sheep) but the sward is highly susceptible to poaching if used in winter. The regular deposition of silt helps maintain fertility but in many cases ploughing and re-seeding with *A. pratensis* and *F. pratensis* (both valuable grasses on moister soils) have been used to increase productivity. The community also occurs more widely as small stands in wet depressions within meadows, when it forms part of an annual hay crop, and on damp road verges. It often grades to the *Holcus-Deschampsia* community on heavy gleyed soils and to the *Festuca-Agrostis-Potentilla* inundation grassland on less well-managed and brackish sites.

It seems likely that some of the river-valley stands of this community have been derived from the vegetation of traditionally managed water-meadows. These have been either neglected during the present century, running to scrub and woodland, or converted by intensive drainage and management to the *Lolio-Cynosuretum*. However, the grass component of the working water-meadow (Fream 1888a, Duffey *et al.* 1974) was very similar to that here and it is possible that where annual inundation by standing water has replaced reg-

ular controlled inundation by running water, this community represents a modified relic of a now defunct style of grassland management. On West Sedgemoor on the Somerset Levels, the community occurs upslope from the *Cynosurus cristatus-Caltha palustris* flood-pasture, the vegetation which perhaps represents the natural precursor of the traditional flood-meadow community (see also O'Sullivan 1968b).

Lolium perenne-Alopecurus pratensis grassland
Lolium perenne-Alopecurus pratensis Nodum   Page 1980

Apart from the absence of *Festuca pratensis*, the floristics of this grassland resemble those of the *Lolium-Alopecurus-Festuca* flood-pasture. *L. perenne* and *A. pratensis* are generally co-dominant as a tall sward with smaller amounts of *D. glomerata*. *H. lanatus*, *Agrostis capillaris*, *A. stolonifera* and *Anthoxanthum odoratum* are occasional. The most frequent dicotyledons are *Taraxacum officinale* agg., *Ranunculus repens*, *Cerastium fontanum*, *Rumex acetosa* and *Trifolium pratense*. Bryophytes are sparse.

This community is commonly treated as hay-meadow, although it is occasionally encountered in damp pastures and on road verges. It is most characteristic of moist and fertile alluvial soils in lowland river valleys where there is less frequent inundation and/or better drainage than is characteristic of the *Lolium-Alopecurus-Festuca* flood-pasture. It is one stage nearer the *Lolio-Cynosuretum* than that community and a sequence of the three vegetation types in relation to soil moisture can be seen at West Sedgemoor on the Somerset Levels.

Lolium perenne-Plantago lanceolata grassland
Lolio-Plantaginetum   (Link 1921) Beger 1930 *emend.* Sissingh 1969

The composition of the somewhat varied swards included here depends largely on age. Young stands are generally dominated by *L. perenne* with some *T. repens*, the two species most commonly included in seed mixtures. With age, the cover of *L. perenne* tends to decline as it is partially replaced by *H. lanatus* and *D. glomerata* and, to a lesser extent, by *F. rubra*, *A. capillaris* and *P. trivialis*. In these older stands, *Plantago lanceolata* becomes constant and it may be abundant, especially where the sward is kept short. *P. major* and *Taraxacum officinale* agg. are also very frequent, the latter being especially conspicuous in spring on verge edges when it is flowering. *Cerastium fontanum*, *Bellis perennis*, *Rumex acetosa*, *Trifolium pratense*, *T. dubium* and *Vicia sativa* ssp. *nigra* are occasional. *Brachythecium rutabulum* and *Eurhynchium praelongum* occur scattered through the sward and the former sometimes attains abundance on damp patches of bare soil.

The community occurs widely on re-seeded verges and lawns which are regularly mown and which receive only moderate trampling. Restriction of mowing to an annual cut and an absence of trampling may permit the invasion of *Arrhenatherum elatius* and the development of an *Arrhenatheretum*. Heavy trampling on verge edges and around gateways opens up the sward and leads to the development of the more weedy communities of the Lolio-Plantaginion. If stands are grazed and fertilised, they can be readily converted to the *Lolio-Cynosuretum*.

Lolium perenne-Poa pratensis grassland
Poo-Lolietum perennis De Vries & Westhoff *apud* Bakker 1965

In general, this vegetation resembles the *Lolio-Plantaginetum* but here the most prominent species are grasses with conduplicate leaves (*L. perenne*, *P. pratensis* and *D.*

glomerata usually co-dominate), rosette dicotyledons resistant to heavy trampling (e.g. *Plantago major* and *Bellis perennis*) and ephemerals of open ground (e.g. *Poa annua*, *Capsella bursa-pastoris* and, especially where dogs urinate, *Hordeum murinum*). Bryophytes are frequent with *Brachythecium rutabulum* and *Eurhynchium praelongum* the most common species.

P. pratensis is not a widely-used agricultural grass but it is commonly included in seed mixtures for banks, verges and coarse recreational swards where there is substantial trampling pressure. The vegetation can survive fairly heavy use but in wet weather the sward may be broken up with severe poaching. It is also somewhat sensitive to drought, when foliage destroyed by trampling cannot quickly be replaced. *Poa annua* is readily able to exploit bare patches which originate in these ways and, without remedial action, weedy Lolio-Plantaginion vegetation may develop. The community is also commonly found as a transitional zone between such vegetation in gateways and intact pasture of the *Lolio-Cynosuretum* beyond.

Floristic table MG7

	a	b	c	d	e	f
Lolium perenne	V (1–9)	V (2–9)	IV (2–7)	V (4–8)	V (2–8)	V (3–9)
Phleum pratense pratense	II (1–5)	IV (1–6)	I (3–5)	I (3)	I (3–4)	I (2–7)
Poa trivialis	I (3–4)	V (3–8)	III (2–6)	II (3–6)	II (2–5)	I (3)
Lolium multiflorum	II (2–6)					
Alopecurus pratensis		I (3)	V (2–8)	V (2–7)	I (1)	
Festuca pratensis	I (2)	I (3–5)	V (1–8)			
Ranunculus acris	I (2–4)	I (2–3)	III (2–5)	II (3–5)	I (3)	I (1–3)
Rumex acetosa		I (1)	III (1–4)	III (1–4)	II (2)	I (1–4)
Agrostis stolonifera		I (4–5)	II (2–8)	II (2–5)	I (3–4)	I (1–2)
Anthoxanthum odoratum	I (8)		II (3–6)	II (4–5)		
Lathyrus pratensis			II (2–3)	I (4)		
Plantago lanceolata	I (3)	I (3)	II (1–4)	II (2–4)	V (2–5)	III (2–4)
Poa pratensis		I (1–2)	I (3)	II (2–4)	I (2)	V (3–9)
Plantago major	I (5)	I (1–5)			III (1–4)	II (1–4)
Bellis perennis	I (3)	I (2–4)	I (3)	I (3)	II (2–6)	II (1–5)
Poa annua	I (3–5)	I (3–5)	I (1–2)	I (2–4)	II (2–4)	II (2–4)
Achillea millefolium	I (4)		I (3)	I (3–4)	II (2–5)	II (1–9)
Vicia sativa nigra	I (5)	I (3)	I (3)		II (1–3)	II (1–4)
Trifolium dubium			I (3)		II (1–4)	II (2–8)
Dactylis glomerata	III (1–7)	IV (1–8)	III (1–5)	V (1–5)	V (1–8)	IV (2–6)
Trifolium repens	III (2–7)	V (2–8)	III (2–8)	II (2–4)	III (2–5)	III (1–6)
Taraxacum officinale agg.	II (1–3)	II (3–4)	III (1–5)	IV (2–4)	V (1–5)	III (1–4)
Holcus lanatus	I (4–6)	III (2–5)	III (1–6)	III (2–5)	IV (3–8)	III (1–5)
Festuca rubra	I (3)	I (2–6)	II (2–5)	II (2–6)	III (1–8)	II (2–7)
Cerastium fontanum	I (2–4)	III (1–4)	III (1–4)	II (2–3)	III (1–3)	III (2–3)
Agrostis capillaris	I (3)	II (2–7)	III (3–5)	III (3–5)	III (4–6)	II (3–7)
Ranunculus repens	II (3–4)	III (1–8)	II (1–7)	III (3–8)	II (1–4)	II (2–6)
Trifolium pratense	II (3–7)	I (2–3)	II (1–4)	II (2–3)	II (4–5)	II (3–5)
Cirsium arvense	I (3–6)	II (1–5)	II (1–5)	I (2–3)	I (1–4)	I (1)
Bromus hordeaceus hordeaceus	I (1)	II (3–8)	I (3–7)	I (3–6)	I (8)	I (3–7)

Floristic table MG7 (*cont.*)

	a	b	c	d	e	f
Stellaria media	I (2–3)	I (1–2)	I (1–3)	I (3)		I (2)
Rumex crispus	I (2)	I (2)	I (4)	I (3)		I (1–3)
Brachythecium rutabulum		II (1–6)	II (2–3)	I (1–3)	II (3–8)	II (2–6)
Eurhynchium praelongum		II (1–5)	II (2–4)	I (2)	I (1–4)	II (1–7)
Rumex obtusifolius		II (1)	I (2–4)	I (5)	I (6)	I (1–3)
Cerastium glomeratum		I (1)	I (1)		I (4)	I (2–4)
Ranunculus bulbosus			I (3)	I (2–6)	I (1)	II (3–5)
Elymus repens			I (1–2)	I (5–7)	I (4)	I (5)
Cardamine pratensis	I (3)		I (2)			I (1)
Capsella bursa-pastoris	I (3)					I (2–4)
Rumex conglomeratus	I (4)				I (1)	I (2–4)
Cirsium vulgare		I (1)		I (4)		I (1)
Hypochoeris radicata			I (6)		I (3–5)	I (1–3)
Juncus inflexus		I (2)	I (1–5)			
Leontodon autumnalis			I (2)			I (3)
Vicia cracca			I (4)		I (2)	
Arrhenatherum elatius			I (2–3)		I (2)	
Number of samples	19	19	19	14	14	26
Number of species/sample	5 (2–8)	8 (4–14)	11 (4–19)	9 (3–14)	10 (5–15)	10 (4–17)

a *Lolium perenne-Trifolium repens* leys
b *Lolium perenne-Poa trivialis* leys
c *Lolium perenne-Alopecurus pratensis-Festuca pratensis* grassland
d *Lolium perenne-Alopecurus pratensis* grassland
e *Lolio-Plantaginetum*
f *Poo-Lolietum perennis*

MG8
Cynosurus cristatus-Caltha palustris grassland

Synonymy
Water meadows Fream 1888*a*, Duffey *et al.* 1974, Ratcliffe 1977, all *p.p.*; *Senecioni-Brometum racemosi* R.Tx. & Preising 1951.

Constant species
Anthoxanthum odoratum, Caltha palustris, Cerastium fontanum, Cynosurus cristatus, Festuca rubra, Holcus lanatus, Leontodon autumnalis, Poa trivialis, Ranunculus acris, Rumex acetosa, Trifolium repens.

Physiognomy
The *Cynosurus-Caltha* community is a species-rich and varied grassland with no single species consistently dominant. Grasses generally account for most of the cover and all five of the constant species may be abundant. A variety of other grasses may be present but these are much less frequent and only occasionally prominent: *Festuca pratensis, Lolium perenne, Agrostis stolonifera, A. capillaris, Dactylis glomerata* and *Briza media*. There are almost always some sedges in the sward, although some of these are smaller species with far-creeping rhizomes and easily under-estimated in sampling. *Carex panicea* and *C. disticha* are the most frequent and abundant with occasional records for *C. flacca, C. nigra, C. demissa, C. ovalis* and *C. hirta*. Junci are uncommon in this vegetation and never dominate.

Dicotyledons are generally well represented and certain species are sometimes sufficiently abundant to give a distinctive stamp to the physiognomy. Notable among these are *Caltha palustris* which is unpalatable to stock and whose large leaves are often prominent in the sward and *Filipendula ulmaria* which, when flowering, may protrude above the other herbage. *Ranunculus acris, R. repens, Trifolium repens* and *T. pratense* are frequent and sometimes abundant with usually smaller amounts of *Leontodon autumnalis, Cerastium fontanum, Bellis perennis* and *Plantago lanceolata*. Among the occasionals are species characteristic of relatively unimproved meadows (e.g. *Sanguisorba officinalis, Lathyrus*

pratensis, Leucanthemum vulgare, Centaurea nigra, Alchemilla glabra and *A. xanthochlora*) and a wide range of poor-fen species (e.g. *Cardamine pratensis, Achillea ptarmica, Lotus uliginosus, Lychnis flos-cuculi, Angelica sylvestris, Valeriana dioica, Galium palustre* and *G. uliginosum*). Other notable species at low frequency are *Geum rivale, Senecio aquaticus, Myosotis scorpioides* and, to the north and west, *Crepis paludosa* and *Trollius europaeus*. Although not recorded in the sampling, the rather local *Bromus racemosus* occurs in this community.

Bryophytes are somewhat patchy but *Calliergon cuspidatum* may be conspicuous with, less frequently, *Brachythecium rutabulum, Eurhynchium praelongum* and *Plagiomnium rostratum*.

Habitat
The community is characteristic of periodically inundated land which has been treated in traditional fashion, usually as pasture. It is most frequent and extensive on the flat or slightly sloping ground by rivers and streams which show seasonal flooding but it also occurs as more fragmentary stands below springs, flushes and seepage lines which produce a trickle of moderately calcareous water. Such sites typically carry gleyed brown earths, gleyed brown calcareous earths or surface-water gleys, of rather silty texture above and sometimes with a humose topsoil.

The soils are naturally enriched by an input of salts from deposited silt or moving water and they do not generally seem to have been improved by anything other than organic manures. The community is characteristically managed as pasture, being grazed by cattle and horses, though not usually by sheep for which the land is often too wet. Poaching may be a severe problem if the vegetation is heavily grazed while the ground is wet. Occasionally a hay crop is taken.

This community seems to be the naturally occurring vegetation which was managed in the past as water-meadow. This type of land use was developed in the sixteenth and seventeenth centuries as a means of providing a much-needed supplement to spring grazing for the increasing numbers of sheep on the southern Chalk. In its primitive form, a water-meadow comprised a sloping hillside along whose contours was dug a series of channels or leats fed from a damned stream. Water was released from the leats through sluices and trickled down over the sward. In time, more sophisticated arrangements extended the system to the gently-sloping land of valley bottoms. In some places, wide valleys were subject to a controlled inundation by 'drowning' or 'floating upwards' where damned water was forced back over entire meadows. More usually, the leat system was developed into a complex series of graded channels running along the tops of ridges through the fields. The aim in 'floating downwards' was to ensure a continuous movement of water along the full channels, down over the gently-sloping sides of the ridges and back along the furrows into the stream or river. The neglected physical remains of these systems, which reached the peak of their development between 1700 and 1850, are widely distributed in southern chalkland valleys and, less extensively, in the Midlands (Taylor 1975).

The value of the water-meadow system was that it provided a supply of water to warm and enrich the soils in spring and so stimulate an early bite from the sward. The vegetation was usually laboriously hand-weeded to remove any of the coarser and unpalatable species and this selected for a grass-dominated sward which was highly productive and palatable. A comparison between the list provided from water-meadows by Fream (1888a in Tansley 1939) and the vegetation included here shows a close correspondence except that certain grasses, notably *Festuca pratensis*, *F. arundinacea* and their hybrids with *Lolium perenne*, are not normal constituents of the *Cynosurus-Caltha* community, although they still occur in the vegetation of some surviving water-meadow stands. Nowadays, these species are much more frequently encountered in other kinds of flood-pasture.

Zonation and succession
Enclosed stands of the community in riverside pastures may be subject to uniform treatment and show no zonations to other vegetation types. In some cases, however, improvement and management have been restricted by difficult topography and an inability to prevent flooding. Here, and around the more fragmentary spring and flush stands, there may be zonations related to soil moisture conditions. A usual pattern is for the *Cynosurus-Caltha* community to give way to inundation communities on the bare substrates of river banks or pool-sides or to small-sedge mires of the Caricion davallianae in base-rich flushes. On drier ground, there may be a transition to the *Lolium-Alopecurus-Festuca* flood-pasture or, where there has been extensive improvement, directly to *Lolio-Cynosuretum*. Such a sequence can be seen on the drained levels of West Sedgemoor in Somerset.

Artificial drainage can mediate a successional sequence from the *Cynosurus-Caltha* community to drier pasture types but neglect permits a fairly rapid invasion of *Salix* spp. and other shrubs and trees of wet woodland.

Distribution
The community has a widespread but rather local distribution throughout the British lowlands. Most water-meadow stands have been either totally neglected or drained and improved and water-meadows worked in the traditional fashion are now very rare although a few remain in chalkland valleys in Wiltshire, Dorset and Hampshire (Ratcliffe 1977).

Affinities
The *Cynosurus-Caltha* community has attracted attention in British descriptive accounts almost exclusively in the modified form of the vegetation which survives in water-meadows. Even in its more natural form, it is, however, a distinctive vegetation type with its combination of Cynosurion pasture species and poor-fen dicotyledons. Other broadly similar communities can generally be distinguished from the *Cynosurus-Caltha* grassland by the physiognomic dominance of *Deschampsia cespitosa* (*Holcus-Deschampsia* community) or Junci (*Holco-Juncetum*).

Vegetation of this kind has been described from West Germany and The Netherlands as the *Senecioni-Brometum racemosi* (Tüxen & Preising 1951, Westhoff & den Held 1969, Ellenberg 1978) and placed in the Calthion. This is a rather ill-defined alliance on the Atlantic fringe of western Europe (O'Sullivan 1965, Werger 1973) and the two original character species of the *Senecioni-Brometum*, *Bromus racemosus* and *Senecio aquaticus* are of restricted distribution in Britain and are not confined to the *Cynosurus-Caltha* community. Nonetheless, this vegetation type remains quite well defined among the more mesotrophic swards and rushy vegetation of the Molinietalia in the constancy of *Caltha palustris* and the preferential occurrence here of *Carex disticha* (Van Schaik & Hogeweg 1977). With further sampling among mire vegetation, it would be worth examining the relationship between the community and the grassier assemblages included within the *Juncus subnodulosus-Cirsium palustre* fen-meadow.

Floristic table MG8

	8	x			
			Crepis paludosa	I (1–2)	+
			Agrostis capillaris	I (4–7)	+
Cynosurus cristatus	V (2–5)	+	*Alchemilla glabra*	I (2–3)	
Caltha palustris	V (1–4)	+	*Bromus hordeaceus hordeaceus*	I (2)	+
Festuca rubra	V (2–7)	+	*Carex flacca*	I (2–3)	
Holcus lanatus	V (1–6)	+	*Carex nigra*	I (2–6)	
Ranunculus acris	V (2–4)	+	*Galium palustre*	I (1–2)	+
Trifolium repens	V (1–7)	+	*Luzula campestris*	I (1–2)	
Cerastium fontanum	IV (1–3)	+	*Succisa pratensis*	I (1–3)	
Poa trivialis	IV (1–6)	+	*Avenula pubescens*	I (2)	+
Rumex acetosa	IV (1–4)	+	*Achillea ptarmica*	I (1–3)	
Anthoxanthum odoratum	IV (2–5)	+	*Angelica sylvestris*	I (2)	
Leontodon autumnalis	IV (1–3)	+	*Cirsium palustre*	I (1–2)	+
			Galium uliginosum	I (2)	
Bellis perennis	III (1–3)	+	*Juncus effusus*	I (2)	
Plantago lanceolata	III (2–4)	+	*Lotus uliginosus*	I (2–3)	+
Ranunculus repens	III (2–5)		*Trollius europaeus*	I (2)	
Trifolium pratense	III (1–5)	+	*Geum rivale*	I (2–3)	+
Carex panicea	III (3–4)		*Myosotis scorpioides*	I (2)	+
Filipendula ulmaria	III (1–9)	+	*Leucanthemum vulgare*	I (2)	+
Agrostis stolonifera	II (2–5)	+	*Alchemilla xanthochlora*	I (3–4)	
Euphrasia officinalis agg.	II (2–3)		*Glyceria declinata*	I (2)	
Lolium perenne	II (2–4)	+	*Mentha aquatica*	I (2–4)	
Taraxacum officinale agg.	II (1–2)	+	*Senecio aquaticus*	I (2)	+
Cardamine pratensis	II (1–3)	+	*Brachythecium rutabulum*	I (1–4)	
Festuca pratensis	II (1–3)	+	*Dactylis glomerata*	I (1–3)	
Rhinanthus minor	II (1–4)		*Briza media*	I (2–3)	+
Sanguisorba officinalis	II (3–4)		*Equisetum arvense*	I (1–2)	
Carex disticha	II (4–5)		*Centaurea nigra*	I (3–4)	
Eleocharis palustris	II (2–8)	+	*Lotus corniculatus*	I (3–4)	+
Juncus articulatus	II (1–4)		*Eurhynchium praelongum*	I (2)	
Leontodon hispidus	II (1–4)	+	*Carex demissa*	I (2–3)	
Prunella vulgaris	II (1–3)	+	*Plagiomnium rostratum*	I (1–4)	
Veronica chamaedrys	II (1–2)	+	*Carex ovalis*	I (3–4)	
Calliergon cuspidatum	II (2–7)		*Juncus acutiflorus*	I (3)	+
Lychnis flos-cuculi	I (1–2)	+			
Equisetum palustre	I (1–2)		Number of samples	15	
Valeriana dioica	I (1–2)	+	Number of species/sample	26 (15–41)	

8   *Cynosurus cristatus-Caltha palustris* grassland
x   Water meadows (Fream 1888*a*)

MG9
Holcus lanatus-Deschampsia cespitosa grassland

Synonymy

Deschampsietum caespitosae Horvatic 1930 *p.p.*; *Arrhenatheretum elatioris deschampsietosum caespitosae* LeBrun *et al.* 1949; Tussocky neutral grassland Duffey *et al.* 1974; *Juncus-Carex hirta-Deschampsia cespitosa* nodum Wheeler 1975 *p.p.*; Moist *Deschampsia-Dactylis* grassland Rahman 1976; Ordinary wet meadows Ratcliffe 1977; *Festuco-Alopecuretum pratensis deschampsietosum caespitosae* Page 1980.

Constant species

Deschampsia cespitosa ssp. *cespitosa*, *Holcus lanatus*.

Physiognomy

The *Holcus lanatus-Deschampsia cespitosa* community has a coarse sward dominated by *D. cespitosa* and other large tufted or tussocky grasses, *Holcus lanatus*, *Dactylis glomerata* and *Arrhenatherum elatius*. The physiognomy and composition of the vegetation are somewhat varied but depend largely on the number, size and disposition of tussocks of *D. cespitosa* and on their pattern of aerial growth through the year. In closed, stable swards like the vegetation of this community, *D. cespitosa* maintains itself by the vegetative expansion of clonal individuals, each of which may come to comprise several thousand tillers arranged in a densely-caespitose tussock. Although the tillers themselves are short-lived, with a continual turnover and gradual accumulation of dead material, whole tussocks may attain a considerable age, perhaps 30 years or more. As they grow, the tussocks do not extend their ground cover greatly but they are able to produce an increasing spread of aerial shoots each year and these may attain a height of 50–200 cm in older, well-grown individuals. There is a maximum proportion of live material in a tussock in August, after which there is extensive die-back although many leaves remain winter-green and there may also be some slow new growth during the winter (Davy 1980).

The vegetation is generally a mosaic built around the *D. cespitosa* tussocks. Where these are scattered, there is between them a sward, often cropped short by preferential grazing, in which *Holcus lanatus*, *Festuca rubra*, *Agrostis stolonifera*, *A. capillaris*, *Poa trivialis*, *Dactylis glomerata*, *Lolium perenne* and *Alopecurus pratensis* may each be locally abundant. Dicotyledons are numerous and varied but none is constant. The most frequent species are *Ranunculus repens*, *R. acris*, *Cirsium arvense*, *Rumex acetosa*, *Cerastium fontanum*, *Plantago lanceolata*, *Lathyrus pratensis* and *Centaurea nigra*. The particular combination of these grasses and dicotyledons in the sward between the tussocks is frequently a reflection of the type of grassland in which *D. cespitosa* has become established. Among other occasionals, poor-fen species may be conspicuous: *Juncus effusus* and *J. inflexus* sometimes occur as scattered clumps and *Filipendula ulmaria*, *Cardamine pratensis*, *Angelica sylvestris*, *Carex hirta*, *Lotus uliginosus* and *Achillea ptarmica* may be encountered at low frequency.

Where the *D. cespitosa* tussocks are close and especially where large tussocks produce a continuous canopy of extended leaves in summer, there is a reduction in diversity in the vegetation. Some species, such as *Dactylis* (Rahman 1976: see below), seem able to compete effectively with *D. cespitosa* under certain soil conditions, but many succumb to the increased lack of ground space and of light. In early spring in such stands, and particularly where the *D. cespitosa* has been grazed hard back over the winter, there may be considerable areas of bare soil and some grasses (e.g. *H. lanatus*, *A. stolonifera*, *L. perenne*, *Alopecurus pratensis*) may show a temporary abundance before the leaves of *D. cespitosa* extend. Between fully-expanded tussocks, however, there is often a very sparse ground cover of spindly *P. trivialis* and *A. stolonifera* with a little *Ranunculus repens*. Certain sprawlers, such as *Lathyrus pratensis* and *Lotus uliginosus*, and some taller dicotyledons, such as *Filipendula ulmaria*, *Centaurea nigra*, *Angelica sylvestris* and *Rumex crispus*, are able to grow up among the tussocks but, in many cases, such stands are extremely species-poor and uncomprisingly dominated by *D. cespitosa*. Where such vegetation is inundated by floodwater, there may be largely bare silty runnels between the tussocks with scattered *Potentilla anserina* and *Mentha aquatica*.

Bryophytes are rather infrequent in the community although *Brachythecium rutabulum* and *Eurhynchium praelongum* are occasionally abundant on bare soil and decaying litter.

Sub-communities

***Poa trivialis* sub-community:** *Deschampsietum caespitosae typicum* and *Festuco-Alopecuretum pratensis deschampsietosum caespitosae* Page 1980. *D. cespitosa* is generally the most abundant species in this sub-community and it may be overwhelmingly dominant. Usually, however, there is a rather open sward in which *P. trivialis*, *Festuca pratensis*, *Alopecurus pratensis*, *Ranunculus acris*, *Trifolium repens* and *T. pratense* are preferentially frequent with occasional *Juncus effusus* and *Filipendula ulmaria* and a wide range of grassland and poor-fen species at low frequency. Bryophytes are slightly more frequent and abundant here and *Calliergon cuspidatum* is preferential.

***Arrhenatherum elatius* sub-community:** *Arrhenatheretum elatioris deschampsietosum caespitosae* LeBrun *et al.* 1949; *Deschampsietum caespitosae arrhenatheretosum elatioris* Page 1980. Here, *D. cespitosa*, *A. elatius*, *D. glomerata* and *H. lanatus* are usually co-dominant in a coarse tussocky sward. Some taller dicotyledons, such as *Centaurea nigra* and *Rumex crispus*, are preferential and, with the thick accumulation of litter, bryophytes tend to be less frequent but, otherwise, the floristics resemble those of the community as a whole.

Habitat

The *Holcus-Deschampsia* community is highly characteristic of permanently moist, gleyed and periodically inundated circumneutral soils throughout the British lowlands. It occurs patchily or as extensive stands on level to moderately steeply sloping ground in pastures and meadows, in woodland rides and clearings, on road verges and in churchyards, on river levees and at fen margins and around the upper limit of inundation by pools, lakes and reservoirs.

The maintenance of the community in such a diverse range of habitats depends largely upon the ability of *D. cespitosa* to survive and become dominant on mineral soils which are often anaerobic and oligotrophic and therefore inhospitable to many other neutral grassland species. *D. cespitosa* has a tolerance of, rather than a need for, high levels of soil moisture (Davy & Taylor 1974*a*). This tolerance may be partly due to its well-developed root aerenchyma which, by reducing oxygen requirements within the tissues, could permit oxidation of the immediate root environment and so mitigate the toxic effect of certain reduced, e.g. ferrous, ions (Martin

1968, Rahman 1976). *D. cespitosa* also has an internal system of nutrient cycling, from older to developing tillers, and this could assist in its survival on oligotrophic soils (Davy & Taylor 1974*b*, 1975, Davy 1980). Further, it has a broad tolerance of soil base-status. Thus, although it tends to grow most vigorously in this community on mesotrophic, periodically inundated soils, the *Holcus-Deschampsia* grassland occurs wherever *D. cespitosa* has a competitive advantage within lowland swards on gleyed brown earths, gleyed brown calcareous earths and surface- and ground-water gleys, including alluvium. Many stands have developed by the establishment and spread of *D. cespitosa* in gleyed hollows within, for example, *Lolio-Cynosuretum* and *Lolium-Alopecurus-Festuca* pastures and *Alopecurus-Sanguisorba* meadows, or where such communities come into contact with a fluctuating water-table by streams and pools.

The growth of *D. cespitosa* in such circumstances is generally vegetative and slow (see above) but some stands of the community have arisen by the explosive spread of *D. cespitosa* on to disturbed moist soils such as occur in rides and new clearings in woods on heavy clays. Vast numbers of seeds can be produced by well-grown individuals in favourable circumstances and the light, plumed caryopses are readily dispersed by wind, sometimes over hundreds of metres (Davy 1980). There is also a persistent soil seed-bank (Grime 1979).

Grazing may accentuate the competitive advantage of *D. cespitosa*. Although it is eaten by cattle (even in the lowlands, cf. Davy 1980), horses, sheep, deer and rabbits, the leaves have a high silica content and are very rough and unpalatable and, in pasture, tussocks are generally avoided if alternative herbage is available. This tends to accentuate the mosaic structure of the vegetation, although heavy trampling by cattle can damage and even flatten small tussocks. On occasion, grazing-sensitive species, such as *Filipendula ulmaria*, may become established on top of the *D. cespitosa* tussocks. When there is little bite to be had from the sward, as in winter, tussocks may be grazed hard back.

The *Poa* sub-community comprises grazed stands on various soil types and those ungrazed stands on moister soils where *D. cespitosa* has become the sole dominant. The *Arrhenatherum* sub-community is always ungrazed and tends to occur on drier soils. Here, *D. cespitosa* is unable to maintain its competitive advantage over other grasses and, in the absence of grazing, *A. elatius* and *D. glomerata* can become co-dominant. This sub-community is particularly common on road verges and in churchyards.

Zonation and succession

The community occurs very frequently as part of zonations related to differences in soil moisture status. The *Poa* sub-community is common in moist hollows within pastures and on grazed verges where there is gleying and it commonly gives way, with a decrease in the cover of *D. cespitosa*, to *Lolio-Cynosuretum* or the *Lolium-Alopecurus-Festuca* community on better-drained soils. Around permanently wet hollows, pools and lakes, the community regularly forms part of an ecotone to some type of fen or swamp or, where there is a frequent rise and fall of water level, as around reservoirs, to a sequence of inundation communities. In many cases, it is possible to discern a range of intermediates between the *Holcus-Deschampsia* community and its neighbours and, in zonations to fen-meadows, there is, or has been in the past, the further complication of mowing and grazing effects.

Such transitional swards may also figure in seral changes that ensue in badly-managed or abandoned pastures and meadows. Once established, *D. cespitosa* cannot be totally eradicated by continued grazing and, if drainage deteriorates, it will slowly spread to dominate in the *Poa* sub-community. Even where drainage is improved, preferential grazing of other species may prevent *D. cespitosa* being ousted by possible competitors. Where there is a reduction of grazing on better-drained land, the *Arrhenatherum* sub-community may develop from pasture swards. Although a switch to mowing is probably the best form of control of *D. cespitosa* (Davy 1980), the silting up of ditches within meadows may precipitate a spread of the community.

If grazing continues, the *Holcus-Deschampsia* grassland is maintained as a plagioclimax and, even when it ceases, the densely-tussocky character of the vegetation often severely hinders any establishment of seedling shrubs and trees. Where succession to woodland does occur, it is usually to the *Deschampsia* sub-communities of the *Alnus-Fraxinus-Lysimachia* or *Fraxinus-Acer-Mercurialis* woodlands. Stands of the *Holcus-Deschampsia* grassland can often be found in rides and clearings in tracts of these kinds of woodlands or planted replacements of them in forests on heavy, clay soils with considerable continuity among the herb flora. In such situations, the *Holcus-Deschampsia* grassland can provide an important seed source for the spread of *D. cespitosa* beneath the trees when felling or coppicing increases the light.

Distribution
The community is virtually ubiquitous in suitable sites throughout the lowlands. At higher altitudes, it is replaced by sub-montane and montane grasslands in which *D. cespitosa* is represented by ssp. *alpina*.

Affinities
Floristically dull grasslands of low agricultural value or little apparent interest for conservation such as the *Holcus-Deschampsia* community have figured rarely in the descriptive literature. However, this vegetation type forms part of a well-defined series of grasslands dominated by the various subspecies of *D. cespitosa*, some elements of which have been described previously from moist soils at higher altitudes and latitudes, e.g. the *Deschampsietum caespitosae alpinum* of McVean & Ratcliffe (1962) and the *Deschampsia-Festuca-Agrostis* grassland of King & Nicholson (1964).

Lowland *D. cespitosa* grasslands have been described from various parts of the Continent (e.g. Horvatic 1930, Tüxen 1937, LeBrun *et al.* 1949, Sissingh & Tideman 1960, Westhoff & den Held 1969), although they have not always been characterised as distinct kinds of *Deschampsietum*. This reflects the fact that, apart from the dominance of *D. cespitosa* and the consequent reduction in the frequency and cover of other grassland species, many stands are very like the various communities in which the grass has but a minor hold. The drawing of boundaries between vegetation types in such cases is obviously difficult, but it seems sensible to retain a distinct unit for those stands where the species is most prominent and where it occurs with a sprinkling of poor-fen plants in a clearly defined range of habitats with a fluctuating water-table. Where lowland communities of this kind have been characterised, authorities differ as to whether they are best placed within the Calthion (with the *Holco-Juncetum*, say) or in a separate alliance, the Deschampsion, within the Arrhenatheretalia.

Floristic table MG9

	a	b	9
Deschampsia cespitosa	V (2–9)	V (1–8)	V (1–9)
Holcus lanatus	IV (2–8)	IV (2–7)	IV (2–8)
Poa trivialis	IV (2–6)	II (2–5)	III (2–6)
Ranunculus acris	II (1–4)	I (2–4)	II (1–4)
Festuca pratensis	II (1–7)	I (2–4)	II (1–7)
Anthoxanthum odoratum	II (2–6)	I (2–3)	II (2–6)

Filipendula ulmaria	II (2–5)	I (2–3)	II (2–5)
Juncus effusus	II (2–7)	I (1–2)	II (1–7)
Taraxacum officinale agg.	II (1–3)	I (1–3)	I (1–3)
Trifolium pratense	II (1–4)	I (2–5)	I (1–5)
Trifolium repens	II (1–6)	I (3–4)	I (1–6)
Brachythecium rutabulum	II (2–7)	I (2–6)	I (2–7)
Eurhynchium praelongum	II (2–6)	I (2–5)	I (2–6)
Calliergon cuspidatum	II (1–7)		I (1–7)
Carex panicea	I (2–5)		I (2–5)
Cirsium palustre	I (1–4)		I (1–4)
Conopodium majus	I (1–4)		I (1–4)
Rumex sanguineus	I (2)		I (2)
Hypericum perforatum	I (1–2)		I (1–2)
Briza media	I (2–3)		I (2–3)
Senecio jacobaea	I (1–2)		I (1–2)
Senecio aquaticus	I (1)		I (1)
Glyceria fluitans	I (1–2)		I (1–2)
Phalaris arundinacea	I (1–5)		I (1–5)
Carex nigra	I (1–4)		I (1–4)
Rumex obtusifolius	I (3)		I (3)
Trisetum flavescens	I (1)		I (1)
Arrhenatherum elatius	I (2)	V (1–8)	II (1–8)
Dactylis glomerata	II (1–7)	V (2–5)	III (1–7)
Centaurea nigra	I (1–4)	III (2–4)	II (1–4)
Rumex crispus	I (1–3)	II (2–4)	I (1–4)
Festuca rubra	II (3–6)	III (2–5)	III (2–6)
Agrostis stolonifera	III (2–7)	III (2–9)	III (2–9)
Ranunculus repens	III (1–7)	II (1–4)	III (1–7)
Agrostis capillaris	II (3–9)	II (2–7)	II (2–9)
Cirsium arvense	II (1–6)	II (2–5)	II (1–6)
Rumex acetosa	II (1–5)	II (1–4)	II (1–5)
Cerastium fontanum	II (1–3)	II (1–3)	II (1–3)
Alopecurus pratensis	II (1–8)	II (3–7)	II (1–8)
Lolium perenne	II (1–6)	II (2–5)	II (1–6)
Plantago lanceolata	II (1–5)	II (1–5)	II (1–5)
Lathyrus pratensis	II (1–5)	II (1–3)	II (1–5)
Juncus inflexus	II (2–5)	II (2–5)	II (2–5)
Cardamine pratensis	I (2–3)	I (3)	I (2–3)
Potentilla erecta	I (2–4)	I (2–3)	I (2–4)
Rumex conglomeratus	I (3–4)	I (4)	I (3–4)
Angelica sylvestris	I (3–4)	I (2–5)	I (2–5)
Primula veris	I (2–3)	I (2)	I (2–3)
Carex hirta	I (3–4)	I (1–3)	I (1–4)
Holcus mollis	I (1–7)	I (2–4)	I (1–7)
Leontodon autumnalis	I (1–3)	I (3)	I (1–3)
Potentilla anserina	I (1–4)	I (3)	I (1–4)
Urtica dioica	I (1–2)	I (2)	I (1–2)
Rubus fruticosus agg.	I (2–4)	I (2–6)	I (2–6)
Poa pratensis	I (3)	I (3)	I (3)
Vicia cracca	I (1–6)	I (2)	I (1–6)

Floristic table MG9 (*cont.*)

	a	b	9
Pulicaria dysenterica	I (1–2)	I (2–3)	I (1–3)
Lotus uliginosus	I (1–4)	I (2–4)	I (1–4)
Achillea millefolium	I (2–3)	I (3)	I (2–3)
Phleum pratense pratense	I (4)	I (3)	I (3–4)
Ranunculus ficaria	I (4–5)	I (4)	I (4–5)
Lotus corniculatus	I (3–4)	I (2–4)	I (2–4)
Juncus articulatus	I (2)	I (3)	I (2–3)
Mentha aquatica	I (1)	I (2–4)	I (1–4)
Danthonia decumbens	I (1)	I (3)	I (1–3)
Succisa pratensis	I (1–3)	I (3)	I (1–3)
Achillea ptarmica	I (3–5)	I (3)	I (3–5)
Rhytidiadelphus squarrosus	I (1–2)	I (3)	I (1–3)
Cynosurus cristatus	I (1–5)	I (4)	I (1–5)
Prunella vulgaris	I (1)	I (3)	I (1–3)
Festuca arundinacea	I (1–5)	I (1–5)	I (1–5)
Galium verum	I (2)	I (2)	I (2)
Silaum silaus	I (3–5)	I (1–4)	I (1–5)
Leucanthemum vulgare	I (2)	I (1–2)	I (1–2)
Dactylorhiza fuchsii	I (2)	I (1)	I (1–2)
Agrimonia eupatoria	I (1)	I (3)	I (1–3)
Hordeum secalinum	I (2)	I (3)	I (2–3)
Heracleum sphondylium	I (1–6)	I (4–5)	I (1–6)
Phleum pratense bertolonii	I (2–5)	I (3)	I (2–5)
Cirsium vulgare	I (1)	I (3)	I (1–3)
Number of samples	33	19	52
Number of species/sample	15 (7–36)	18 (11–34)	16 (7–36)

a *Poa trivialis* sub-community

b *Arrhenatherum elatius* sub-community

9 *Holcus lanatus-Deschampsia cespitosa* grassland (total)

Floristic table MG10

	a	b	c	10
Juncus effusus	V (2–8)	III (1–4)	IV (2–6)	V (1–8)
Holcus lanatus	V (3–8)	IV (4–7)	IV (3–7)	V (3–8)
Agrostis stolonifera	IV (3–9)	V (5–8)	IV (3–8)	IV (3–9)
Ranunculus repens	III (3–7)	V (3–5)	IV (3–6)	IV (3–7)
Phleum pratense pratense	I (2–5)			I (2–5)
Angelica sylvestris	I (2–4)			I (2–4)
Carex panicea	I (3–5)			I (3–5)
Pulicaria dysenterica	I (2–5)			I (2–5)
Prunella vulgaris	I (4)			I (4)
Juncus articulatus	I (3)			I (3)
Juncus inflexus		V (3–8)		II (3–8)
Carex hirta	I (4)	III (3–4)		I (3–4)
Iris pseudacorus			V (1–7)	II (1–7)
Alopecurus pratensis			II (2–6)	I (2–6)
Filipendula ulmaria	I (4)		II (2–4)	I (2–4)
Phalaris arundinacea	I (5)		II (3–7)	I (3–7)
Lotus uliginosus	I (5)	I (2)	II (3)	I (2–5)
Glyceria fluitans	I (3)	I (3)	II (2–5)	I (2–5)
Myosotis scorpioides	I (3)		II (2)	I (2–3)
Urtica dioica		I (2)	II (2–3)	I (2–3)
Lychnis flos-cuculi			I (3)	I (3)
Poa trivialis	III (2–5)	IV (4–6)	IV (2–5)	III (2–6)
Cardamine pratensis	II (1–4)	III (2–4)	III (1–3)	III (1–4)
Ranunculus acris	III (2–5)	III (3–4)	II (2)	III (2–5)
Trifolium repens	II (3–9)	III (2–3)	I (1)	II (1–9)
Lolium perenne	II (2–4)	II (3–4)	II (3–4)	II (2–4)
Rumex crispus	I (2–4)	III (3–4)	III (2–3)	II (2–4)
Alopecurus geniculatus	I (1–4)	II (3–4)	III (1–3)	II (1–4)
Rumex acetosa	II (2–4)	I (2)	II (3–4)	II (2–4)
Festuca pratensis	I (2–3)	II (2–4)	II (4–5)	II (2–5)
Rumex obtusifolius	I (3–4)	II (2–3)	I (2)	II (2–4)
Plantago lanceolata	I (2–4)	I (6)	II (1–3)	II (1–6)
Potentilla anserina	II (2–5)		II (1–3)	II (1–5)
Cerastium fontanum	II (3)	I (2)	I (3)	II (2–3)
Cirsium arvense	I (1)	II (3–4)	I (2–4)	I (1–4)
Festuca rubra	I (4–5)	II (3–5)		I (3–5)
Rumex conglomeratus	I (2–3)	II (3–6)		I (2–6)
Calliergon cuspidatum	I (5–6)	I (5)	I (3)	I (3–6)
Equisetum arvense	I (1–2)	I (4)	I (1)	I (1–4)
Dactylis glomerata	I (4–7)	I (5)	I (6)	I (4–7)
Trifolium pratense	I (4)	I (4)	I (2)	I (2–4)
Polygonum hydropiper	I (2)	I (4)	I (3)	I (2–4)
Poa pratensis	I (3–5)		I (1)	I (1–5)
Anthoxanthum odoratum	I (3)		I (2)	I (2–3)
Caltha palustris	I (2)		I (4)	I (2–4)

Floristic table MG10 (*cont.*)

	a	b	c	10
Stellaria alsine	I (3)		I (3)	I (3)
Oenanthe crocata	I (4)		I (3)	I (3–4)
Eurhynchium praelongum	I (3)		I (3)	I (3)
Cynosurus cristatus	I (1–6)	I (3–5)		I (1–6)
Agrostis capillaris	I (5–9)	I (5–6)		I (5–9)
Taraxacum officinale agg.	I (1–2)	I (4)		I (1–4)
Plantago major	I (2)	I (1)		I (1–2)
Cirsium palustre	I (3)	I (2)		I (2–3)
Equisetum palustre	I (1)	I (4)		I (1–4)
Mentha aquatica	I (2–3)	I (1)		I (1–3)
× *Festulolium loliaceum*	I (1)	I (3)		I (1–3)
Vicia cracca		I (3)	I (3)	I (3)
Senecio aquaticus		I (4)	I (3)	I (3–4)
Number of samples	17	15	8	40
Number of species/sample	12 (6–20)	15 (8–24)	13 (9–16)	13 (6–24)

a　Typical sub-community
b　*Juncus inflexus* sub-community
c　*Iris pseudacorus* sub-community
10　*Holco-Juncetum effusi* (total)

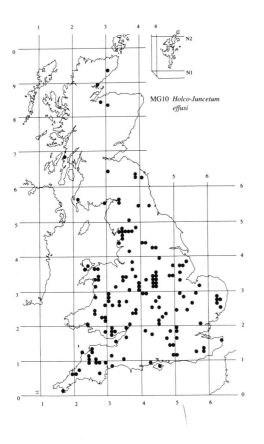

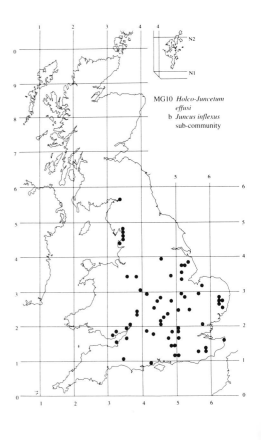

	a	b	c	11
Agrostis capillaris	II (3–7)		II (3)	I (3–7)
Plantago maritima	I (2–3)	II (2–3)	II (1–3)	I (1–3)
Cirsium vulgare	I (1–2)	I (1)	I (1–2)	I (1–2)
Galium aparine	I (2–3)	I (2–4)	I (1–3)	I (1–4)
Cochlearia officinalis	I (2)	I (2)	I (1–3)	I (1–3)
Bellis perennis	I (1–4)	I (3)	I (1–3)	I (1–4)
Hypochoeris radicata	I (2)	I (3)	I (1)	I (1–3)
Alopecurus geniculatus	I (3)	I (2)	I (1)	I (1–3)
Poa annua	I (2–3)	I (3)	I (3)	I (2–3)
Juncus articulatus	I (5)	I (4)	I (1)	I (1–5)
Leontodon autumnalis	I (3–4)	I (2–3)	I (1)	I (1–4)
Brachythecium rutabulum	I (2–4)		I (1)	I (1–4)
Festuca arundinacea	I (2–4)	I (2–5)		I (2–5)
Juncus gerardi	I (3–5)	I (2–5)		I (2–5)
Urtica dioica	I (2–6)	I (2–3)		I (2–6)
Triglochin maritima	I (2–3)	I (2–5)		I (2–5)
Carex distans	I (2–3)	I (1–6)		I (1–6)
Poa trivialis	I (1–7)	I (2–6)		I (1–7)
Plantago major	I (1–3)	I (3)		I (1–3)
Parapholis strigosa	I (2–3)	I (2)		I (2–3)
Arrhenatherum elatius	I (4)	I (3–4)		I (3–4)
Glaux maritima	I (2–4)	I (4)		I (2–4)
Chamomilla suaveolens	I (2)	I (3)		I (2–3)
Bromus hordeaceus hordeaceus	I (3–4)	I (5)		I (3–5)
Odontites verna	I (2–3)	I (3)		I (2–3)
Juncus bufonius	I (3)	I (4)		I (3–4)
Number of samples	50	27	17	94
Number of species/sample	11 (4–25)	10 (5–14)	13 (7–29)	11 (4–29)

a   *Lolium perenne* sub-community
b   *Atriplex prostrata* sub-community
c   *Honkenya peploides* sub-community
11   *Festuca rubra-Agrostis stolonifera-Potentilla anserina* grassland (total)

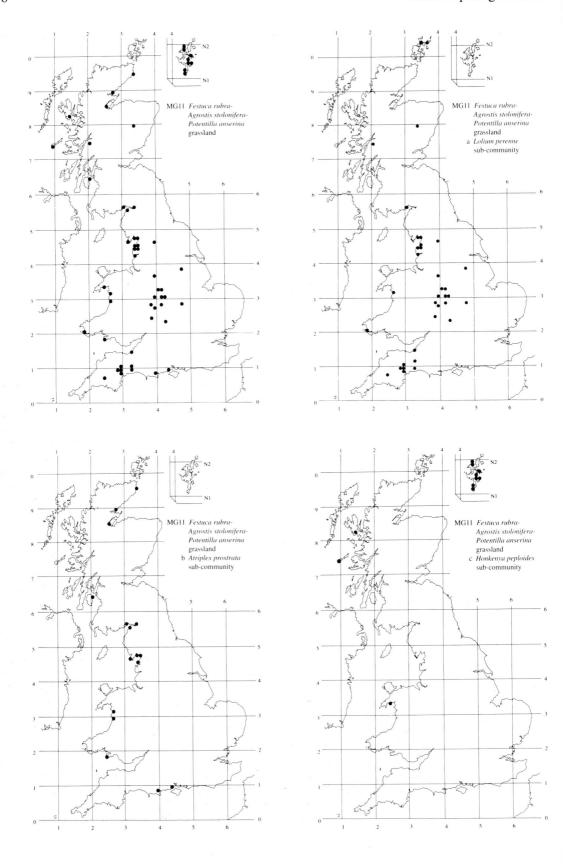

MG11 *Festuca rubra-*
Agrostis stolonifera-
Potentilla anserina
grassland

MG11 *Festuca rubra-*
Agrostis stolonifera-
Potentilla anserina
grassland
a *Lolium perenne*
sub-community

MG11 *Festuca rubra-*
Agrostis stolonifera-
Potentilla anserina
grassland
b *Atriplex prostrata*
sub-community

MG11 *Festuca rubra-*
Agrostis stolonifera-
Potentilla anserina
grassland
c *Honkenya peploides*
sub-community

MG12
Festuca arundinacea grassland
Potentillo-Festucetum arundinaceae   Nordhagen 1940

Synonymy
Lolium perenne-Matricaria suaveolens Ass. R.Tx. 1937 *p.p.*; *Festuca arundinacea-Dactylis glomerata* Ass. R.Tx. 1950; *Festuceto-Dactyletum* Oberdorfer 1957; *Agrostis stolonifera-Festuca arundinacea* nodum Adam 1976; includes *Lolio-Agrostetum stoloniferae festucetosum arundinaceae* Page 1980.

Constant species
Agrostis stolonifera, Festuca arundinacea, F. rubra.

Physiognomy
The *Potentillo-Festucetum arundinaceae* is a coarse grassland usually dominated by large tussocks of *F. arundinacea* with often abundant *F. rubra* and *A. stolonifera* and generally smaller amounts of *Holcus lanatus*. Apart from *Elymus repens* and *Poa pratensis*, which are occasional throughout, other grasses are rare. Salt-marsh stands in both sub-communities frequently have *Carex distans, C. otrubae* and *Juncus gerardi.*

Potentilla anserina and *Trifolium repens* are the most frequent and abundant dicotyledons with occasional *Cirsium arvense, Vicia cracca, Lotus corniculatus, Trifolium pratense* and *Plantago lanceolata.*

Bryophytes are rather infrequent, although *Eurhynchium praelongum* is sometimes abundant.

Sub-communities

***Lolium perenne-Holcus lanatus* sub-community:** *Agrostis stolonifera-Festuca arundinacea* nodum Adam 1976 *p.p.*; *Potentillo-Festucetum arundinaceae* Nordhagen 1940 *sensu* Page 1980 and *Lolio-Agrostetum stoloniferae festucetosum arundinacaea* Page 1980. In this sub-community *H. lanatus* and/or *L. perenne* are abundant among the *F. arundinacea, A. stolonifera* and *F. rubra* and sometimes share dominance. *Anthoxanthum odoratum* and *Arrhenatherum elatius* occur occasionally and each may be prominent in particular stands. *Ranunculus acris, Lotus corniculatus, Trifolium pratense, Plantago*

lanceolata and *Cerastium fontanum* are preferential here but halophytes are poorly represented, although salt-marsh stands frequently have some *Carex distans, C. otrubae* and *Juncus gerardi.*

***Oenanthe lachenalii* sub-community:** *Festuceto-Caricetum distantis* Duvigneaud 1967; *Potentillo-Festucetum arundinaceae ranunculetosum acer* Krisch 1974; *Agrostis stolonifera-Festuca arundinacea* nodum Adam 1976 *p.p.* Although *H. lanatus* and a number of species characteristic of inland mesotrophic grasslands are represented here, this sub-community is strongly distinguished by the frequency of salt-marsh species such as *O. lachenalii, J. gerardi, Glaux maritima, Juncus maritimus* and *Triglochin maritimus. Sonchus arvensis, Odontites verna, Atriplex prostrata, Leontodon autumnalis* and *Hypochoeris radicata* are also preferential and there are occasional records for *Phragmites australis, Eleocharis uniglumis* and *Triglochin palustris.*

Habitat
The community is characteristic of moist but usually free-draining soils in coastal sites which receive frequent inundation by brackish water, occasional tidal inundation or small amounts of salt-spray. It occurs along the banks of tidal rivers, on the upper salt-marsh and occasionally on slumping clay sea cliffs. It is generally ungrazed.

The differences between the sub-communities are probably related to the amount of sea-salts, particularly sodium chloride, in the flood-waters or spray. The *Lolium-Holcus* sub-community occurs typically alongside brackish streams, rivers and ditches (where it often forms fragmentary strips on the steeply-sloping embankments), towards the upper limit of salt marshes and on sea cliffs which receive relatively small amounts of spray. The *Oenanthe* sub-community is confined to salt-marshes and occasionally forms extensive stands on sites with rare tidal inundation.

Although accessible stands alongside coastal farmland are sometimes grazed or included within a hay crop, the community is generally untreated. In some places, growth may be controlled by periodic burning.

Zonation and succession
The *Lolium-Holcus* sub-community generally shows abrupt zonations to other vegetation types, sharpened by treatment where stands abut on to agricultural land or by sudden topographic differences where stands lie on steep banks alongside ditches. Occasionally, there are more gradual zonations to the *Festuca-Agrostis-Potentilla* community with increased grazing pressure or, on the upper salt-marsh, to the *Oenanthe* sub-community on more saline soils. Although this latter sub-community sometimes occurs as extensive pure stands, it is frequently encountered in a mosaic with some form of *Juncus maritimus* salt-marsh which replaces it on more frequently submerged ground.

Distribution
The *Potentillo-Festucetum* is exclusively a coastal community recorded from estuaries and salt-marshes on the south and west coasts of England and Wales and on Arran in Scotland and from clay cliffs in Dorset, Kent and North Yorkshire.

Affinities
The community shows very close affinities with the *Festuca-Agrostis-Potentilla* community and, like that vegetation type, has a good representation of species considered characteristic of the Elymo-Rumicion crispi, into which similar communities described from the Continent have been placed (e.g. Nordhagen 1940, Duvigneaud 1967).

Floristic table MG12

	a	b	12
Festuca arundinacea	V (1–9)	V (1–7)	V (1–9)
Agrostis stolonifera	V (4–7)	V (4–7)	V (4–7)
Festuca rubra	V (3–7)	V (4–8)	V (3–8)
Lolium perenne	IV (2–8)		II (2–8)
Holcus lanatus	IV (1–7)	III (2–4)	III (1–7)
Ranunculus acris	III (2–4)	I (1–3)	II (1–4)
Lotus corniculatus	II (2–4)	I (2–4)	II (2–4)
Trifolium pratense	II (3–5)	I (3)	II (3–5)
Anthoxanthum odoratum	II (3–5)		I (3–5)
Arrhenatherum elatius	II (1–7)		I (1–7)
Plantago lanceolata	II (2–3)	I (1–3)	II (1–3)
Cerastium fontanum	II (2–3)	I (2)	I (2–3)
Lathyrus pratensis	I (2–4)		I (2–4)
Cynosurus cristatus	I (3–4)	I (2)	I (2–4)
Dactylis glomerata	I (2–6)		I (2–6)
Juncus acutiflorus	I (2–4)		I (2–4)
Festuca pratensis	I (2–4)	I (2)	I (2–4)
Torilis japonica	I (2–3)		I (2–3)
Agrostis capillaris	I (4–5)		I (4–5)
Oenanthe lachenalii	I (2–4)	V (1–5)	III (1–5)
Juncus gerardi	II (2–3)	IV (2–5)	III (2–5)
Glaux maritima		IV (2–3)	II (2–3)
Carex otrubae	II (2–3)	IV (2–6)	III (2–6)
Sonchus arvensis		III (2–3)	II (2–3)
Phragmites australis	I (2–6)	III (2–7)	II (2–7)
Eleocharis uniglumis	I (3)	III (2–5)	II (2–5)
Leontodon autumnalis	I (3)	III (2–3)	II (2–3)
Elymus pycnanthus	I (2)	III (2–5)	II (2–5)

CALCICOLOUS GRASSLANDS

INTRODUCTION TO CALCICOLOUS GRASSLANDS

The sampling of calcicolous grasslands

Grasslands in which calcicoles are a prominent feature have long attracted attention among British ecologists. Some of these swards were the subject of classic, early descriptions of our vegetation (Moss 1907, 1911, 1913, Tansley & Rankin 1911, Tansley & Adamson 1925, 1926) and the behaviour of plants in what are often rich and delightful assemblages has proved an abiding fascination. Huge losses among these grasslands and their continuing vulnerability to either agricultural improvement or neglect have also greatly enhanced the nature conservation interest of surviving stands. The contribution of these swards to impressive open landscapes with their associations of settled pastoralism is much valued, too: the more so with the proximity of many tracts to centres of dense population like the south-east of England. Yet, in some ways, these very qualities have hindered the development of a universally accepted classification of calcicolous grasslands.

For one thing, much scientific study has been preoccupied with particular regions or sites. This has been enormously informative about the relationships between their different swards and the subtle interplay of climatic factors, soil conditions and treatment history which have influenced them in those particular localities. At the same time, the very wealth of detail has been difficult to accommodate within a balanced overview of floristic variation among all these grasslands and its relation to broader environmental trends. Rather, there has grown up a series of regional perspectives from, for example, the southern English Chalk, the Peak District, Upper Teesdale and the Breadalbane Mountains, each viewpoint acquiring some sort of determinative authority for interpreting the rest. From the start of this survey, we were resolved to sample as widely as possible among these grasslands so as to set local insights within a general framework of understanding and to test the status of accepted distinctions between vegetation types that sometimes seemed to owe too much to particular enthusiasms.

Closely associated with the commitment to understanding and safeguarding the swards of particular localities, there has developed from the earliest days a persistent habit of speaking about 'Chalk grassland', 'Carboniferous Limestone grassland' and so on, as if the nature of the bedrock were the controlling factor in determining the character of the vegetation and the best basis for classifying it. We did not ourselves wish to begin with this assumption and our sampling of these grasslands was part of the much wider survey of vegetation, on and off Chalk, other kinds of limestone and calcareous superficials throughout the country. As always, samples were located simply by the homogeneity of composition and structure in the swards we encountered and, though details of geology were recorded along with the other environmental data, habitat features were not used in the disposition of quadrats. In this way, we hoped for sufficient breadth of coverage for the major lines of floristic variation to become evident in the processing of the samples, and to be able to test the meaning of these in relation to geological and other environmental variables.

We also tried to avoid any preoccupation with more species-rich or rarer kinds of sward. There are some striking examples of these among British calcicolous grasslands, but they can so easily emerge as too sharply defined from the rest and become the standard against which all else is judged as impoverished. While striving to ensure inclusion of the complete range of variation among the grasslands, we did not therefore concentrate our own sampling on more unusual assemblages, nor make any special effort to place quadrats around rare plants. Much survey was carried out among more species-poor and nondescript grasslands and the sampling also extended on to the vegetation of artificial calcareous habitats like chalk pits, limestone quarries, ballast and spoil heaps. We hoped that this would enable us to define and understand better the richer, rare and more venerable swards and provide an accurate indication of the community context of the many species of restricted distribution that find a home among these grasslands. By sampling widely and within the context of the broader survey, we also hoped to give a clear definition to boundaries and transitions between calcicolous grasslands and other vegetation types.

Generally speaking, it was not too difficult to detect homogeneity among the grasslands included here. Quadrats of 2 × 2 m were usually sufficient to provide a representative sample of the vegetation, with 4 × 4 m being occasionally necessary among more grossly structured swards, like those dominated by the tussocky grasses *Bromus erectus*, *Brachypodium pinnatum*, *Festuca rubra* or *Avenula pubescens*. Patchy prominence of large individuals or small clones of such plants would be treated as an integral part of the variegation in the grasslands being sampled, but more extensive clumps would be sampled in their own right. With more complex mosaics involving patterning among greater numbers of species, the elements of the mosaic would be sampled separately and notes made on the way in which the vegetation types related to one another. In such situations, stands of peculiar shape would be sampled using rectangular or other quadrats of the requisite area, while stands smaller than 2 × 2 m could be recorded in their entirety. This kind of flexibility was especially helpful in sampling the vegetation of cliff ledges and broken rock outcrops.

As always, recording of the floristic data involved assigning Domin scores to all vascular plants, bryophytes and macrolichens present in the sample plot. Special care was taken with the identification of grasses and sedges although, in very closely grazed swards, of which there were many here, it was often difficult to make a sure diagnosis of every nibbled shoot. Separation of *Festuca ovina* and *F. rubra* could be particularly troublesome in this respect, but results indicate that it is always worth attempting this distinction because of the differing ecological amplitude of the two grasses: their relative abundance can be very informative about the character of the grassland environment. *Agrostis* spp. were much less of a problem although, compared with some earlier surveys of southern English swards, it was noticeable how very much less common *A. stolonifera* was among our samples: this may be attributed largely to the decreased influence of rabbits, but again there may have been some errors in identification. Among dicotyledons, *Thymus pulegioides* has almost certainly been under-recorded because of confusion with *T. praecox* and, in the drier summers when we were sampling, we may have missed some winter annuals which were, by then, shrivelled to indistinguishable remains. We always attempted to separate the subspecies of *Cerastium diffusum*, *Phleum pratense* and *Alchemilla filicaulis*, but generally recorded *Euphrasia officinalis*, *Taraxacum officinale*, *Arctium minus* and

Rosa canina to the aggregate, and *Hieracium* (apart from *H. pilosella*) to the section. Some bryophytes were also especially difficult to identify and we often recorded *Weissia* to the genus.

In addition to these quantitative floristic data, we also often made notes on the structure of the vegetation, detailing any kind of dominance, other fine patterning in the sward or the various contributions by different life forms. The context of the stand and its relation to neighbouring vegetation types were recorded, with information on the character of the boundaries and transitions, together with any evidence from the floristics and physiognomy of the grassland that successional processes were in train.

The usual environmental data were recorded for each sample and supplemented where appropriate with details of the habitat. With these grasslands, for example, it was often helpful to know, not simply the identity of the underlying bedrock or superficial deposit, and the general slope and aspect of the ground, but also the nature of any fine surface relief and the pattern of drainage, and whether there seemed to be any exposure to strong insolation, wind or frost, or any suggestion of snow-lie. Then, where bedrock was exposed, notes would be made on the character of the outcrop, and the influence of any bedding, jointing or erosion features, such as surface sculpturing and the accumulation of rock waste, on the disposition and composition of the vegetation. Where a soil pit was dug and the profile described, particular attention was paid to the nature of the weathering parent material and whether this appeared to include any till, wind-blown sand or loess which muffled the influence of the bedrock and affected the drainage (Figure 11).

Also of great significance for many of the swards included here are biotic influences, particularly the impact of grazing by stock and wild herbivores which is often vital for maintaining these grasslands as plagioclimax vegetation. As with the survey of mesotrophic grasslands, we had to rely almost entirely on field observations here, although some sites we sampled had an outline of treatment history and informal conversations with farmers and reserve wardens were sometimes very informative about current practice. Details of the intensity and timing of grazing, and of the kinds of stock and other herbivores involved, were especially helpful, together with notes of any particular visible effects of cropping, manuring and trampling, each of these being variously influential on the different elements in the swards. Information about the impact of rabbits was especially interesting because these have been enormously important as grazers of calcicolous grasslands in Britain, particularly those in the lowlands, and their almost total demise with myxomatosis after the mid-1950s had a startling effect on the appearance of many swards. We are still living with, and trying to understand, the heritage of those changes, although it is clear that the return of rabbits in many places, sometimes in huge numbers again, is rendering remarks that we have made in this volume themselves out of date. As to the local, though sometimes striking, influence of biota like mound-building ants, we have relied almost entirely on the observations of previous workers.

Other features were sometimes worthy of note. For example, many tracts of calcicolous grasslands have been subject to agricultural improvement of their soils or swards to provide more productive pasturage, and it is important to understand the precise effects of this. Fertilising with cattle dung and urine is itself more enriching than manuring from sheep, such that the vegetation of the 'cow downs' of Chalkland villages is often noticeably mesophytic, but it was also informative to know of any use of artificial fertilisers or of any top-sowing in the past, or drift of seed or dressings from neighbouring enclosures. Evidence of past ploughing, as in wartime campaigns to increase arable production, was also noted, as was the suspicion that the ground had gone undisturbed for long periods of time: in this latter respect, the calcicolous swards sampled from ancient hill

Figure 11. Completed sample cards from calcicolous grasslands.

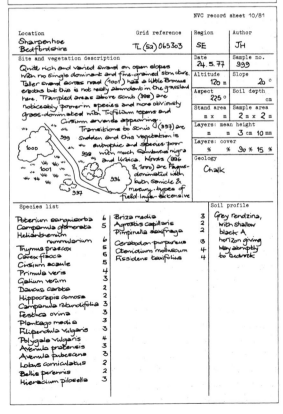

forts and burial mounds were especially interesting. More recent human activity was often evident, too, in the impact of visitors using the grasslands for various kinds of recreation. Signs of trampling, disturbance and cultural eutrophication were thus frequently to be seen around car parks and picnic spots and along tracks.

In addition to the records of such floristic and environmental data in the samples we ourselves collected, we were very fortunate, as far as these grasslands were concerned, in the ready availability of a wealth of existing information and experience. Among the larger data sets to which we were kindly given access was a survey of grasslands on English limestones conducted by Mr Terry Wells of the then Nature Conservancy (Wells 1975), and a study of calcicolous swards, mostly from northern England, western England and Wales and the Midlands, by Dr David Shimwell prior to his involvement with the project (Shimwell 1968a, 1971a, b). Then, there were various smaller surveys of Upper Teesdale by Pigott (1956a), Ratcliffe (1965) and Jones (1973) and of the Derbyshire Dales (Pigott unpublished, Shimwell 1968b) and, begun during the tenure of this project, of West Yorkshire (West Yorkshire Biological Data Bank 1983), the North York Moors (Atherden 1983) and Durham (Graham 1988). Some NCC reports on particular sites yielded small numbers of samples and, although most of the grassland surveys of the England Field Unit, under the leadership of Dr Tim Bines and, later, Dr Kevin Charman, came too late to supply data, they were of great value in confirming the shape of the emerging classification and providing further details of distribution. For Scotland, we again had the benefit of data from McVean & Ratcliffe (1962 and unpublished) and the Macaulay Institute (Birse & Robertson 1976, Birse 1980, 1984).

Where data in existing surveys were fully compatible with our own, we were able to include them for analysis and finally assembled just under 2500 samples from which to erect the classification. Geographical spread of the samples across Britain was extensive (Figure 12) and coverage of the range of floristic variation good, although we were not able to supplement existing data from the Scottish Highlands as much as we had wished.

Data analysis and the description of calcicolous grassland communities

As always, data analysis used only the floristic records from the samples for erecting the classification, with environmental information being held back for testing the ecological meaning of the distinctions recognised between the vegetation types. The quantitative scores for all vascular plants, bryophytes and macrolichens were employed and no special weight given to any reputed indicators of particular kinds of grassland or habitat conditions.

In all, fourteen communities have been included in this section of the work: swards in which mixtures of grasses, sedges and dicotyledonous herbs usually dominate, with a generally strong contingent of calcicoles among the vascular plants and cryptogams. We have also thought it sensible to deal with most British vegetation with *Dryas octopetala* here, but readers will have to look elsewhere for calcicolous mires with a prominent grassy element, for calcicolous types of mesotrophic sward, maritime grasslands and sand-dune vegetation, grassy transitions to calcicolous scrub and woodland, and fern-dominated vegetation of base-rich rock exposures and spoil. As for those distinctive mixtures of calcicoles and calcifuges which have been termed 'Chalk heath' or 'Limestone heath', these have emerged for the most part as mosaics or transitions between calcicolous grasslands and ericoid heaths: separate communities have not been described, although the composition and ecology of these assemblages are dealt with both in this volume and among the heaths.

As one might expect, the distribution of the fourteen communities presents a reasonable approximation to a map of pervious calcareous bedrocks in Britain (Figure 13). Beyond this general relationship, however, the

Figure 12. Distribution of samples available from calcicolous grasslands.

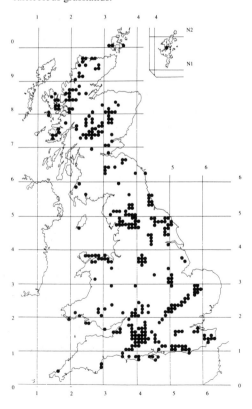

Figure 13. Exposures of lime-rich bedrocks and important locations mentioned in the text.

a	Durness
b	Inchnadamph
c	Skye
d	Caenlochan–Glen Clova
e	Breadalbane
f	Moor House/Upper Teesdale
g	Craven District/Yorkshire Dales
h	Morecambe Bay
i	Great Orme
j	Derbyshire Dales
k	Yorkshire–Durham Magnesian Limestone
l	North York Moors
m	Yorkshire Wolds
n	Chilterns
o	Cotswolds
p	Breckland
q	Gower
r	Mendips
s	North Downs
t	South Downs
u	Hampshire, Dorset & Wiltshire Downs
v	Torbay

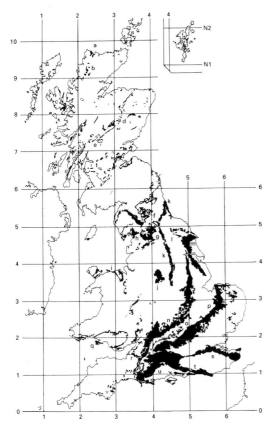

lithology of the substrate exerts relatively little influence on the floristic composition and range of the vegetation types. It is true that there are some striking geographical associations between geological formations and particular kinds of calcicolous grassland, but most of this effect is attributable to the fortuitous disposition of the different kinds of calcareous rock in relation to the climatic zones in this country. It is variations in climate, particularly differences in regimes of temperature and precipitation, which appear to be of prime importance in determining the composition and distribution of the communities. Some of these effects of climate act directly upon the plants themselves, influencing the pace and timing of vegetative growth and sexual reproduction of individual species across their range and their ability to compete against one another in the swards. But of fundamental importance, too, is the indirect impact which climate has through the processes of soil development, in the formation of rendzinas and their maturation from the raw, lithomorphic state, and the balance of water and nutrients in the various kinds of profile. Then, there is the third general influence of treatments, pre-eminently grazing, because, with few notable exceptions, it is not severe climatic or edaphic conditions which maintain these vegetation types as grassy swards, but their continual defoliation by herbivores.

Calcicolous grasslands of the south-east lowlands

Among the fourteen communities, there is a major floristic distinction between grasslands of the south-east of Britain and those of the north and west, separated geographically by a line running roughly from Durham, down through Derbyshire to the Mendips and then skirting the seaboard of Wales (Figures 14 and 15). In environmental terms, this boundary corresponds approximately to the 1000 mm annual isohyet (*Climatological Atlas* 1952) or the 160 wet days yr^{-1} line (Ratcliffe 1968). In the drier south and east, the representation of calcicoles in the swards is strong and related essentially to the occurrence of rendziniform soils on pervious calcareous bedrocks, free of any influence which detracts from the excessively-draining, base-rich and oligotrophic character of the immature profiles. In Britain, there is in these grasslands a background of quite mesophytic Arrhenatherion herbs, such as *Galium verum*, *Plantago lanceolata*, *Trifolium repens*, *Leontodon hispidus*, *Lotus corniculatus* and, where there is some relief from grazing, *Arrhenatherum elatius* and *Avenula pubescens*. Much more distinctive, however, is the variety and abundance of plants characteristic of the less arid of the Festuco-Brometea grasslands, more particularly those of the Mesobromion alliance: *Avenula pratensis*, *Briza media*, *Koeleria macrantha*, *Carex flacca*, *Sanguisorba minor*, *Helianthemum nummularium*,

Scabiosa columbaria, *Filipendula vulgaris* and, in less heavily grazed situations, *Bromus erectus* and *Brachypodium pinnatum*, together with the bryophytes *Fissidens cristatus*, *Homalothecium lutescens* and *Weissia* spp.

The core of this kind of vegetation is represented by the *Festuca ovina-Avenula pratensis* grassland (CG2), the most widespread of the heavily-grazed Mesobromion swards in the British lowlands (Figure 16). Over the warmest part of its range, where mean daily August maxima reach more than 20°C (Pigott 1970*b*), the flora of the community is further enriched by a group of calcicoles with a Continental distribution in Europe (Matthews 1955). Most common among these plants are *Cirsium acaule*, *Hippocrepis comosa* and *Asperula cynanchica*, along with a number of nationally rare plants like *Polygala calcarea*, *Senecio integrifolius*, *Thesium humifusum*, *Euphrasia pseudokerneri*, *Orchis ustulata* and *Herminium monorchis*. The climatic zone where these species are concentrated in Britain happens to include most of our exposures of Chalk, and it is this richer kind of sward that has traditionally been described as 'Chalk grassland' (Tansley 1939, Ratcliffe 1977, Smith 1980), though it is not strictly confined to the Chalk, nor are

other communities of calcicolous grassland absent from this formation (Figure 17). Towards the north, with the shift to cooler, moister conditions, this sort of sward becomes increasingly confined to south-facing slopes and, particularly on the northern Chalk of the Yorkshire Wolds and the Carboniferous Limestone of Derbyshire and the Mendips, we see a floristic shift in the *Festuca-Avenula* grassland towards its counterpart in the upland fringes (Hope-Simpson & Willis 1955, Perring 1960, Pigott 1968, Shimwell 1968*b*).

Even within southern Britain, however, other trends are visible. First, towards the southern and western coastal fringe, where Chalk, Devonian and Carboniferous limestones crop out in the Isle of Wight, around Torbay, in the Mendips and in south and north Wales, there is a local attenuation of the Mesobromion sward with the loss of some mesophytes and calcicoles demanding of less parched conditions. Pauciennial plants which, generally speaking, find no more than an occasional niche in the plush *Festuca-Avenula* turf, are also better represented in this more open kind of grassland, with frequent records for *Carlina vulgaris*, *Blackstonia perfoliata*, *Medicago lupulina*, *Crepis capillaris*

Figure 14. Upland and lowland suites of calcicolous grasslands in relation to 26°C mean annual maximum isotherm and cooler (stippled).

Figure 15. Upland and lowland suites of calcicolous grasslands in relation to 160 wet days per year and wetter (stippled).

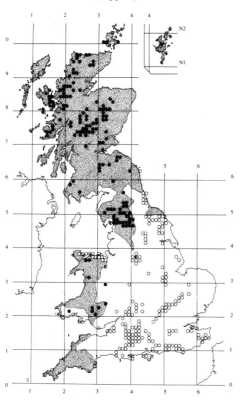

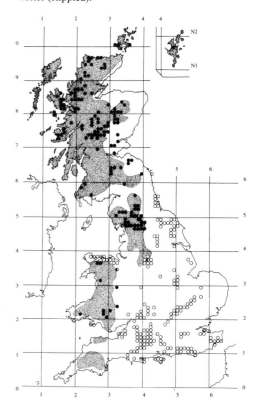

and winter annuals like *Centaurium erythraea*, *Bromus hordeaceus* ssp. *ferronii*, *Desmazeria rigida* and the rare *Cerastium pumilum* and *Gentianella anglica*. Within this warm, oceanic zone with its very mild winters and long growing season, such assemblages are strictly confined to west- and south-facing slopes where the effects of summer insolation are maximised, and to harder exposures which weather to produce open and rocky, but essentially stable, slopes (Figure 18). It is in this *Festuca ovina-Carlina vulgaris* grassland (CG1) that we find the best representation in Britain of most of our Xerobromion rarities – species such as *Koeleria vallesiana*, *Helianthemum canum*, *H. apenninum* and *Trinia glauca*, which have perhaps found a congenial refuge in these hot, unshaded and barren situations right through periods of post-Glacial forest spread (Pigott & Walters 1954, Proctor 1958).

The second trend among the southern swards is best seen towards the east, particularly over the Chalk of East Anglia, very little of which now carries calcicolous grassland. In this region, there is a marked local shift away from dominance by grasses and hemicryptophytes towards an abundance of chamaephytes and thero-

phytes. *Thymus praecox* and, often here, *T. pulegioides*, become very prominent, together with *Hieracium pilosella* and, more locally, *Sedum acre*, while, in the numerous open patches among the tussocky sward, a variety of coarse weeds and pauciennials come and go: *Senecio jacobaea*, *Potentilla reptans*, *Fragaria vesca*, *Rumex acetosa*, *Erigeron acer*, *Arenaria serpyllifolia*, *Aphanes arvensis*, *Myosotis ramosissima* and *Veronica arvensis*. Bryophytes, notably *Hypnum cupressiforme* var. *lacunosum*, *Pseudoscleropodium purum* and *Homalothecium lutescens*, can be very abundant and there may be plentiful *Cladonia* spp. and some unusual encrusting lichens. This *Festuca ovina-Hieracium pilosella-Thymus praecox/pulegioides* grassland (CG7) is always associated with very dry and shallow rendzinas and its open character is accentuated also by extreme nutrient poverty and often some kind of disturbance, as on occupied ant-hills and abandoned ploughland, in old quarries and around rabbit warrens (Lloyd 1964, Lloyd & Pigott 1967, Wells *et al.* 1976, King 1977*a*, *b*, *c*, Smith 1980). It is, however, strongly concentrated in areas with the most extreme kind of continental climate in Britain, as around Breckland where there is less than 600 mm of rain annually with frequent droughts, much wind-blow and frost-heave in the bitter winters. There, in the kinds

Figure 16. Distribution of Continental Mesobromion species in CG2 *Festuca-Avenula* grassland in relation to mean daily maximum August temperatures.

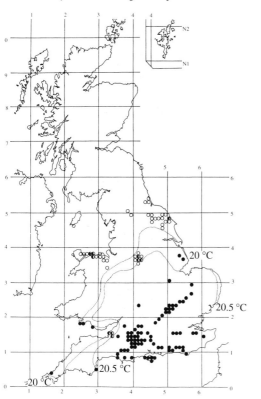

Figure 17. The constituents of 'Chalk Grassland'.

CG1	*Festuca-Carlina* grassland
CG2	*Festuca-Avenula* grassland
CG3	*Bromus* grassland
CG4	*Brachypodium* grassland
CG5	*Bromus-Brachypodium* grassland
CG6	*Avenula pubescens* grassland
CG7	*Festuca-Hieracium-Thymus* grassland
MG1	*Arrhenatheretum elatioris*
MG5	*Centaureo-Cynosuretum*

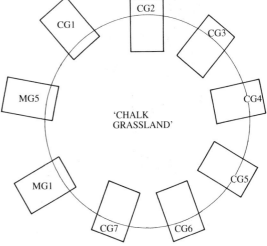

of *Festuca-Hieracium-Thymus* grassland which Watt (1940) described in his Grasslands A and B, we see an abundance of the Continental Northern *Astragalus danicus* and the sporadic occurrence of rarities like *Phleum phleiodes*, *Silene otites* and *Medicago sativa* ssp. *falcata*, which are most characteristic in mainland Europe of the calcicolous steppe-grasslands of the Festucion vallesiacae or Koelerio-Phleion.

In their most extreme forms, the *Festuca-Carlina* and *Festuca-Hieracium-Thymus* grasslands may constitute herbaceous climax vegetation but, in less harsh climatic and edaphic conditions, it is grazing that maintains these lowland calcicolous swards as plagioclimax pastures: sheep and rabbits were most widely influential in the past, with mixtures of sheep and cattle now increasingly important, and rabbits once again becoming locally abundant. Where the *Festuca-Avenula* grassland extends some way on to rather more nutrient-rich and less drought-prone brown calcareous earths, derived from softer and more argillaceous limestones, downwash or light-textured drift, such pasture can take on a distinctly mesophytic aspect and transitions to the *Centaureo-Cynosuretum* (MG5) can be especially hard to

define. This, though, is now a much-declining kind of traditional agricultural sward and often the use of artificial fertilisers and re-seeding of more tractable ground has converted calcicolous swards into the *Lolio-Cynosuretum* (MG6). Zonations through various grades of improved sward have become a common feature of almost every tract of calcicolous grassland in the southern lowlands of Britain.

Equally apparent in many areas are the results of the neglect of grazing or, as most dramatically after myxomatosis, of perturbations in wild herbivore predation. In such conditions, responses among the various elements in the swards can be complex, but most obvious is the spread of coarse Mesobromion or Arrhenatherion grasses, previously held in check by grazing, and the demise of many of the more diminutive light-demanding associates of the close-cropped pasture. We have distinguished four communities of such ranker, tussocky swards dominated by various combinations of *Bromus erectus*, *Brachypodium pinnatum*, *Festuca rubra* and *Avenula pubescens* (CG3, CG4, CG5, CG6) with an abundance of *Arrhenatherum elatius* providing a link with more calcicolous *Arrhenatheretum* (MG1). Exactly what factors control the response of the different species to the relaxation of grazing is still unclear (Hope-Simpson 1940*b*, Wells *et al.* 1976) but, through most of southern Britain, such swards eventually converge, with the invasion of shrubs and trees, into calcicolous *Crataegus-Hedera* scrub (W21). This is itself eventually succeeded by *Fraxinus-Acer-Mercurialis* woodland (W8) or its analogues dominated by beech (W12) or yew (W13).

Calcicolous grasslands of the north-west uplands

Beyond the 1000 mm annual isohyet, such grassland calcicoles as extend outside the warmer continental south-eastern lowlands of Britain are fighting a losing battle against the surface leaching that the higher rainfall induces, even in profiles derived from drift-free, lime-rich bedrocks. In fact, beyond the extensive Carboniferous Limestone deposits of the Craven Pennines, such parent materials are mostly of rather local occurrence through north-west Britain, with few stretches of sedimentary rocks and but small exposures of calcareous igneous and metamorphic rocks. Many outcrops are also mantled by superficial deposits of one sort or another, such as till, aeolian sand or loess, often of an inherently lime-poor and/or impermeable character. In this region, then, the less drought-prone and lime-saturated nature of the soils is often marked by the frequency of mesophytic plants like *Festuca rubra* and *Viola riviniana*, and of such calcifuges as *Agrostis capillaris*, *Anthoxanthum odoratum*, *Nardus stricta*, *Carex pilulifera*, *Luzula campestris*, *Potentilla erecta*, *Galium saxatile*, *Alchemilla alpina* and the mosses *Hylocomium splendens* and *Rhytidiadelphus squarrosus*. Among such

Figure 18. Distribution of CG1 *Festuca-Carlina* grassland in relation to warm oceanic climate (stippled).

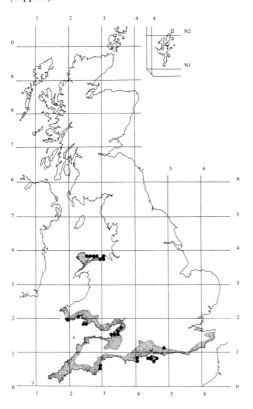

Grass-heath vegetation with combinations of above species absent

> **CG11** *Festuca ovina-Agrostis capillaris-Alchemilla alpina* grass-heath 41

41   Five or more of the following present: *Selaginella selaginoides, Ranunculus acris, Campanula rotundifolia, Carex pulicaris, C. panicea, Plantago lanceolata, Viola palustris, Anemone nemorosa, Veronica serpyllifolia, Juncus squarrosus*

> **CG11** *Festuca ovina-Agrostis capillaris-Alchemilla alpina* grass-heath
> *Carex pulicaris-Carex panicea* sub-community

Around springs and flushes this vegetation type can grade into small-sedge mire communities but the rise to prominence of Caricion davallianae sedges and bryophytes will usually effect a separation.

Combinations of the above species rare and three or more of the following present: *Dicranum scoparium, Pleurozium schreberi, Ptilidium ciliare, Trollius europaeus, Peltigera canina, Cladonia uncialis, Rumex acetosa, Vaccinium vitis-idaea, Blechnum spicant*

> **CG11** *Festuca ovina-Agrostis capillaris-Alchemilla alpina* grass-heath
> Typical sub-community

42   Five or more of the following present: *Trifolium repens, Galium saxatile, Luzula campestris, Rhytidiadelphus squarrosus, Dicranum scoparium, Veronica officinalis, Cerastium fontanum, Achillea millefolium, Carex caryophyllea*

> **CG10** *Festuca ovina-Agrostis capillaris-Thymus praecox* grassland

Trifolium repens-Luzula campestris sub-community

In the enclosed land around upland farms, agricultural improvement tends to blur the distinctions between this vegetation type and its calcifugous counterpart, the *Festuca ovina-Agrostis capillaris-Galium saxatile* grassland, and reseeded derivatives of both of them that have developed into *Lolio-Cynosuretum*.

Combinations of above species infrequent but three or more of the following present: *Carex panicea, C. pulicaris, C. flacca, Linum catharticum, Succisa pratensis, Alchemilla glabra, Selaginella selaginoides, Ctenidium molluscum* 43

43   Three or more of the following present: *Saxifraga aizoides, Bellis perennis, Tortella tortuosa, Briza media, Polygonum viviparum, Saxifraga oppositifolia, Pinguicula vulgaris, Deschampsia cespitosa*

> **CG10** *Festuca ovina-Agrostis capillaris-Thymus praecox* grassland
> *Saxifraga aizoides-Ditrichum flexicaule* sub-community

Combinations of above species absent

> **CG10** *Festuca ovina-Agrostis capillaris-Thymus praecox* grassland
> *Carex pulicaris-Carex panicea* sub-community

Around springs and flushes, both these vegetation types can grade to small-sedge mire communities but the rise to prominence there of Caricion davallianae sedges and bryophytes will usually effect a separation.

COMMUNITY DESCRIPTIONS

CG1
Festuca ovina-Carlina vulgaris grassland

Synonymy

Natural pasture Moss 1907 *p.p.*; Carboniferous Limestone Grassland Hope-Simpson & Willis 1955 *p.p.*; *Helianthemum apenninum* localities Proctor 1956, 1958; *Helianthemum canum* localities Proctor 1956 *p.p.*; *Helianthemum canum-Thymus drucei* Association Proctor 1958 *p.p.*; *Arabis stricta* localities Pring 1961; Carboniferous Limestone Grassland Gittins 1965*a* *p.p.*; *Thymo-Festucetum* Williams & Varley 1967 *p.p.*; *Poterio-Koelerietum vallesianae* Shimwell 1968*a*; *Helianthemum apenninum-Euphorbia portlandica* Nodum Shimwell 1968*a*; *Helianthemo-Koelerietum cristatae* Shimwell 1968*a* *p.p.*; *Draba aizoides* localities Kay & Harrison 1970 *p.p.*; Carboniferous Limestone Grassland Ratcliffe 1977 *p.p.*; Devonian Limestone Grassland Ratcliffe 1977 *p.p.*; Typical Limestone Grassland 3a South Gower Coast Report 1981.

Constant species

Carlina vulgaris, Dactylis glomerata, Festuca ovina, Hieracium pilosella, Lotus corniculatus, Plantago lanceolata, Sanguisorba minor, Thymus praecox.

Rare species

Arabis stricta, Aster linosyris, Bupleurum baldense, Carex humilis, Cerastium pumilum, Draba aizoides, Euphorbia portlandica, Gentianella anglica, Helianthemum apenninum, H. canum, H. × sulfureum, Hypochoeris maculata, Koeleria vallesiana, Potentilla tabernaemontani, Scilla autumnalis, S. verna, Sedum forsteranum, Senecio cineraria, Veronica spicata, Scorpiurium circinnatum, Tortella nitida.

Physiognomy

The *Festuca-Carlina* grassland has a characteristically short and open tussocky turf which, even where it forms extensive stands, is usually interrupted by fractured rock outcrops and small patches of bare soil and which is often disposed in more fragmentary fashion over narrow ledges and in crevices. *Festuca ovina* (or, in one of the sub-communities, *F. rubra*) is almost always an abundant component of the sward and either *Koeleria macrantha* or *K. vallesiana* is also very frequent throughout, though usually in smaller amounts. There are often some rather small tufts of *Dactylis glomerata*, perhaps of the form variously described as var. *abbreviata*, var. *maritima* or ssp. *hispanica* (see Tutin *et al.* 1980). Generally, however, it is not perennial grasses but the woody chamaephytes *Thymus praecox* and *Helianthemum* spp., together with the deep-rooted hemicryptophyte *Sanguisorba minor*, which give the vegetation its distinctive stamp. Each of these can be abundant and combinations of them frequently co-dominate with either *F. ovina* or *F. rubra*.

Hieracium pilosella and *Lotus corniculatus* are constant throughout and these, too, may be abundant; *Anthyllis vulneraria*, though its occurrence is rather more uneven, can also be locally prominent. There are very frequently some scattered rosettes of *Plantago lanceolata* and, less commonly, *Leontodon taraxacoides*, *Galium verum* and *Daucus carota* (including some records for ssp. *gummifer*) occur in small quantities. Although many stands occur close to the sea, strictly maritime species are rare in the community as a whole.

There are two further general features of this vegetation. The first is the frequency of those ephemeral species which are characteristic of more calcicolous open habitats. The most frequent and showy of these (though it is rarely abundant) is the biennial *Carlina vulgaris*. Less common, but sometimes present in abundance on the fine patchwork of areas of bare soil, are some of the following annuals or pauciennials: *Euphrasia officinalis* agg., *Centaurium erythraea*, *Blackstonia perfoliata*, *Medicago lupulina*, *Crepis capillaris*, *Arenaria serpyllifolia* and the grasses *Bromus hordeaceus* ssp. *ferronii* and *Desmazeria rigida*; the national rarities *Cerastium pumilum* and *Gentianella anglica* also occur

more slow-growing species (like *Carlina vulgaris*) requiring more space than the smaller annuals. In some cases, however, other factors may account for the relative rarity of these pauciennials: in some dune vegetation, for example, the abundance of *C. vulgaris* seems to be strongly influenced by seed predation by small mammals (Grieg-Smith & Sagar 1981).

Although the openness of this vegetation is probably primarily controlled by the physical characteristics of the habitat, grazing of the herbage by sheep and rabbits is also likely to be of importance in maintaining the cover of grasses at relatively low levels and so favouring the abundance of chamaephytes and the relatively high frequency of pauciennials. The former, though they may be sometimes nibbled (e.g. *Thymus praecox*: Pigott 1955) or have their seed-set limited by the removal of inflorescences (e.g. *Helianthemum canum*: Griffiths & Proctor 1956), are probably partly protected by their very prostrate habit. The latter perhaps benefit by the grazing-out of grass tussocks and the creation of scuffed areas of bare soil. These effects are likely to be of greater significance around the less rocky fringes of stands of the community where access is easier: on narrow ledges and in remoter crevices, the vegetation is probably out of reach of sheep at least. On these margins, heavy grazing may help maintain the community under more mesic edaphic conditions than are usual: on the Great Orme in Gwynedd, for example, the *Helianthemum canum* sub-community occurs in stands of quite substantial size under very heavy sheep-grazing over somewhat deeper soils than normal (Proctor 1958).

In some localities, especially where the community occurs close to coastal resorts, where large summer populations are attracted to the fine views afforded by the cliffs over which it grows or to the pleasing 'rock-garden' effect of the vegetation itself, trampling can be considerable. The unevenness of the terrain can afford some protection to craggier stands and some sensitive species, such as *Draba aizoides* and *Arabis stricta*, are generally confined to such situations. Others, such as the rare *Helianthemum* spp. and *Thymus praecox*, are resistant to a certain amount of trampling and will persist in more accessible stands. *Cynosurus cristatus*, a grass which is usually no more than very occasional in the community, is markedly frequent and sometimes abundant in stands of certain sub-communities where trampling is considerable, as in the *Trinia* sub-community on the Mendips around Weston-super-Mare and the *Festuca-Scilla* sub-community on the south Gower coast. Heavy trampling has undoubtedly contributed to the now rather scruffy appearance of some stands of the *Scilla-Euphorbia* sub-community around Torquay and Brixham in Devon.

Some of the floristic variations between the different sub-communities are perhaps related to present-day differences in these climatic, topographic and biotic variables. As outlined above, there is a trend among the vegetation types included here from the more extreme (the *Carex humilis* and the *Scilla-Euphorbia* sub-communities) to swards which are very similar to the *Festuca ovina-Avenula pratensis* grassland (the *Koeleria* sub-community) in which an increase in Mesobromion species may be partly related to the occurrence of somewhat deeper soils over less rocky terrain in slightly damper and less sunny climates. In this respect, the slight ecological differences between *H. apenninum* and *H. canum* are of interest. The former is very much confined in Britain to the more extreme kinds of sward in this community; the latter has a wider geographical and phytosociological distribution, occurring not only in more mesic swards here but also in the *Sesleria-Galium* grassland in the considerably cooler, wetter and duller climate of the Pennine Carboniferous Limestone. Its topographic vicarism with *H. nummularium* is also somewhat less marked than that of *H. apenninum*, although this may be affected by grazing intensity (Proctor 1956, 1958). Then, again, in the *Carex humilis*, and, especially, the *Festuca-Scilla* sub-community, there is a more obvious maritime influence where the occasional occurrence of species such as *Plantago coronopus*, *P. maritima* and *Armeria maritima* may reflect a greater input of salt-spray in stands very close to the sea. Only detailed comparative studies of the physical environments of the different sub-communities can elucidate the extent of such influences.

It seems, however. very probable that some of the floristic variation within the community, and especially that involving the coincidental occurrence in the various sub-communities of different members of the group of Oceanic Southern and Continental Southern rarities, is related, not only to present-day ecological variation, but also, and perhaps primarily, to the accidents of historical survival. The habitat of the community as a whole is one of those kinds of environment that Pigott & Walters (1954) saw as likely to have retained its open and calcareous character, by virtue of its topographic and edaphic features, in the face of the forest advance and soil leaching that occurred as the Post-glacial progressed. Indeed, it is not impossible that the localities where the community occurs, or perhaps more restricted rockier areas within them, have remained in largely the same physiographic form as they are today, since the Late-glacial. On this hypothesis, they could have provided refuges for vegetation which, with the close of the Glacial, spread to become much more widely distributed than now, but which, with the later advance of a tree cover and the development of deeper and less calcareous soils over less extreme topographies, became progressively more restricted. Such a view could account not only for the present confined and disjunct distribution of the community and the rarity in Britain of some of its most distinctive species, but also for the somewhat anomalous pattern of presence and absence of these species in moving from one locality to another. Which of these rarities survived where, may be the largely fortuitous result of an uneven process of forest advance and irregular isolation of remaining areas of open ground. Some support for this proposal is given by the fact that, in the equivalent kinds of vegetation in Continental Europe, species which here seem to replace one another in different sites occur together more frequently and consistently in communities that are both richer and more widely distributed than the *Festuca-Carlina* grassland. Such Continental communities also sometimes have consistent records for species which still survive in Britain but which here are not confined to the *Festuca-Carlina* community, but also occur in other kinds of disjunctly-distributed open vegetation (e.g. *Veronica spicata* and *Potentilla tabernaemontani* (part of the European *P. verna* agg.)) or are wholly restricted to the latter (e.g. *Phleum phleoides* and *Scleranthus perennis*). What we seem to have in this community is therefore part of the remnants of a flora which is still widespread in parts of the unglaciated area of Europe but which has been sorted and fragmented through history in Britain and is now confined to a small number of scattered sites which remain ecologically suitable.

We cannot reconstruct the detailed history of the localities where the community survives or of the vegetation itself. The pollen record provides little information about the former distribution of its preferentials or differentials: although *Helianthemum* pollen is very widespread in Late-glacial deposits (Godwin 1975), it is not separable reliably into its species. What does seem clear, however, is that, as forest was cleared for agriculture, there was no large-scale reappearance of habitat conditions suitable for expansion of the community. Grazing, especially where very heavy, may have permitted some extension of stands around the margins of rockier areas but the community still retains its essentially disjunct distribution and strict confinement to steep, warm and dry slopes.

Moreover, even stands which occur in close proximity to one another have not been colonised by species which occur nearby: thus *Carex humilis* remains abundant on Brean down in Somerset, yet totally absent from Uphill, less than two kilometres away; conversely, *Trinia glauca*, which is common on the latter site, is not present on the former. Colonies of some other species remain similarly localised, yet good growth in cultivation has been recorded for certain of them well outside their present natural ranges (e.g. *H. apenninum* (Proctor 1956), *Aster linosyris* (C.D. Pigott, pers. comm.)). In some cases, therefore, there may be hindrances to even small-scale migration which compound the isolation of tracts of the suitable habitat. Certain species, like *H.*

apenninum (Proctor 1956), *H. canum* (Griffiths & Proctor 1956), *Draba aizoides* (Kay & Harrison 1970) and *Arabis stricta* (Pring 1961), though they fruit well, have no special dispersal mechanism: seeds seem to be scattered to within but a short distance of the parent plants and cannot establish in the closed swards that often separate stands of the community. There may also be difficulties in survival for young seedlings of perennials in the parched conditions of the summer following germination. Again, seeds or seedlings may be heavily predated, as in *D. aizoides*, where snails are a major cause of seedling mortality in the first winter (Kay & Harrison 1970). It would be wrong to presume a common reason for slow migration of these species and only detailed programmes of experimentation, with perhaps seeding-in or transplantation of species into stands of the community from which they are absent, could provide a firm answer to their behaviour.

An additional complexity here is that some of the distinctive species of the community show ecotypic responses to what must now be very long periods of isolation of their populations. *Helianthemum apenninum*, for example, though very uniform in its British localities, retains in cultivation a prostrate habit that differs markedly from the erect form of this species in its nearest Continental locality along the Seine valley in France (Proctor 1956, 1958). Other species vary from one British locality to another, though differences may be hard to discern in the often exposed conditions in which the plants grow in the wild. Plants of *H. canum* from different British sites show a certain amount of this kind of variation and they also differ in their somewhat hairier leaves from the *H. canum* which survives in the Teesdale locality of the *Sesleria-Galium* grassland (forms sometimes described as var. *vineale* Pers. or ssp. *levigatum* Griffiths & Proctor 1956, Proctor 1957). *Aster linosyris* from the Mendips in Somerset and Berry Head in Devon has, in cultivation, a sub-erect form with the plants attaining heights of 30–35 cm; erect and taller plants, 40–45 cm high, are produced from the populations on the Great Orme in Gwynedd (and Humphrey Head in Cumbria where this species occurs in the *Sesleria-Galium* grassland) (C.D. Pigott, pers. comm.). The Humphrey Head and some, at least, of the Great Orme populations are also known to be self-incompatible clones, though they readily produce fertile achenes when plants propagated from root cuttings are brought together with other populations (Pigott 1977). By contrast, the Berry Head population fruits well in the wild, being either self-fertile or composed of a group of strains of mixed incompatibilities.

Veronica spicata is another species which shows considerable variation, both between and within its populations in stands of this community and of those other vegetation types in which it survives, such that some of the more robust and larger-leaved western forms which can be found in the *Festuca-Carlina* grassland have been considered as the separate species, *V. hybrida* L. A more satisfactory solution, proposed by Pigott & Walters (1954; see also Tutin *et al.* 1972), seems to be to regard such differences as largely ecotypic responses within a single species whose populations have been widely separated in habitats which have a generally open and calcareous character but which perhaps differ slightly in certain features. As with the community as a whole, a great deal remains to be known about the interplay between historical and ecological factors that have produced such distinctive patterns of distribution.

Zonation and succession

Stands of the *Festuca-Carlina* grassland generally occur as relatively small islands of open, rocky vegetation within more extensive tracts of closed swards or patch-works of grassland, heath and scrub. Much of this variation is explicable in terms of an edaphic transition in moving from the limestone exposures with their often fragmentary and highly calcareous soils to areas of deeper and moister soils which, in the absence of heavy grazing, become invaded by ericoids or other shrubs.

In many places, and especially where exposures of bedrock are low and much fractured, the community may cover entire outcrops. In some sites, however, the cliffs are more substantial and here the *Festuca-Carlina* grassland may become fragmented in the crevices and on the ledges of the rock face and give way over the more massive exposures to fern-dominated stands of the *Asplenium trichomanes-Fissidens cristatus* community in narrow fissures. Some of the species of the *Festuca-Carlina* grassland, such as certain of the pauciennials and *Draba aizoides*, can survive in this more chasmo-phytic vegetation. Elsewhere, on more unstable areas of bare limestone, where there is clitter and shifting accumulations of wind-blown organic and mineral detritus, there can be a transition to the more ephemeral vegetation of the Thero-Airion.

On grazed sites (which are the majority), the most common transition around stands of the community is to some form of the *Festuca ovina-Avenula pratensis* grassland, frequently, on these warm, south-facing slopes, the *Helianthemum nummularium-Filipendula vulgaris* variant of the Typical sub-community. As the soil cover becomes more continuous and slightly deeper (though remaining calcareous and base-rich), the sward thickens up and there is a switch to a more obviously grass-dominated turf with an increase in Mesobromion dicotyledons and a reduction in the frequency of pauci-ennials. Often, such junctions are marked by a strong vicarism between either *H. apenninum* or *H. canum* and *H. nummularium*, patterns which have been strikingly depicted by Proctor (1958). In those sub-communities where *Koeleria vallesiana* occurs (the *Carex humilis* and *Trinia glauca* sub-communities), there may be a similar

abrupt switch to *K. macrantha*; elsewhere, this latter species runs right through the grasslands to even the most strongly parched stands of the *Festuca-Carlina* community.

Variations in the sub-surface configuration of the limestone or local accumulations of loess or head can complicate such transitions. Then, more calcifuge com-munities developed over the deeper and initially less calcareous or superficially leached and acidified soils, can occur in close contact with the *Festuca-Carlina* grassland. This kind of mosaic is very characteristic where the *Festuca-Scilla* sub-community occurs on the Gower cliffs in West Glamorgan (South Gower Coast Report 1981).

In the more maritime sites, as in Gower and on Brean Down, there is the additional complexity imposed by gradients, running more or less normal to the coastline, of salt-spray deposition. Here, stands of the *Festuca-Scilla* and *Carex humilis* sub-communities may have occasional records for certain maritime species and give way towards the sea to the *Festuca-Armeria* grassland or directly to the crevice vegetation of the *Crithmo-Spergu-larietum* on sea-splashed rocks. Likewise, transitions to Thero-Airion and heath communities further up the cliffs may involve more maritime vegetation types like the *Armeria-Cerastium* therophyte community or the *Calluna-Scilla* heath.

It is doubtful whether grazing is necessary to maintain the openness of the more extreme habitats of the community, where topographic and climatic conditions are probably sufficient to prevent the development of a closed sward, subsequent invasion of shrubs and the formation of a closed woody canopy. However, bird-sown seed of shrubs can germinate in rock crevices and, if grazing is relaxed on the slopes where the community occurs, stands can be encroached upon by scrub relati-vely quickly: thus, Proctor (1958) noted a marked change in some stands of the *Scilla-Euphorbia* sub-community on Anstey's Cove in Devon in only three years following the demise of rabbits from myxomato-sis. *Ligustrum vulgare* seems to be the most characteristic early invader, but *Crataegus monogyna*, *Rubus frutico-sus* agg. and, at some sites, the Himalayan introduction *Cotoneaster microphylla*, have also been recorded. In the sunny gaps within this patchy cover, tussocks of *Brachy-podium sylvaticum* can grow up together with prominent plants of *Teucrium scorodonia*, *Geranium sanguineum* and *Centaurea scabiosa* with, in the shade beneath the shrubs, sprawls of *Hedera helix*. In time, such vegetation probably develops into the kind of wind-cut spinose scrub characteristic of many ungrazed sea cliffs. In some cases, where such scrub is encroaching more slowly from areas of deeper soil, it may grade to the *Festuca-Carlina* grassland through an intermediate zone of rank *Avenula pubescens* grassland with its characteristic mixture of Mesobromion herbs, tall grasses and dicotyledons and seedling shrubs or, in more maritime situations, through the *Festuca rubra-Daucus carota* ssp. *gummifer* grassland (South Gower Coast Report 1981). Some of the chamaephytes and pauciennials of the *Festuca-Carlina* grassland seem to be able to persist for some time in more open areas but they are soon overwhelmed by the taller herbage.

Distribution

The community is characteristically limited to scattered sites on harder limestones around the southern and western coasts of England and Wales. The most widely distributed type is the *Koeleria macrantha* sub-community which occurs over Carboniferous Limestone in north Wales and the Mendips and extends the occurrence of the community on to harder Chalk strata in Dorset, the Isle of Wight and Sussex. All the remaining sub-communities are much more strictly confined to geographically isolated localities: the *Carex humilis* sub-community to Brean Down in Somerset, the *Trinia* sub-community to various sites between Brean and Axbridge along the southern Mendips, the *Festuca-Scilla* sub-community to cliffs along the south Wales coast from St Govan's Head to Gower, the *Helianthemum canum* sub-community to the north Wales coast between Anglesey and the Vale of Clwyd, all these over Carboniferous Limestone, and, over Devonian Limestone, the *Scilla-Euphorbia* sub-community to the area around Tor Bay in Devon.

Affinities

The vegetation types included in this community represent the nearest approach among British calcicolous grasslands to the Continental *pelouses sêches calcaires* and *felsigen Trockenrasen* of the sub-alliance Xerobromion. Open, rocky swards of this kind are found more widely in mainland Europe where limestones (including Chalk) occur in climatic conditions which are attained in this country in but a few scattered localities, but which there characterise whole regions. As in Britain, however, the communities still have their maximum expression on the warmest, south-facing slopes, especially where the annual precipitation exceeds 900 mm (Shimwell 1971a). Vegetation of this kind has been described from northwest France (Allorge 1921–2, Litardière 1928, Hagène 1931, Liger 1952, Bournérias 1968; see also Stott 1970), south-west Germany (Volk 1937, Krause 1940), the Jura (Pottier-Alapetite 1943), the Auvergne (Luquet 1937) and the Alpes Maritimes (Braun-Blanquet 1961). In these regions, those rarities of the *Festuca-Carlina* grassland which are of very sporadic occurrence in Britain become a more consistent feature of the vegetation, providing a strong link with the steppe-grasslands of eastern Europe (in the order Festucetalia vallesiacae) and, to the south, with the calcicolous *garrigue* scrub of the Mediterranean (Proctor 1958, Shimwell 1971a).

Lying, as it does, on a far fringe of the distribution of this kind of vegetation in Europe, the *Festuca-Carlina* community has more in common with the swards of the Mesobromion than have the mainstream Xerobromion associations of mainland Europe and it could be argued that only the more extreme types should be separated off from vegetation of the former alliance. This was the solution proposed by Shimwell (1968a, 1971a), but if his suggestion were to be followed consistently, the *Trinia glauca* sub-community of his *Poterio-Koelerietum* (equivalent to the *Trinia* sub-community here) would belong better alongside the *Helianthemum canum* vegetation of the Welsh localities which he placed in the Mesobromion: apart from the absence of *Trinia glauca* itself, the replacement of *Helianthemum apenninum* by *H. canum* and of *Koeleria vallesiana* by *K. macrantha*, the vegetation is very similar indeed. While recognising the transition within the sub-communities to the Mesobromion (a trend which Shimwell himself acknowledged), it seems preferable to keep these vegetation types together in a single distinct community. All are characterised by some floristic and physiognomic features that separate them from the closed swards of the *Festuca ovina-Avenula pratensis* grassland and all occupy distinctly more extreme habitats. Moreover, grouping them in this way provides a unified framework for considering the effects of those historical factors that have apparently played some part in determining the structure and distribution of each.

Two other floristic affinities deserve mention. First, the *Festuca-Carlina* grassland is replaced on warm, rocky slopes on the more northerly Carboniferous Limestone around Morecambe Bay by the *Helianthemum canum-Asperula cynanchia* sub-community of the *Sesleria-Galium* grassland. Apart from the rather particular conditions of Upper Teesdale, this is the only vegetation type in Britain in which *Sesleria albicans* occurs with any of the Oceanic Southern or Continental Southern rarities, a situation which contrasts strikingly with mainland Europe where *S. albicans* is a constant species in a number of Xerobromion communities. The peculiar British distribution of this grass is taken up below.

Second, the *Festuca-Carlina* community shares with the *Festuca-Hieracium-Thymus* grassland a distinctly open structure and an abundance of pauciennials. In that vegetation type, however, the fragmentary nature of the sward is attributable to very heavy grazing by sheep or (especially in the past) rabbits and the instability of the habitat is reflected by the additional frequency of weed species which are never prominent in the *Festuca-Carlina* community. Such swards can develop on all aspects and, though on south-facing slopes the effects of insolation may accentuate the openness of the cover, they do not depend on topoclimatic conditions for the maintenance of their distinctive physiognomy.

profile and, on cooler, north-facing slopes, by some slight solifluctional churning. Such disturbance may be further enhanced by the trampling of stock and the activities of rabbits. Terracing of the surface, with the development of small-scale mosaics within the turf, is common in *Festuca-Avenula* swards (e.g. Cornish 1954). Although bedrock exposures and screes are uncommon on the Chalk (but see Smith 1980), they occur widely on the older and harder limestones and fragmentation of the soil cover around these may introduce some heterogeneity into the vegetation. In general, however, this grassland is characterised by a closed sward and it gives way to other communities where the stability of the soil cover is markedly disrupted.

In a typical valley-side topography, stands of the community are usually limited above by often more poorly draining and less calcareous superficial deposits (such as till within the limit of the Final Glaciation and, beyond this zone, periglacial material like Clay-with-Flints) and below by accumulations of colluvium. The soils developed from such materials tend, through brown rendzinas, towards calcareous or typical brown earths, generally deeper, moister and more mesotrophic and with some horizon differentiation which can involve the eluviation of any calcium carbonate from the upper layers with a slight fall in pH. Where such soils remain moderately calcareous, the *Festuca-Avenula* grassland may extend on to them in the form of the *Holcus-Trifolium* sub-community. Stands of this vegetation type often mark the upper, gently-sloping brows of limestone slopes where there is some plateau downwash, perhaps augmented by ploughing above. This kind of pattern has been well described from some Derbyshire Dales (e.g. Balme 1953, Pigott 1962, Grime 1963*a, b*, Bryan 1967) and from the Mendips (Shimwell 1968*a*, 1971*b*) and it can also be discerned in data from the Pewsey Vale scarp (Thomas *et al.* 1957: see also Figure 21). Deeper, mesotrophic but still moderately base-rich soils over flat limestone surfaces can also carry this sub-community (e.g. Gittins 1965*a*).

Within these general limits, climatic variation plays a very considerable part in influencing the floristic differences between the various sub-communities and its effects are visible at a variety of levels: over whole regions, between one local area and another and even within individual stands of the vegetation. Geology is important here because of the coincidental way in which the different limestones are disposed over the country and because the physiography of the exposures modifies the impact of climate, most notably through aspect and slope.

The most obvious regional climatic trend reflected in the community is that from the lowland south and east (where the major limestone is the Chalk) to the upland fringes of the north and west (where older limestones occur with the northern Chalk and the Corallian). The

former area is, generally speaking, drier, warmer and sunnier, especially in the growing season. Over much of this region, the annual rainfall is less than 800 mm with usually under 250 mm falling between May and August (*Climatological Atlas* 1952, Chandler & Gregory 1976). Average means of daily maximum August temperatures exceed 20 °C and, throughout the summer, there is much bright sunshine, generally more than $5\frac{1}{2}$ hours daily in August (Pigott 1970*b*). Towards the more elevated northern and western limits of the community, beyond

Figure 21. Simplified vegetation map of Pewsey Down in Wiltshire (after NCC reports).

CG2b *Festuca-Avenula* grassland, *Succisa-Leucanthemum* sub-community
CG2c *Festuca-Avenula* grassland, *Holcus-Trifolium* sub-community
MG1 *Arrhenatheretum elatioris*
MG5b *Centaureo-Cynosuretum, Galium* sub-community
MG6c *Lolio-Cynosuretum, Trisetum* sub-community
MG7 *Lolium perenne* ley
W21a *Crataegus-Hedera* scrub, *Hedera-Urtica* sub-community
W21d *Crataegus-Hedera* scrub, *Viburnum* sub-community

A–A indicates scarp top, B–B scarp bottom.

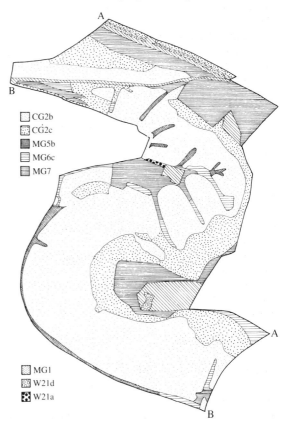

the Humber–Severn line, the annual rainfall is generally higher, often approaching and sometimes exceeding 1000 mm, with up to 350 mm falling in the summer months which are also markedly cooler and cloudier (Pigott 1970b, Chandler & Gregory 1976). Among the different kinds of *Festuca-Avenula* grassland, the *Cirsium-Asperula* and *Succisa-Leucanthemum* sub-communities are almost entirely confined to the former region, the *Dicranum* sub-community to the latter. Two components can be discerned in this floristic response.

The most obvious is the restriction of the Continental element largely to the former sub-communities. This is unlikely to be a simple matter but, for some species at least, it is known to be related to the need for a drier, warmer and sunnier climate to maintain their reproductive capacity. In *Cirsium acaule*, for example, both fruit production and the occurrence of established seedlings decline towards the north-western limit of distribution (Pigott 1968, 1970b). The former is dependent on long daylight hours for the initiation of capitula and on high temperatures for their continuing development and embryo growth after pollination. The stigmas are receptive for but a short period, which must coincide with fine clear weather to maximise the possibility of visits by flying insects, in this case mostly bees. Furthermore, wet weather may allow the capitula to become infected with the fungus *Botrytis cinerea*, which can severely limit fruit production or cause entire heads to rot. Such mature fruits as do survive and disperse do not germinate or grow well in moister conditions, perhaps because of damping-off. Thus, where a late spring gives way to generally cooler, damper and cloudier summers, with a greater unpredictability in the weather, as to the north and west, this species is less able to consolidate its position and more vulnerable to environmental change. Similar climatic influences, perhaps equally complex, have been suggested as playing a role in limiting the distribution of *Thymus pulegioides* (Pigott 1955), *Pulsatilla vulgaris* (Wells & Barling 1971), *Hippocrepis comosa* (Fearn 1973) and *Senecio integrifolius* ssp. *integrifolius* (Smith 1979). Close investigation of other members of the Continental element in this respect would clearly be worthwhile.

The other component in the response works through soil differences. To the north and west, there is a greater tendency for even drift-free soils on gentle slopes to become leached of calcium carbonate in their superficial layers. This is reflected in the *Dicranum* sub-community by its more narrow restriction to steeper slopes than are occupied by the other sub-communities further south and by the occasional presence in the swards of some species indicative of slight surface leaching, species which are, in general, rare in southern *Festuca-Avenula* grasslands except where there is contamination with less lime-rich superficials.

This major floristic distinction among the sub-communities can be seen, writ small, where local topographic variation accentuates or ameliorates the effects of the large-scale climatic trend. This phenomenon was noted in some early studies (e.g. Tansley & Adamson 1925, 1926, Hope-Simpson 1940a, 1941b) but first given neat expression in Perring's (1958, 1959, 1960) attempts to produce a theoretical model to account for floristic variation in *Festuca-Avenula* grasslands over Chalk in relation to climate and topography.

The most obvious effects are felt through the influence of aspect though, as always with this variable, the frequent absence of a complete range of expositions within local areas may make it difficult to perceive its impact. Over the range of the community as a whole, however, two effects are quite clear. First, towards the north and west, the Continental species become increasingly confined to those south and south-west facing slopes where the topoclimate continues to approximate to the regional climate further south. There is thus a narrow zone of overlap between the *Dicranum* and *Cirsium-Asperula* sub-communities (mostly between the 20° and 20.5° August isotherms with an outlier in north Wales) where both vegetation types can be seen on opposed aspects. This is particularly striking in those few Derbyshire Dales which happen to run in an east–west direction and have identical pastoral regimes on both their northern and southern slopes (Pigott 1970b). Conversely, in the south-east of the country, it is on cooler and damper, north-facing slopes that the *Cirsium-Asperula* sub-community most closely approaches the *Dicranum* sub-community in its floristics, with a sometimes marked quantitative reduction in, though rarely a total loss of, various of the Continental species.

The other effect is seen on what are probably the very warmest and sunniest slopes occupied by the *Cirsium-Asperula* sub-community. On some south-west facing slopes through the range of this kind of *Festuca-Avenula* grassland, a good representation of Continental species is further augmented by unusually high frequencies of *Helianthemum nummularium*, *Filipendula vulgaris* and *Sanguisorba minor* in the *Filipendula-Helianthemum* variant, a floristic transition to the Xerobromion swards of the *Festuca-Carlina* grassland.

However, attempting to unravel the effects of climate in the southern part of the range of the community is very difficult because here there is a further climatic trend of major importance. This exerts its influence again partly through temperature but most obviously through precipitation, both its absolute values and its seasonality. Towards East Anglia, the annual rainfall may be as little as 600 mm but there is a slight tendency for a summer maximum which accentuates the continental quality of the climate with its warm summers and cold winters (Chandler & Gregory 1976). Towards the south-west, on the other hand, annual rainfall is mostly in excess of 800 mm and locally (as in the western South Downs (Adamson 1921, Hope-Simpson 1941*b*) and on the western Chalk in Wiltshire and Dorset (Rodda *et al.* 1976, Smith 1980)) as high as 1000 mm. There is, moreover, a strong tendency for a winter maximum which, with the milder temperatures in that season, gives the local climate a distinctly oceanic character.

Many of the typical species of the southern *Festuca-Avenula* swards retain their frequency across this region, though some (e.g. *Polygala calcarea* and the Oceanic West European *Thesium humifusum*) are more abundant to the west. Among other associates, though, some clearer responses can be seen. First, within the *Cirsium-Asperula* sub-community, the *Pseudoscleropodium-Prunella* variant, with its distinctive abundance and variety of bryophytes, tends to be better represented in the wetter, western areas, the Typical variant being somewhat more widespread. As before, local topographic variation can enhance or reduce this effect. The former variant, for example, is especially prominent on north-facing slopes within the wetter areas, most notably in the western South Downs. The fact that most of those few accounts of 'Chalk grassland' bryophytes that we have (e.g. Tansley & Adamson 1925, 1926, Hope-Simpson 1940*a*, Watson 1960) have drawn many of their data from this area is of some importance. It should be remembered, too, that the bryophyte component in these swards is strongly influenced by grazing; and that particular kinds of pastoral treatment have sometimes been concentrated in areas with a distinctive regional climate or where certain aspects predominate, thus compounding these effects (see below). Outside the wetter areas, the Typical variant tends to occur even on the cooler, damper northern slopes, as on the Chiltern scarp, until, on the sand-smeared Chalk of Breckland, it gives way to the more obviously continental vegetation of the *Festuca-Hieracium-Thymus* grassland.

Second, the more oceanic climate of the western Chalk probably plays some part in influencing the distribution and floristics of the *Succisa-Leucanthemum* sub-community. This very distinctive kind of *Festuca-Avenula* grassland is almost entirely restricted to parts of Wiltshire and Dorset, where it is especially well developed (or, at least, now largely remains) over south- and west-facing slopes where higher amounts of rainfall, especially winter rainfall, coincide with a warm and sunny summer topoclimate. Substantial amounts of periglacial and post-glacial loess seem to have been deposited in this area (e.g. Catt *et al.* 1971, Perrin *et al.* 1974, Cope 1976) and incorporated into the soils which, generally, are humic rendzinas, often quite poor in calcium carbonate in their superficial layers. Moreover, both soils and vegetation appear to have remained largely undisturbed for very considerable periods of time, being on the more intractable topographies or responding poorly to earlier forms of arable cultivation, the soils turning to dust when exposed (Cope 1976). There has also been a longer tradition of cattle-, rather than sheep-, grazing in some parts of this region (e.g. Smith 1980: see below). Rewarding work could be done on elucidating how much of this combination of variables is coincidental and which components of it influence the different elements of the peculiar mixture of calcicoles, more mesotrophic pasture species, plants like *Stachys betonica* and *Succisa pratensis*, often indicative of a pH hovering between high and low, and the rarities. Among the last, *Carex humilis* has already attracted particular attention (e.g. Coombe in Pigott & Walters 1954, Wells 1975). It does not occur in all stands, being more strictly confined to south-west facing slopes than the sub-community as a whole, a strong association with aspect that is also seen in the *Festuca-Carlina* grassland (Figure 20).

Under whatever climatic conditions the *Festuca-Avenula* grassland occurs, permanent stands of the community are always plagioclimax vegetation maintained by grazing. Repeated defoliation by stock and wild herbivores is responsible for the characteristic close, tight structure of the sward with its preponderance of perennials, especially hemicryptophytes, which can escape permanent damage by having buds very close to the ground, and chamaephytes, such as *Thymus praecox*, *Helianthemum nummularium* and *Hippocrepis comosa*, which benefit indirectly through the maintenance of high light in the short turf. The most important grazers have traditionally been sheep (with some cattle and, rarely, horses) and rabbits (less commonly, hares). Any pronounced relaxation of grazing by one or more of these components in the herbivore population results in alterations in the sward and, if maintained, can lead to successional change (detailed below). The balance between the different components also influences floristic variation in the community.

Wells (1969, 1971) has warned of the dangers of making facile assumptions about the past uniformity of pastoral practice over the Chalk and, in assessing the impact of grazing on the *Festuca-Avenula* grassland throughout its range, account should always be taken of the possibility of fine variation in factors like the breeds of animals used, stocking rates and grazing regimes, which might lie behind general regional patterns and apparently stable traditions prevailing in particular periods of agricultural history. Close observation on the effects of grazing on individual species which are especially characteristic of the community also indicates the complexity of responses among its components (e.g. Pigott 1955 on *Thymus praecox* and *T. pulegioides*, Proctor 1956 on *Helianthemum nummularium*, Wells

1967 on *Spiranthes spiralis*, Wells & Barling 1971 on *Pulsatilla vulgaris*, Fearn 1973 on *Hippocrepis comosa*, Smith 1979 on *Senecio integrifolius* ssp. *integrifolius*). Unfortunately, apart from the early enclosure studies (Tansley & Adamson 1925, Hope-Simpson 1941*b*), almost all experimental work on the influence of variations in grazing on lowland calcicolous grasslands has been on swards which are not mainstream types of *Festuca-Avenula* grassland (e.g. the Hurley grassland, which seems to be a rough approximation to the *Holcus-Trifolium* sub-community developed over ploughed and re-sown land: Norman 1957, Norman & Green 1958, Kydd 1964) or which already have substantial amounts of *Bromus erectus* (e.g. Morris 1967, 1971*b*, Wells 1971) or *Arrhenatherum elatius* (e.g. Wells 1969). (The work of Farrow (1917) and Watt (1962 *et seq.*) on the Breckland grasslands of Norfolk is considered under the *Festuca-Hieracium-Thymus* community.) The following generalisations should, therefore, be treated with caution.

Until recent decades, the most characteristic grazing stock pastured on *Festuca-Avenula* grasslands have been sheep and, over the Chalk, the traditional style of folding management was an integral part of an extensive, artificial ecosystem which, with its vast expanses of rolling downland, combined the aesthetic delights of small-scale variegation (the 'fairy flora' of Hudson 1900) and airy exhilaration (especially in Wiltshire: e.g. Aubrey 1685, Hudson 1910). With the demise of this kind of low-cost, labour-intensive agriculture, much of this landscape and the larger stands of the community have been lost but, where *Festuca-Avenula* grassland remains, it still often owes its distinctive physiognomy and its attractive floristic diversity to the very close, even grazing that sheep provide. This is especially true of the *Cirsium-Asperula* sub-community which comprises the bulk of the community in many parts of the southern Chalk but it is also true of the *Dicranum* sub-community, even though most of its stands lie in areas like Derbyshire and the Mendips, which have long had a more mixed regional grassland economy with different breeds of sheep and also dairy cattle and which have probably never known the kind of folding pastoralism characteristic of the Chalk.

In moderate numbers, rabbits produce a sward which seems to be physiognomically and floristically very similar to that developing under sheep (e.g. Tansley & Adamson 1925) and they probably often simply supplemented the effect of stock or, as when their increasing numbers compensated for the decline in sheep between the First and Second World Wars, replaced it. Even when relatively sparse, however, they can contribute to sward heterogeneity in *Festuca-Avenula* grassland by avoiding species such as *Helianthemum nummularium* (Tansley 1939) or *Galium verum* (Smith 1980). The more dramatic zonational and successional impacts of heavy rabbit pressure and the results of their decline after myxomatosis are dealt with below.

Even over the southern Chalk, with its traditional devotion to sheep-rearing, many parishes long had their 'cow down' and in some areas, such as parts of Wiltshire (Hosier & Hosier 1951, Cope 1976, Smith 1980), cattle-grazing seems to have been practised on *Festuca-Avenula* grasslands earlier and more systematically than in others. Furthermore, over the past few decades, there has been an increasing tendency to return to grazing these grasslands, after the neglect between the wars, not with sheep alone, but, at least partly and often wholly, with dairy and beef cattle (e.g. Smith 1980). There is a clear association within the community between cattle-grazing and the *Holcus-Trifolium* sub-community which can develop, under this kind of pastoralism, from both the southern *Cirsium-Asperula* and the northern *Dicranum* types. In general, cattle graze less closely and more selectively than sheep and their larger and more persistent faeces lead more readily to the development of avoidance-mosaics. The effects of these differences can be seen in the *Holcus-Trifolium* sub-community in the increase in coarser grasses and dicotyledons and in the development of an often tussocky sward with a fall in abundance, and sometimes a loss, of the more shade-sensitive vascular plants, acrocarpous mosses and less robust pleurocarps. Although there is not always a reduction in species-richness where this sub-community develops under light cattle-grazing, there is a shift from the more fine-textured turf with its range of bright-flowered calcicoles towards a ranker sward with duller mesophytes, and prolonged and/or heavy cattle-grazing may help mediate a transition to other grassland types (see below). In heavily-grazed swards, there is also an increased danger of poaching where turf becomes broken by the heavier hoof pressure of cattle. However, the likelihood is that it is the manurial effects of the stock that are of especial importance in the development of these *Holcus-Trifolium* swards. The same group of mesophytic species increases also where the *Festuca-Avenula* grassland extends on to naturally more mesotrophic soils (see above) or when it, or grasslands approximating to it (e.g. Norman 1956, Smith *et al.* 1971), are enriched by the addition of major nutrients in artificial fertilisers.

Some of the *Holcus-Trifolium* species are also characteristic of the *Succisa-Leucanthemum* sub-community and it is possible that cattle-grazing has also played a part in its development, typical as it is of areas of the Wiltshire Chalk. Interestingly, Wells (1969) noted that grazing by Ayrshires in this region produced a much more even sward than is usual with cattle and this may assist in maintaining the higher diversity of this vegetation.

The general importance of grazing in the maintenance

and into the 1950s when their impact became especially obvious with the decline in numbers of sheep. Then, their vociferous nibbling and scraping, their dunging and urinating on the sward and their excavations in the softer strata and superficial deposits, produced striking and very widespread effects (see, for example, plates 2 and 3 in Tansley 1939 and figure 5 in Hope-Simpson 1940*b*). In some areas, their numbers have risen again as less virulent strains of myxomatosis have stabilised within the population. Their general effect on the *Festuca-Avenula* grassland seems to be to convert it to the kind of vegetation that is included in the *Festuca-Hieracium-Thymus* grassland, the community on which most of the classic studies of the effects of rabbits have been carried out (Farrow 1917, Watt 1940 *et seq.*). Here, there is a marked reduction in the numbers and cover of more palatable grasses, sedges and dicotyledons and a rise to prominence of certain bryophytes, notably *Hypnum cupressiforme*, *Homalothecium lutescens* and *Pseudoscleropodium purum*, in the more open sward. Also, with the greater disturbance, there is an increase in weedy species, not only those therophytes which are characteristic of open areas within the *Festuca-Avenula* grassland, but also grosser species such as *Senecio jacobaea*, *Erigeron acer* and *Fragaria vesca*. It was the local abundance of these, together with patches of, for example, *Urtica dioica*, *Arctium* spp., *Verbascum thapsus*, *Hyoscyamus niger*, *Atropa belladonna* and *Solanum dulcamara*, often attaining a grand stature on the rich burrow spoil, that gave the old rabbit warrens their distinctive, and often highly peculiar, appearance (e.g. Tansley & Adamson 1925, Thomas 1960, 1963).

Variations in grazing intensity in the *Festuca-Avenula* grassland have a very considerable impact, not only on the vegetation itself, but also on its often very characteristic invertebrate populations (e.g. Duffey 1962*a*, *b*, Duffey *et al.* 1974, Morris 1967, 1968, 1969, 1971*a*, *b*, 1973). The complex relationships between the numerous and varied species and the plants are an essential part of the *Festuca-Avenula* grassland environment and make themselves felt through inter-dependencies such as those involved in pollination. Many of these effects are subtle and hidden. One often very obvious impact of the invertebrate fauna on the vegetation, however, is that produced by the mound-building activities of ants, especially *Lasius flavus* (Thomas 1962, Grubb *et al.* 1969, Wells *et al.* 1976, King 1977*a*, *b*, *c*). The hills produced by this species are large and roughly hemispherical and they introduce an element of structural complexity into stands of the *Festuca-Avenula* grassland, especially on the southern Chalk, which long outlasts their occupation; indeed, the ant-hills may even outlast the grassland, persisting after the abandonment of grazing and beyond the development of a woodland cover.

The soil of the ant-hills is structurally finer and less dense than that of the surrounding intact profiles, though it does not seem to be significantly different in its chemical properties. It is also more free-draining and subject to greater fluctuations of temperature, the drying and heating being especially marked on the southern face. The mounds are also subject to various kinds of disturbance by rabbits, which use them as latrines, and various ant-eating or dust-bathing birds. Both sheep and rabbits may graze them preferentially, especially where the surrounding vegetation has grown rank.

These effects combine to produce a complex floristic mosaic within *Festuca-Avenula* swards which is subject to its own distinctive successional changes. In King's studies, the most characteristic species of the ant-hills were of two kinds: those within the immediately surrounding sward which were more readily able to grow through or on to the accumulating soil (notably the chamaephytes *Thymus praecox*, *Helianthemum nummularium*, *Cerastium fontanum* and, to a lesser extent, *Asperula cynanchica*) and therophytes, virtually absent from the intact grassland but able to find a very congenial germination ground on the mounds (*Arenaria serpyllifolia*, *Veronica arvensis*). With ageing, but continued occupation, the vegetation seemed to stabilise with large amounts of *T. praecox* and *H. nummularium* but, on abandonment, other species, especially *Hieracium pilosella*, became conspicuous to form vegetation which appears to resemble that of the *Festuca-Hieracium-Thymus* grassland.

Although hills produced by moles (*Talpa europaea*) are frequent in *Festuca-Avenula* grassland and may be confused with ant-hills (see King 1977*a*), they present a somewhat different disruption of the surface and show a less peculiar flora. Mole-holes are rapidly excavated in a single operation and are quickly eroded by rain and trampling by stock to ground-level. Their soil is not structurally sorted, being simply a disturbed pile of material, and it may contain seeds and vegetative fragments which can readily regenerate (King 1977*a*). Though there may be a temporary flush of therophytes on the mound, the vegetation seems quickly to return to that of the surrounding grassland.

The aesthetic appeal of *Festuca-Avenula* swards has long made them popular for recreation which can entail heavy trampling and sometimes the driving or parking of vehicles on the vegetation. Around picnic spots, there may also be particular kinds of cultural eutrophication. In recent years, a declining number of stands has come under increasing pressure from a growing and more mobile population and some sites, as on the North Downs (Streeter 1971, Dixon 1973) or along the north Wales coast (Rodwell 1974), have become especially vulnerable.

The distinctive springy turf of the community seems

resilient to a certain amount of trampling but, where this is concentrated, as along paths and around viewing-points, clear changes occur in the vegetation. In such places, there is often a local increase in resistant grasses, notably *Cynosurus cristatus*, but in some cases also *Dactylis glomerata*, *Agrostis stolonifera* and *Holcus lanatus*, and in some of the rosette species, such as *Plantago lanceolata*, *Bellis perennis* and *Taraxacum officinale* agg., at the expense of more sensitive plants like the chamaephytes. This gives such swards a some-what similar composition to the *Holcus-Trifolium* sub-community and recreational trampling may play a part in the development of some stands of this kind of *Festuca-Avenula* grassland. Typically, however, the sward lacks the tussocky physiognomy of this vege-tation, being extremely short. Also, the further appear-ance in the turf of *Lolium perenne*, especially where there is some eutrophication, often presages a rapid move towards Lolio-Plantaginion vegetation. Very similar changes to these were observed by Perring (1967) with an increase in horse-galloping pressure on stretches of Newmarket Heath, Suffolk, used for race-horse exercis-ing. Most unusually among *Festuca-Avenula* species, the rarity *Pulsatilla vulgaris* is actually benefited by heavy trampling, which stimulates the formation of new rosettes from deep adventitious root-buds (Wells & Barling 1971). Even where this species is present, how-ever, very heavy pressure may eventually disrupt the sward, allow soil erosion and expose the bedrock.

Distribution
The *Festuca-Avenula* grassland is widely distributed over southern lowland limestones. The *Cirsium-Asper-ula* sub-community characteristically occurs south of the Humber–Severn line where it is largely confined to the Chalk, though it has outliers on the Carboniferous Limestone of south and north Wales and Derbyshire. The *Succisa-Leucanthemum* sub-community is virtually restricted to parts of the Wiltshire and Dorset Chalk. The *Dicranum* sub-community is the northern counter-part of the *Cirsium-Asperula* type and it occurs over Carboniferous Limestone in north Wales, the Mendips and Derbyshire (with a very few samples further north on the Isle of Man, Craven and around Morecambe Bay), on the northern Chalk of the Yorkshire Wolds, the Corallian of the North York Moors and the Magnesian Limestone of Durham. The *Holcus-Trifolium* sub-community occurs in both the north and south, being especially frequent in Wiltshire, Derbyshire, the York-shire Wolds and the North York Moors.

Some idea of the remaining extent of the community over the Chalk was provided by Blackwood & Tubbs (1970), although, in that survey, 'Chalk grassland' also included swards dominated by rank grasses and some small areas of abandoned arable and more mixed vege-tation over superficials.

Affinities
This community is the central kind of stable plagio-climax Mesobromion grassland in Britain. It takes in much of the vegetation included as 'Chalk grassland' in some descriptive accounts (e.g. Tansley & Adamson 1925, 1926, Tansley 1939, Ratcliffe 1977) as well as essentially similar swards from other limestones (notably those included in the *Helictotricho-Caricetum flaccae* of Shimwell 1968a). It thus returns to the early concept of a *Festucetum* enunciated by Tansley & Ran-kin (1911) and recently resurrected by Smith (1980) and cuts across schemes which rely on bedrock type as the basis of classification. It also includes a variety of vegetation types erected around individual species, often rare ones. Similar *Halbtrockenrasen* have been widely described from limestones in north-west Europe (e.g. LeBrun *et al.* 1949, Bornkamm 1960, Westhoff & den Held 1969, Stott 1970, Oberdorfer 1978), though it should be noted that, in moving from the Continent to Britain, there is a shift in the fidelity of species regarded there as characteristic of the different kinds of Brometa-lia grasslands (e.g. Shimwell 1968a, 1971a, b).

The *Festuca-Avenula* grassland is very closely related to both the rank grasslands dominated by such species as *Bromus erectus*, *Brachypodium pinnatum* and *Avenula pubescens* and also the *Festuca-Hieracium-Thymus* grassland. These communities, too, are of the Mesobro-mion type and show floristic transitions to the *Festuca-Avenula* grassland which are mediated largely by grazing.

In Britain, these kinds of grasslands are intermediate between the sub-Mediterranean Xerobromion swards of the *Festuca-Carlina* grassland and the sub-montane and montane grasslands in which Mesobromion plants are eclipsed by calcicoles of cooler, damper climates and Nardo-Galion species. The *Festuca-Avenula* grassland grades to the former through the *Filipendula-Helianthe-mum* variant of the *Cirsium-Asperula* sub-community and to the latter through the *Dicranum* sub-community.

More diffuse relationships can also be seen within the community, through the *Holcus-Trifolium* sub-community, to the more mesophytic swards, both grazed and ungrazed, of the Arrhenatheretalia.

Floristic table CG2

	a	b	c	d	2
Festuca ovina	V (1–9)	V (2–9)	V (2–9)	V (3–8)	V (1–9)
Carex flacca	V (1–8)	V (2–8)	V (1–8)	V (1–9)	V (1–9)
Sanguisorba minor	V (1–8)	V (1–7)	IV (1–8)	IV (1–6)	V (1–8)
Koeleria macrantha	V (1–6)	IV (1–5)	V (1–4)	IV (1–6)	V (1–6)
Plantago lanceolata	IV (1–7)	V (1–7)	V (1–6)	IV (1–5)	IV (1–7)
Briza media	IV (1–9)	V (1–6)	V (1–5)	IV (1–5)	IV (1–9)
Lotus corniculatus	IV (1–6)	V (1–7)	V (1–6)	IV (1–5)	IV (1–7)
Avenula pratensis	IV (1–7)	V (1–6)	IV (1–6)	IV (1–8)	IV (1–8)
Leontodon hispidus	IV (1–7)	V (1–7)	IV (1–7)	III (1–5)	IV (1–7)
Linum catharticum	III (1–4)	IV (1–4)	IV (1–5)	IV (1–5)	IV (1–5)
Hieracium pilosella	IV (1–7)	III (1–4)	III (1–5)	IV (1–7)	IV (1–7)
Scabiosa columbaria	III (1–5)	V (1–5)	IV (1–5)	III (1–6)	IV (1–6)
Thymus praecox	IV (1–7)	III (1–4)	III (1–7)	V (1–6)	IV (1–7)
Cirsium acaule	IV (1–7)	V (1–6)	II (1–7)	I (1–7)	III (1–7)
Asperula cynanchica	III (1–7)	IV (1–4)	I (1–3)	I (1)	II (1–7)
Hippocrepis comosa	II (1–6)	III (1–8)	I (3)	I (1)	II (1–8)
Polygala calcarea	I (1–4)	II (1–4)	I (1–2)		I (1–4)
Onobrychis viciifolia	I (1–4)	I (2–4)	I (3)		I (1–4)
Thesium humifusum	I (1–4)	I (1–4)	I (1)		I (1–4)
Senecio integrifolius integrifolius	I (1–3)	I (1)	I (1)		I (1–3)
Thymus pulegioides	I (1–5)	I (1)	I (1)		I (1–5)
Hypochoeris maculata	I (1–3)	I (1–2)			I (1–3)
Gentianella anglica	I (1–2)	I (1)			I (1–2)
Astragalus danicus	I (1–3)			I (1)	I (1–3)
Phyteuma tenerum	I (1–5)				I (1–5)
Trifolium pratense	I (1–5)	IV (1–6)	IV (1–6)	I (1–4)	III (1–6)
Carex caryophyllea	III (1–7)	IV (1–5)	III (1–4)	III (1–5)	III (1–7)
Prunella vulgaris	III (1–4)	IV (1–4)	III (1–4)	II (1–3)	III (1–4)
Dactylis glomerata	II (1–5)	V (1–6)	III (1–6)	III (1–4)	III (1–6)
Plantago media	II (1–4)	V (1–5)	III (1–5)	I (1–5)	III (1–5)
Succisa pratensis	I (1–6)	IV (1–6)	II (1–5)	I (1–3)	II (1–6)
Centaurea nigra	I (1–5)	III (1–4)	II (1–4)	II (1–6)	II (1–6)
Avenula pubescens	II (1–6)	III (1–5)	II (1–8)	I (1–4)	II (1–8)
Leucanthemum vulgare	I (1–3)	III (1–3)	I (1–3)	I (1–3)	I (1–3)

Floristic table CG2 (*cont.*)

	a	b	c	d	2
Campanula glomerata	I (1–4)	II (1–3)	I (1–2)	I (1)	I (1–4)
Serratula tinctoria	I (1–5)	II (1–5)	I (1–2)	I (2)	I (1–5)
Stachys betonica	I (1–3)	II (1–5)	I (1–7)	I (1–4)	I (1–7)
Carex humilis	I (3–8)	II (1–8)	I (5)		I (1–8)
Holcus lanatus	I (1–5)	I (1–7)	IV (1–6)	II (1–6)	II (1–7)
Trifolium repens	I (1–4)	I (1–4)	IV (1–7)	I (1–5)	II (1–7)
Medicago lupulina	I (1–5)	III (1–5)	III (1–5)	II (1–5)	II (1–5)
Agrostis stolonifera	I (1–3)	II (1–4)	III (1–5)	I (2–3)	I (1–5)
Cynosurus cristatus	I (1–5)	II (1–5)	III (1–6)	I (1–4)	I (1–6)
Trisetum flavescens	I (1–5)	I (1–2)	III (1–5)	II (1–6)	I (1–6)
Achillea millefolium	I (1–5)	I (1–3)	II (1–4)	II (1–5)	I (1–5)
Crepis capillaris	I (1–5)	I (1–2)	II (1–3)	I (1–5)	I (1–5)
Phleum pratense bertolonii	I (1–2)	I (1–3)	II (1–4)	I (1–2)	I (1–4)
Arrhenatherum elatius	I (1–6)	I (1–4)	II (1–7)	I (1–5)	I (1–7)
Pulsatilla vulgaris	I (1–4)		II (1–3)	I (1–3)	I (1–4)
Dicranum scoparium	I (1–4)		I (1–3)	III (1–4)	I (1–4)
Leontodon taraxacoides	I (1–4)	I (1–5)	I (1–5)	II (1–4)	I (1–5)
Agrostis capillaris	I (2–4)	I (1)	II (1–7)	II (1–5)	I (1–7)
Brachypodium sylvaticum	I (1–5)	I (1–2)	I (1–7)	II (1–9)	I (1–9)
Anthoxanthum odoratum	I (1–5)	I (1)	I (1–3)	III (1–5)	I (1–5)
Ctenidium molluscum	I (1–8)	I (1–4)	I (1)	II (1–4)	I (1–8)
Fissidens cristatus	I (1–8)	I (1–4)		II (1–2)	I (1–5)
Galium sterneri	I (1–5)	I (1–4)	I (1–2)	I (1–5)	I (1–5)
Campanula rotundifolia	III (1–5)	III (1–4)	III (1–4)	III (1–4)	III (1–5)
Helianthemum nummularium	III (1–8)	III (1–6)	II (1–7)	IV (1–7)	III (1–8)
Pseudoscleropodium purum	III (1–8)	III (1–8)	III (1–4)	II (1–3)	III (1–8)
Euphrasia officinalis agg.	II (1–6)	III (1–4)	II (1–3)	II (1–5)	II (1–6)
Ranunculus bulbosus	II (1–4)	III (1–4)	II (1–3)	II (1–3)	II (1–4)
Pimpinella saxifraga	II (1–4)	III (1–4)	III (1–4)	II (1–4)	II (1–4)
Galium verum	II (1–4)	III (1–7)	III (1–5)	III (1–5)	III (1–7)
Bromus erectus	II (1–4)	II (1–3)	II (1–4)	I (1–3)	II (1–4)
Gentianella amarella	II (1–4)	II (1–3)	I (1–3)	II (1–4)	II (1–4)
Filipendula vulgaris	II (1–6)	II (1–4)	I (1–5)	I (1–5)	II (1–6)
Homalothecium lutescens	II (1–8)	II (1–4)	I (1–3)	II (1–5)	II (1–8)

Species				
Primula veris	I (1-4)	II (1-4)	II (1-5)	II (1-5)
Viola hirta	II (1-5)	I (1-3)	II (1-3)	I (1-5)
Polygala vulgaris	II (1-4)	I (1)	II (1-3)	I (1-4)
Senecio jacobaea	I (1-3)	II (1-3)	II (1-3)	I (1-3)
Anthyllis vulneraria	I (1-7)	I (1-3)	I (1-7)	I (1-7)
Festuca rubra	I (2-9)	I (1-5)	II (1-5)	I (1-9)
Bellis perennis	I (1-4)	I (1-5)	I (1-5)	I (1-5)
Taraxacum officinale agg.	I (1-3)	I (1-3)	I (1-3)	I (1-3)
Calliergon cuspidatum	I (1-7)	I (1-3)	I (1-4)	I (1-7)
Centaurea scabiosa	I (1-4)	I (1-3)	I (1-5)	I (1-5)
Blackstonia perfoliata	I (1-5)	I (1-5)	I (1-3)	I (1-5)
Galium mollugo	I (1-4)	I (1-3)	I (1-2)	I (1-5)
Weissia cf. *microstoma*	I (1-3)	I (1-5)	I (1-3)	I (1-3)
Danthonia decumbens	I (1-6)	I (1)	I (1-7)	I (1-7)
Campylium chrysophyllum	I (1-4)	I (1-7)	I (1-3)	I (1-4)
Picris hieracioides	I (1-2)	I (2)	I (1-2)	I (1-4)
Rhytidiadelphus squarrosus	I (1-3)	I (1-4)	I (1-3)	I (1-4)
Daucus carota	I (1-4)	I (1-4)	I (1)	I (1-4)
Festuca arundinacea	I (1-6)	I (1-2)	I (1-4)	I (1-6)
Carlina vulgaris	I (1-5)	I (1-5)	I (1-3)	I (1-5)
Poa pratensis	I (1-3)	I (1-5)	I (1-3)	I (1-3)
Cerastium fontanum	I (2-3)	I (1-3)	I (1-3)	I (1-3)
Luzula campestris	I (1-2)	I (1)	I (1-3)	I (1-3)
Brachypodium pinnatum	I (1-4)	I (1-3)	I (3)	I (1-4)
Cirsium vulgare	I (1)	I (1-3)	I (1-3)	I (1-3)
Leontodon autumnalis	I (1-3)	I (3)	I (1)	I (1-3)
Rhinanthus minor	I (1-5)	I (1)	I (1)	I (1-5)
Agrimonia eupatoria	I (1-2)	I (1-3)	I (1)	I (1-3)
Crataegus monogyna sapling	I (1-3)	I (1)	I (3)	I (1-3)
Ulex europaeus	I (1-4)	I (2-3)	I (1)	I (1-4)
Ophrys apifera	I (1-3)	I (1-3)	I (1)	I (1-3)
Dactylorhiza fuchsii	I (1-3)	I (1)	I (1)	I (1-3)
Cirsium arvense	I (1-3)	I (1)	I (1-2)	I (1-3)
Carduus nutans	I (1-6)	I (1)	I (1-3)	I (1-6)
Veronica chamaedrys	I (1)	I (1)	I (1-2)	I (1-2)
Lolium perenne	I (1-3)	I (1-3)	I (1-3)	I (1-3)
Rhytidiadelphus triquetrus	I (1)	I (1-3)	I (1)	I (1-3)
Ononis repens	I (1-5)	I (1-6)	I (1-3)	I (1-6)
Tragopogon pratensis	I (1-3)	I (1)	I (1-2)	I (1-3)

Floristic table CG2 (*cont.*)

	a	b	c	d	2
Urtica dioica	I (1)	I (1)	I (1)		I (1)
Coeloglossum viride	I (1–3)	I (1)	I (1–3)		I (1–3)
Centaurium erythraea	I (1–4)		I (1–3)	I (1–3)	I (1–4)
Hypochoeris radicata	I (2–4)		I (1–3)	I (1)	I (1–4)
Sedum acre	I (1–3)		I (1)	I (1)	I (1–3)
Fragaria vesca	I (1–4)		I (1–3)	I (1–4)	I (1–4)
Teucrium scorodonia	I (2–3)		I (1)	I (1–3)	I (1–3)
Arenaria serpyllifolia	I (2–3)		I (1)	I (1–3)	I (1–3)
Rumex acetosa	I (1)		I (1)	I (3)	I (1–3)
Hylocomium splendens	I (1)		I (1)	I (1)	I (1)
Deschampsia cespitosa	I (2)		I (1)	I (1–3)	I (1–3)
Lathyrus pratensis	I (1)		I (1)	I (1)	I (1)
Hypnum cupressiforme	I (1–7)	I (1–2)		I (1–3)	I (1–7)
Vicia cracca	I (1–3)	I (1)		I (3)	I (1–3)
Neckera complanata	I (1–3)	I (1)		I (1)	I (1–3)
Gymnadenia conopsea	I (1–2)	I (1–3)		I (1–4)	I (1–4)
Cirsium palustre		I (1)	I (1)	I (1–3)	I (1–3)
Inula conyza	I (1–3)			I (1)	I (1–3)
Geranium sanguineum	I (1)			I (1–3)	I (1–3)
Phleum pratense pratense	I (3)		I (1–3)		I (1–3)
Viola riviniana			I (1–4)	I (1–5)	I (1–5)
Cladonia rangiformis			I (1)	I (1–5)	I (1–5)
Hypericum montanum			I (1–3)	I (3)	I (1–3)
Potentilla erecta			I (1–5)	I (1–5)	I (1–5)
Cirsium eriophorum			I (1–3)	I (1)	I (1–3)
Number of samples	343	177	167	169	856
Number of species/sample	25 (7–45)	30 (18–45)	30 (16–44)	26 (5–47)	27 (5–47)

a   Cirsium acaule-Asperula cynanchica sub-community
b   Succisa pratensis-Leucanthemum vulgare sub-community
c   Holcus lanatus-Trifolium repens sub-community
d   Dicranum scoparium sub-community
2   Festuca ovina-Avenula pratensis grassland (total)

Floristic table CG2a, variants

	ai	aii	aiii
Festuca ovina	IV (3–9)	V (2–9)	V (1–9)
Carex flacca	V (1–5)	V (2–8)	V (1–8)
Sanguisorba minor	V (1–8)	V (2–8)	IV (1–7)
Koeleria macrantha	V (1–6)	IV (1–5)	V (1–6)
Plantago lanceolata	IV (1–7)	IV (1–5)	V (1–5)
Briza media	IV (1–6)	IV (1–7)	V (1–9)
Lotus corniculatus	IV (1–5)	IV (1–6)	V (1–6)
Avenula pratensis	IV (1–6)	IV (1–6)	V (1–7)
Leontodon hispidus	IV (1–4)	IV (1–5)	IV (1–7)
Linum catharticum	II (1–4)	III (1–3)	IV (1–4)
Hieracium pilosella	IV (1–7)	III (1–5)	V (1–6)
Scabiosa columbaria	III (1–4)	IV (1–5)	III (1–5)
Thymus praecox	III (1–5)	III (1–6)	IV (1–6)
Cirsium acaule	III (1–6)	IV (1–7)	IV (1–7)
Asperula cynanchica	III (1–5)	III (1–4)	IV (1–7)
Hippocrepis comosa	II (1–4)	II (1–6)	II (1–6)
Helianthemum nummularium	IV (1–6)	III (2–8)	II (1–6)
Filipendula vulgaris	V (1–6)	I (1–4)	I (1–6)
Festuca rubra	III (2–9)	I (3–8)	I (1–9)
Pseudoscleropodium purum	III (1–8)	III (1–7)	IV (1–8)
Prunella vulgaris	III (1–4)	I (1–3)	IV (1–4)
Homalothecium lutescens	II (1–6)	I (1–5)	III (1–8)
Ranunculus bulbosus	I (1–2)	I (1–3)	III (1–4)
Bellis perennis	I (2–3)	I (1)	II (2–4)
Senecio jacobaea	I (1–3)	I (1–3)	II (1–3)
Ctenidium molluscum	I (1–4)	I (1–8)	II (1–6)
Fissidens cristatus	I (1–4)	I (1–5)	II (1–3)
Trifolium pratense	I (1–4)	I (1–5)	II (1–4)
Medicago lupulina	I (1–4)	I (1–3)	II (1–5)
Rhytidiadelphus triquetrus	I (5)		I (1–4)
Neckera complanata			I (1–3)
Hylocomium splendens			I (1–3)
Deschampsia cespitosa			I (2)
Ulex europaeus			I (1–4)
Campanula rotundifolia	III (1–5)	III (1–3)	III (1–3)
Carex caryophyllea	III (1–7)	II (1–4)	III (1–5)
Galium verum	III (1–4)	II (1–4)	II (1–4)
Pimpinella saxifraga	II (1–4)	III (1–4)	II (1–3)
Plantago media	II (1–4)	II (1–4)	II (1–4)
Euphrasia officinalis agg.	II (1–3)	II (1–5)	II (1–6)
Avenula pubescens	II (1–6)	II (1–4)	II (1–4)
Bromus erectus	II (2–4)	II (1–4)	I (1–4)
Polygala vulgaris	II (1–4)	I (2–4)	II (1–3)
Gentianella amarella	I (1–3)	II (1–3)	II (1–4)

Floristic table CG2a, variants (*cont.*)

	ai	aii	aiii
Viola hirta	II (1–4)	I (1–4)	II (1–5)
Dactylis glomerata	I (1–3)	II (1–3)	II (1–5)
Centaurea nigra	I (2–4)	II (1–5)	I (1–4)
Anthyllis vulneraria	I (1–5)	II (1–4)	I (1–7)
Succisa pratensis	I (1–4)	I (1–6)	I (1–5)
Cynosurus cristatus	I (2–3)	I (1–2)	I (1–5)
Trisetum flavescens	I (1–4)	I (1–5)	I (1–5)
Achillea millefolium	I (2–3)	I (1–5)	I (1–3)
Primula veris	I (1–3)	I (1–4)	I (1–4)
Campanula glomerata	I (1–3)	I (1–4)	I (1–3)
Calliergon cuspidatum	I (1–2)	I (2–5)	I (1–7)
Daucus carota	I (1–3)	I (1–4)	I (1–3)
Campylium chrysophyllum	I (1–3)	I (1–3)	I (2–4)
Danthonia decumbens	I (1–4)	I (1–6)	I (1–5)
Leucanthemum vulgare	I (1–2)	I (1–3)	I (1–3)
Leontodon taraxacoides	I (1–3)	I (3–4)	I (1–4)
Galium mollugo	I (1–3)	I (1–3)	I (1–4)
Crataegus monogyna sapling	I (1–3)	I (1–3)	I (1–2)
Thesium humifusum	I (1–3)	I (3–4)	I (1–2)
Hypnum cupressiforme	I (1–7)	I (3–7)	I (1–4)
Centaurea scabiosa	I (1–4)	I (1–4)	I (1–4)
Polygala calcarea	I (1–4)	I (1–4)	I (1–4)
Thymus pulegioides	I (1–5)	I (1–4)	I (2–5)
Onobrychis viciifolia	I (3–4)	I (1–4)	I (4)
Blackstonia perfoliata	I (2)	I (1–5)	I (1–3)
Carex humilis	I (5)	I (3–7)	I (3–8)
Arrhenatherum elatius	I (3)	I (1–4)	I (1–6)
Taraxacum officinale agg.	I (1–2)	I (1)	I (1–3)
Fragaria vesca	I (1–3)	I (2–4)	I (1–4)
Anacamptis pyramidalis	I (1–3)	I (1–3)	I (1–3)
Phyteuma tenerum	I (3–5)	I (1–3)	I (2–4)
Carlina vulgaris	I (1–3)	I (1–5)	I (1–4)
Weissia cf. *microstoma*	I (1–2)	I (1–3)	I (1–3)
Crepis capillaris	I (1–5)	I (2)	I (1–2)
Brachypodium sylvaticum	I (1)	I (1–5)	I (2–5)
Hypochoeris radicata	I (2–3)	I (3–4)	I (3)
Cerastium fontanum	I (2–3)	I (2)	I (2–3)
Senecio integrifolius integrifolius	I (1–2)	I (1)	I (1–3)
Centaurium erythraea	I (2–3)	I (3)	I (1–4)
Agrostis stolonifera	I (2)	I (1–2)	I (1–3)
Astragalus danicus	I (1)	I (1–3)	I (1–4)
Agrostis capillaris	I (3–4)	I (3)	I (2–4)
Anthoxanthum odoratum	I (1–2)	I (3)	I (1–5)
Picris hieracioides	I (2)	I (1–2)	I (1–2)
Trifolium repens	I (1)	I (2)	I (1–4)
Phleum pratense bertolonii	I (1)	I (1)	I (1–2)

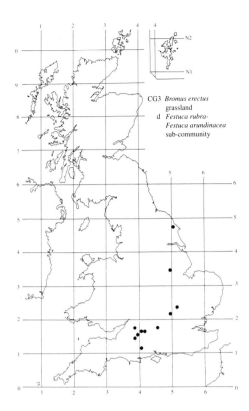

CG3 *Bromus erectus*
grassland
d *Festuca rubra-*
Festuca arundinacea
sub-community

CG4
Brachypodium pinnatum grassland

Synonymy
Chalk grassland *auct. angl. p.p;*; *Brachypodium pinnatum* grassland *auct. angl.*; *Cirsio-Brometum brachypodietosum* Shimwell 1968a.

Constant species
Brachypodium pinnatum, Carex flacca, Festuca ovina.

Rare species
Astragalus danicus, Herminium monorchis.

Physiognomy
This community includes all swards in which *B. pinnatum* exceeds 10% cover in the virtual absence of other bulky tussock grasses such as *Bromus erectus* and *Avenula pubescens*. Like the *Bromus* grassland, it takes in vegetation which is similar to the *Festuca-Avenula* grassland as well as some much ranker swards but it is somewhat poorer overall in *Festuca-Avenula* species than the *Bromus* grassland, its general floristics being shifted somewhat towards more mesotrophic grasslands. Only two associates attain constancy throughout, *Carex flacca* and *Festuca ovina* and, of the frequent species, only *Sanguisorba minor, Linum catharticum* and *Briza media* are stricter Mesobromion plants. As well as being often too rank to offer a congenial site for more diminutive rarities, many stands occur outside the range of the Continental element in British calcicolous grasslands.

Sub-communities

***Avenula pratensis-Thymus praecox* sub-community.**
Although *B. pinnatum* is almost always the most abundant species here, sometimes growing as conspicuous tussocks, its cover is often patchy and, between the clumps, less coarse grasses, such as *Festuca ovina, Avenula pratensis, Briza media* and *Koeleria macrantha*, and chamaephytes, notably *Thymus praecox* and *Helianthemum nummularium*, occur frequently and sometimes with local abundance. *Cirsium acaule* and *Asperula*

cynanchica are preferential to this sub-community. Many of the swards included here, especially those from colonised quarry spoil, are also somewhat open and on patches of bare soil there are occasional records for a variety of pauciennials, such as *Euphrasia officinalis* agg., *Carlina vulgaris, Blackstonia perfoliata, Inula conyza, Centaurium erythraea* and annual Hieracia, and bryophytes like *Weissia* cf. *microstoma, Fissidens cristatus, Ctenidium molluscum, Campylium chrysophyllum* and *Pseudoscleropodium purum. Anacamptis pyramidalis* is sometimes conspicuous in this kind of vegetation.

***Centaurea nigra-Leontodon hispidus* sub-community.**
Some Mesobromion dicotyledons, such as *Helianthemum nummularium, Sanguisorba minor* and *Linum catharticum*, remain frequent in this sub-community but the taller and ranker swards are generally more overwhelmingly dominated by *B. pinnatum* and the most obvious features of the vegetation are the frequency of the taller hemicryptophytes *Leontodon hispidus, Centaurea nigra* and *Knautia arvensis* and the increasing prominence of more mesophytic species such as *Plantago lanceolata, Bellis perennis, Trifolium pratense, Trisetum flavescens* and *Holcus lanatus. Ononis repens* occurs occasionally.

***Holcus lanatus* sub-community:** *Cirsio-Brometum brachypodietosum* Shimwell 1968a *p.p.* In this sub-community more mesophytic species virtually eclipse Mesobromion associates in usually very rank and grassy swards. *B. pinnatum* is generally dominant but *Trisetum flavescens* and *Holcus lanatus* become constant and there are also occasional records for *Cynosurus cristatus* and *Arrhenatherum elatius*. Among the dicotyledons, *Prunella vulgaris, Cruciata laevipes, Veronica chamaedrys, Cerastium fontanum, Lathyrus pratensis, Trifolium repens, Rumex acetosa* and *Cirsium vulgare* are all preferential at low frequencies.

Habitat
As with *Bromus erectus*, the dominance of *Brachypodium pinnatum* is essentially associated with an

absence of or a relaxation of grazing in predominantly calcicolous swards. However, though it has a roughly similar distributional limit in Britain to that of *B. erectus*, *Brachypodium* does not have so narrow a Continental range through Europe and it seems to attain prominence here under climatic and edaphic conditions somewhat different to those favoured by that species. That being said, and despite the all too familiar occurrence of these grasslands in certain areas, we still know very little about the ecology of *B. pinnatum* or the detailed floristic and physiognomic consequences of its expansion.

Like *B. erectus*, *Brachypodium* does occur as an occasional within calcicolous swards (as in the *Festuca-Avenula* grassland) over shallow and free-draining grey rendzinas on limestones in the warmer and drier southeast. With a reduction in grazing pressure, such swards can develop into the kind of vegetation included here as the *Avenula-Thymus* sub-community. Such grasslands are, however, of much more local occurrence than their *Bromus* counterparts, being largely absent, for example, from much of the Chalk of the Chilterns and East Anglia. It may be that, in these more Continental areas, *B. erectus* is more readily able to invade or expand because it is better fitted to the climatic conditions though even in some of the wetter western areas of the Chalk, *Brachypodium*-dominated grasslands are distinctly local.

In general, however, it seems to be where there is some regional or local topographic amelioration of these more extreme climatic and edaphic conditions that *B. pinnatum* rises to prominence as a dominant in ungrazed or lightly-grazed grasslands. The bulk of the swards included here are characteristic of cooler and damper sites towards the western and northern fringe of the Chalk and over the Oolite. In these regions, the community can occur as the more calcicolous *Avenula-Thymus* sub-community over shallow rendzinas (for example, an Oolite quarry spoil) but, more frequently, it is found as the more mesophytic *Centaurea-Leontodon* and *Holcus* sub-communities over deeper and moister, though still often base-rich, soils which tend towards brown rendzinas or calcareous brown earths.

Although many stands of the community now carry stock and/or rabbits, it is reduction in grazing pressure that is, above all, responsible for the development of these swards. In general, *B. pinnatum* is a highly unpalatable grass: it may be eaten by cattle when there is no alternative herbage (Hope-Simpson 1940*b*) and sheep and rabbits may nibble younger shoots which appear in spring (Duffey *et al.* 1974), but otherwise it seems hardly to be grazed at all by these species. They may, however, keep it in check by grazing hard and close around established tussocks, by inhibiting litter accumulation and by trampling damage (Hope-Simpson 1940*b*, Green

1973). If such pressure is lessened, *B. pinnatum* responds vigorously, its tussocks expanding by growth of creeping rhizomes and eventually coalescing (Hall & Russell 1911, Hope-Simpson 1940*b*, Duffey *et al.* 1974: see especially their plates 13–15) and its coarse litter forming a thick layer which decays but slowly, perhaps because of the high silica content of the herbage (Elton 1966, Duffey *et al.* 1974). As in the *Bromus* grassland, it is the consequent reduction in light penetration to the sward that is perhaps primarily responsible for the decline in competitive ability of many of the typical *Festuca-Avenula* species and their eventual demise here. And again, as there, differences in the grazing history, in the regimes and kind of stock employed, may play some part in determining the floristics of the different sub-communities. For example, there is some evidence in the data for an association, in the period of sampling, between sheep-grazing and the more calcicolous *Avenula-Thymus* sub-community, and cattle-grazing and the more mesophytic *Centaurea-Leontodon* and *Holcus* sub-communities, though this may not, of course, reflect long-established differences in pastoral treatment. Burning also helps maintain the more mixed, species-rich and calcicolous swards of the former vegetation type (see below).

Without detailed site histories, it is impossible to define precisely the balance of such edaphic or treatment influences on the composition of the sub-communities or to assess how much of their floristic character is inherited from the antecedent grasslands and how much is consequent upon the expansion of the dominant. Neither do we know whether any of these swards have developed from ploughed and abandoned ground. Like *B. erectus*, *Brachypodium* can produce prolific crops of large, awned fruits and it will invade open soil and spoil but it, too, may face problems of dispersal and establishment.

Zonation and succession

The various kinds of *Brachypodium* grassland, like those dominated by *Bromus erectus*, commonly occur in zonations which reflect seral changes related to grazing intensity, forming patchworks with *Festuca-Avenula* grassland, other rank swards and scrub as on the Wye and Crundale Downs in Kent (Green 1973), at Castor Hanglands on the Huntingdon/Peterborough Oolite (Duffey *et al.* 1974) and over the Chalk of the Yorkshire Wolds (NCC Yorkshire Chalk Grassland Survey 1985). More locally, the community occurs with more open weed vegetation and scrub on quarry spoil (e.g. Shimwell 1968*a*).

Because of its unpalatability, the expansion of *B. pinnatum* seems to be reversible only with difficulty. Grazing with sheep alone, especially just in winter, is ineffective and even all year round grazing involving cattle may do no more than simply restrict further spread and reduce the pronounced tussocky character of the sward (Hope-Simpson 1940b, Green 1973). Likewise, burning alone, though cheaper than grazing, destroys accumulated litter but may not reduce the cover of *B. pinnatum*: indeed, it can encourage further expansion, though twice-yearly burns in alternate years may prevent this (Green 1973). More effective seem to be combinations of these treatments, for example a burn in very early spring followed by very hard grazing, preferably by cattle first, then sheep (Hope-Simpson 1940b, Green 1973). Sprinkling salt on the tussocks may encourage stock to eat the herbage (Hope-Simpson 1940b).

The assiduous care needed in such treatments to avoid deleterious side effects such as poaching of the soil, always a danger where cattle are grazing steeper slopes, or loss of invertebrate populations with burning, and the difficulties of integrating them with recreational use of the grasslands, makes mowing an attractive alternative. Mowing with removal of the cuttings has been shown to be effective in maintaining less tussocky and more varied *Brachypodium* swards and reduction in the height of the herbage may induce rabbits to graze (Green 1973).

The further development of ungrazed *Brachypodium* grasslands has not been monitored in detail. As with *Bromus erectus*, the dense tussocky nature of some of the swards and the high population of voles which this encourages, may hinder scrub invasion. Though *B. pinnatum* is actually palatable to *Microtus agrestis* (Godfrey 1953, 1955, Chitty et al. 1968), it is doubtful whether voles make great inroads on the cover of the species.

Distribution

The community occurs locally over the Chalk of the North and South Downs and more commonly in Dorset where it is also found on the Corallian Limestone. It is more frequent on the Oolite of the Cotswolds and Northamptonshire (where mixed *Bromus-Brachypodium* swards are very common) and very characteristic of the northern Chalk of the Yorkshire Wolds.

Affinities

As defined here, the *Brachypodium* grassland takes in some of the vegetation traditionally included within a compendious 'Chalk grassland' community or its partial phytosociological equivalent, the *Cirsio-Brometum* of Shimwell (1968a, 1970b), as well as ranker swards which have sometimes been separately treated as some kind of '*Brachypodietum*' (e.g. Wells 1975, Smith 1980). Under *Brachypodium* dominance, it brings together grasslands which show a range of variation roughly analogous to that included within the *Festuca-Avenula* grassland, running from more calcicolous to more mesophytic. Although the representation of Mesobromion calcicoles in the community as a whole is very much weaker than in the *Festuca-Avenula* grassland (and less, too, than in the *Bromus*-dominated counterparts), the general affinities of this vegetation type are with that alliance and, for the most part, the swards seem to originate from mainstream plagioclimax Mesobromion vegetation whose stability has been disrupted by a reduction in grazing pressure.

B. pinnatum plays a major part, often with *Bromus erectus*, in Mesobromion grasslands in mainland Europe and grasslands similar to those included here have been described from France (e.g. Allorge 1921–2), The Netherlands (Westhoff & den Held 1969) and Germany (e.g. Ellenberg 1978, Oberdorfer 1978) where communities sometimes take in more mesophytic kinds of vegetation like those included here. Like *B. erectus*, *Brachypodium pinnatum* is also present on the Continent in the drier, rocky swards of the Xerobromion and in some of the steppe-grasslands of the Festucetalia valesiacae.

Floristic table CG4

	a	b	c	4
Brachypodium pinnatum	V (5–9)	V (5–9)	V (5–9)	V (5–9)
Festuca ovina	IV (1–7)	V (3–8)	V (4–8)	V (1–8)
Carex flacca	IV (2–6)	IV (2–6)	III (2–5)	IV (2–6)
Helianthemum nummularium	IV (1–7)	III (2–5)	I (2–4)	II (1–7)
Thymus praecox	IV (1–6)	II (1–7)	I (2–4)	II (1–7)
Avenula pratensis	IV (1–4)			I (1–4)
Koeleria macrantha	III (1–4)	I (3–5)	I (2–5)	I (1–5)
Cirsium acaule	III (1–8)	I (4)		I (1–8)
Carex caryophyllea	III (1–4)	I (3)		I (1–4)
Weissia cf. *microstoma*	III (1–3)			I (1–3)

Euphrasia officinalis agg.	II (1–2)	I (2–3)	I (1–4)	I (1–4)
Taraxacum officinale agg.	II (1–3)	I (1–3)	I (1–2)	I (1–3)
Carlina vulgaris	II (1–4)	I (1–2)	I (2–3)	I (1–4)
Viola hirta	II (1–4)	I (2–3)		I (1–4)
Blackstonia perfoliata	II (1–3)	I (1)		I (1–3)
Festuca rubra	II (2–6)		I (2–4)	I (2–6)
Anthoxanthum odoratum	II (1–4)		I (1)	I (1–4)
Hieracium sect. *Vulgata*	II (1–3)		I (1)	I (1–3)
Inula conyza	II (1–3)			I (1–3)
Asperula cynanchica	II (1–3)			I (1–3)
Anacamptis pyramidalis	II (1–3)			I (1–3)
Origanum vulgare	I (4–5)			I (4–5)
Centaurium erythraea	I (1–3)			I (1–3)
Hippocrepis comosa	I (2–4)			I (2–4)
Pimpinella saxifraga	I (1–3)			I (1–3)
Fissidens cristatus	I (1–2)			I (1–2)
Campylium chrysophyllum	I (1–5)			I (1–5)
Ctenidium molluscum	I (1–5)			I (1–5)
Rosa canina agg.	I (2–3)			I (2–3)
Hypochoeris radicata	I (2–3)			I (2–3)
Lotus corniculatus	III (2–7)	IV (2–5)	III (2–7)	III (2–7)
Campanula rotundifolia	III (1–3)	IV (1–4)	III (1–3)	III (1–4)
Sanguisorba minor	III (2–6)	IV (2–6)	II (2–6)	III (2–6)
Leontodon hispidus	II (2–7)	IV (2–4)	I (1–3)	II (1–7)
Centaurea nigra	I (3–4)	III (2–5)	II (2–4)	II (2–5)
Knautia arvensis	I (1)	III (1–3)	II (1–3)	II (1–3)
Trifolium pratense	I (1)	II (1–5)	I (1–4)	I (1–5)
Ononis repens		II (2–4)	I (2–4)	I (2–4)
Trisetum flavescens	I (1)	III (3–5)	IV (2–6)	III (1–6)
Holcus lanatus	I (2–3)	II (2–5)	IV (2–8)	III (2–8)
Prunella vulgaris	I (1–2)	I (2–4)	II (1–4)	II (1–4)
Cynosurus cristatus	I (1–3)	I (3–5)	II (3–5)	II (1–5)
Arrhenatherum elatius	I (4)	I (2–4)	II (2–7)	II (2–7)
Cruciata laevipes	I (2–3)	I (1)	II (1–3)	I (1–3)
Veronica chamaedrys	I (1–3)	I (1–3)	II (1–4)	I (1–4)
Cerastium fontanum	I (1)	I (1)	II (1–3)	I (1–3)
Lathyrus pratensis	I (1)	I (1)	II (1–3)	I (1–3)
Trifolium repens		I (2–3)	II (2–5)	I (2–5)
Rumex acetosa		I (1–2)	II (1–3)	I (1–3)
Cirsium vulgare		I (1)	II (2–4)	I (1–4)
Linum catharticum	III (1–4)	III (1–3)	II (1–3)	III (1–4)
Briza media	III (1–5)	III (3–6)	II (2–5)	III (1–6)
Hieracium pilosella	III (1–8)	III (1–4)	I (2–3)	II (1–8)
Plantago lanceolata	I (2–6)	III (1–8)	III (1–6)	III (1–8)
Galium verum	II (1–4)	II (1–4)	II (1–4)	II (1–4)
Bellis perennis	I (2)	II (1–3)	II (1–3)	II (1–3)
Crataegus monogyna sapling	I (2–3)	II (1–3)	II (1–5)	II (1–5)
Senecio jacobaea	I (3)	II (1–3)	II (2–3)	II (1–3)
Pseudoscleropodium purum	II (1–4)	II (1–5)	I (1–5)	I (1–5)

Floristic table CG4 (*cont.*)

	a	b	c	4
Achillea millefolium	II (2–3)	I (1–3)	II (1–4)	I (1–4)
Dactylis glomerata	II (1–4)	I (2–4)	II (2–6)	I (1–6)
Medicago lupulina	I (2)	I (1–4)	I (2–3)	I (1–4)
Centaurea scabiosa	I (3–4)	I (1–4)	I (3)	I (1–4)
Scabiosa columbaria	I (1–4)	I (2–4)	I (2)	I (1–4)
Campanula glomerata	I (3)	I (2–4)	I (3)	I (2–4)
Polygala vulgaris	I (2–4)	I (1–2)	I (2–3)	I (1–4)
Crepis capillaris	I (2)	I (2–3)	I (1–3)	I (1–3)
Primula veris	I (3–4)	I (1–2)	I (2–4)	I (1–4)
Leucanthemum vulgare	I (1)	I (2–3)	I (2–3)	I (1–3)
Fraxinus excelsior sapling	I (1–3)	I (1–3)	I (1)	I (1–3)
Dicranum scoparium	I (1)	I (1–4)	I (3–4)	I (1–4)
Ulex europaeus	I (1–3)	I (1)	I (1)	I (1–3)
Daucus carota	I (2–5)	I (2)	I (2)	I (2–5)
Danthonia decumbens	I (2–4)	I (3–4)	I (4)	I (2–4)
Filipendula vulgaris	I (2–5)	I (2–3)	I (3)	I (2–5)
Succisa pratensis	I (1)	I (2–3)	I (3)	I (1–3)
Urtica dioica	I (1)	I (1)	I (1–4)	I (1–4)
Galium mollugo	I (3)	I (2–3)	I (2–5)	I (2–5)
Ranunculus repens	I (2)	I (1)	I (1–3)	I (1–3)
Phleum pratense bertolonii	I (1)	I (4)	I (2–4)	I (1–4)
Agrostis stolonifera	I (2)	I (4)	I (2–4)	I (2–4)
Vicia cracca	I (1)	I (1)	I (1–3)	I (1–3)
Tragopogon pratensis	I (1)	I (1–3)		I (1–3)
Anthyllis vulneraria	I (2)	I (1)		I (1–2)
Agrostis capillaris	I (4–6)		I (4–5)	I (4–6)
Avenula pubescens	I (2–6)		I (2–4)	I (2–6)
Teucrium scorodonia	I (2–3)		I (2–4)	I (2–4)
Hypericum hirsutum	I (1–3)		I (3)	I (1–3)
Lolium perenne	I (1)		I (1–3)	I (1–3)
Ranunculus bulbosus	I (1–3)		I (1)	I (1–3)
Sambucus nigra sapling	I (1)		I (1)	I (1)
Cirsium arvense	I (2)		I (1–4)	I (1–4)
Luzula campestris		I (1)	I (1–3)	I (1–3)
Rhytidiadelphus squarrosus		I (2–3)	I (2–3)	I (2–3)
Hylocomium splendens		I (1)	I (1)	I (1)
Gentianella amarella		I (1–3)	I (2)	I (1–3)
Number of samples	34	96	101	231
Number of species/sample	21 (12–33)	15 (8–25)	16 (7–28)	17 (7–33)

a   *Avenula pratensis-Thymus praecox* sub-community
b   *Centaurea nigra-Leontodon hispidus* sub-community
c   *Holcus lanatus* sub-community
4   *Brachypodium pinnatum* grassland (total)

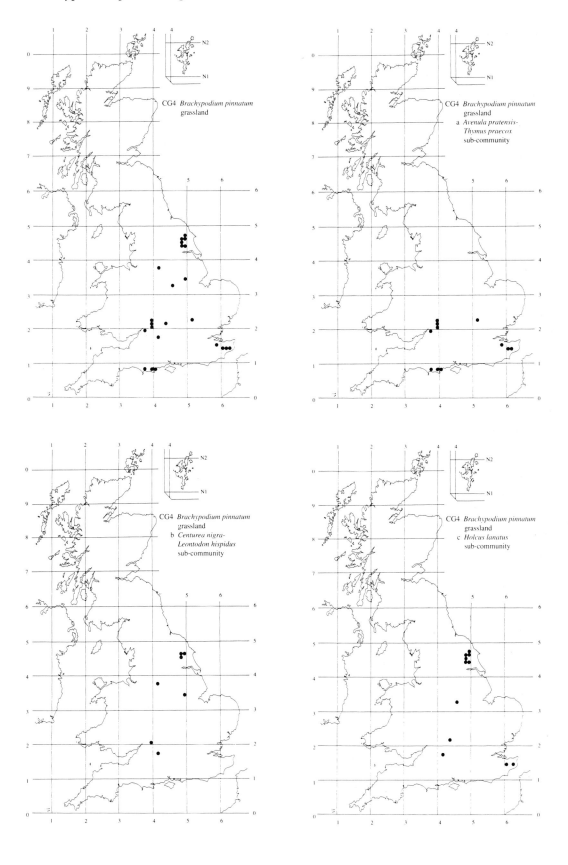

CG5
Bromus erectus-Brachypodium pinnatum grassland

Synonymy

Inferior Oolite grassland Tansley 1939; Unaltered Oolitic Limestone grassland Hepburn 1942; *Cirsio-Brometum typicum* and *astragaletosum* Shimwell 1968a; Cotswold Limestone grassland Wells & Morris 1970; *Pulsatilla vulgaris* stands Wells & Barling 1971 *p.p.*; Jurassic Limestone grassland Ratcliffe 1977 *p.p.*

Constant species

Brachypodium pinnatum, Briza media, Bromus erectus, Carex flacca, Cirsium acaule, Festuca ovina, Helianthemum nummularium, Hieracium pilosella, Leontodon hispidus, Lotus corniculatus, Sanguisorba minor, Thymus praecox.

Rare species

Aceras anthropophorum, Astragalus danicus, Carex ericetorum, Galium pumilum, Herminium monorchis, Phyteuma tenerum, Polygala calcarea, Pulsatilla vulgaris, Thesium humifusum, Thymus pulegioides.

Physiognomy

This community comprises open or closed, sometimes rank and tussocky, swards in which complementary proportions of *Bromus erectus* and *Brachypodium pinnatum* are dominant. In some cases, small tussocks of both species occur together in quite intimate mixtures; in others, more extensive patches of each make up more coarse-grained mosaics, as at Barnack in Northamptonshire, where *Bromus* tends to be more prominent over the 'hills' of spoil and *Brachypodium* in the 'holes' between (e.g. Hepburn 1942). *Carex flacca* and the finer grasses *Festuca ovina* and *Briza media* are constant and each can be locally abundant. Together, and frequently with some smaller amounts of *Avenula pratensis* and *Koeleria macrantha*, they often make up the bulk of the remainder of the sward, giving a generally grassy feel to the vegetation.

Some dicotyledons are, however, a constant feature of the community and may be locally abundant. The chamaephytes *Hieracium pilosella, Thymus praecox* and *Helianthemum nummularium* may all be patchily prominent, together with sprawls of *Lotus corniculatus* and, less commonly, *Anthyllis vulneraria, Hippocrepis comosa* and *Asperula cynanchica*. *Sanguisorba minor, Cirsium acaule* and *Leontodon hispidus* are constant and their rosettes occasionally abundant; other frequent hemicryptophytes are *Scabiosa columbaria, Pimpinella saxifraga* and *Campanula rotundifolia*.

Although the distribution of the community is very much towards the limit of the ranges of many of the characteristic Continental rarities of British calcicolous grasslands, certain of these species do occur here. The most frequent is *Pulsatilla vulgaris* which now survives in this country mostly in these ranker swards and in similar vegetation dominated by *B. erectus* (Wells 1968, Wells & Barling 1971). The largest British colony of this species, comprising some 30,000 individuals, occurs in a stand of this community at Barnsley Warren in Gloucestershire (Wells & Morris 1970, Ratcliffe 1977). Other rarities recorded here include the robust orchids *Aceras anthropophorum* and *Herminium monorchis, Galium pumilum, Polygala calcarea, Phyteuma tenerum* and the Oceanic West European *Thesium humifusum*.

Towards the north-eastern edge of the range of the community, two Continental Northern species also occur in this vegetation: *Carex ericetorum*, on the Magnesian Limestone of Derbyshire and Yorkshire and *Astragalus danicus*, largely on the Oolite of Lincolnshire and Northamptonshire. From stands where the distribution of the latter species overlapped with occurrences of *Pulsatilla vulgaris* and *Aceras anthropophorum* (and of the more widespread *Genista tinctoria* and *Serratula tinctoria*), Shimwell (1968a, 1971b) characterised a distinct type of *Bromus-Brachypodium* grassland, his *Cirsio-Brometum astragaletosum*. Although it is certainly the case that this combination of species is very much associated with *Bromus-Brachypodium* swards over Oolite rather than *Bromus* grasslands over Chalk (Wells & Barling 1971), their coincidental occurrence within

monest of these are *Potentilla reptans*, which can produce locally prominent pads of stoloniferous rosettes, *Tragopogon pratensis* and the scrambler *Vicia cracca*. Less frequent, but sometimes giving a pronounced individual character to particular stands, are *Knautia arvensis*, *Cynoglossum officinale*, *Erigeron acer*, *Reseda luteola*, *R. lutea*, *Linaria vulgaris*, *Cirsium arvense*, *Pastinaca sativa* and *Crepis capillaris*.

Habitat

The *Avenula pubescens* grassland typically occurs over moister and more mesotrophic calcareous soils on flat or gently-sloping sites on a variety of lowland limestones where there is sometimes a history of gross disturbance, such as ploughing, but little or no grazing.

Festuca rubra occurs with somewhat uneven frequency and prominence throughout the *Festuca-Avenula* grassland but it seems to be able to attain prominence, in the absence of grazing, where there is some amelioration of the dry and oligotrophic soil conditions characteristic of lowland grey rendzinas. An adequate supply of available nitrogen appears to be of especial importance (e.g. Elston 1963, Mirghani 1965, Bunting & Elston 1966, Smith *et al.* 1971 on the Swyncombe Down grasslands in Oxfordshire and Lloyd 1964, Lloyd & Pigott 1967 on the Pulpit Hill Field in the Buckinghamshire Chilterns). The other major distinctive grass of this community, *Avenula pubescens*, though little investigated, seems to respond to similar conditions.

There are two particular situations over limestones where such conditions are met. The first is towards the foot of valley sides where the accumulation of colluvium (often over solifluctional head) gives rise to deep and moist, though usually free-draining, colluvial rendzinas or calcareous brown earth soils (e.g. Green & Fordham 1973, Jarvis 1973, Cope 1976, Smith 1980, *Soil Survey* 1983). It is in such sites, and especially towards the bottom of north-facing slopes, where the low insolation reduces water loss from the soils, that the *Dactylis-Briza* sub-community is characteristically found (e.g. Duffey *et al.* 1974, Smith 1980). This vegetation is often lightly pastured, though usually by cattle rather than sheep, and their more selective grazing helps maintain the typically patchy sward in which some light-demanding or more diminutive calcicoles can survive.

The second situation, where tracts of deeper, moister and more mesotrophic soils occur over flat limestone surfaces which are not exploited for agriculture, is now much rarer. Virtually all the available samples from this kind of habitat originate from Porton Down where, within the bounds of the Ministry of Defence ranges, there occur extensive stands of vegetation, long ungrazed by stock and now available only to fluctuating populations of rabbits and hares and military personnel.

It is here that the *Potentilla-Tragopogon* sub-community was first described by Wells *et al.* (1976) from areas where there was no record of ploughing for at least 50 years and sometimes evidence of continuous grassland cover for more than 130 years. Under such conditions, it was suggested, this kind of vegetation had developed, with an absence of pasturing, over soils which had slowly accumulated a reasonable level of fertility. The colonisation of ant-hills, initially by *F. rubra* and then, after abandonment, by other coarse grasses, was thought to have given a head start in some areas to the development of the pronounced tussock physiognomy so characteristic of this vegetation. Different levels of rabbit-grazing were invoked to explain varying degrees of dominance by the rank grasses within stands of the sub-community.

Wells *et al.* (1976) noted that many tussocks consisted of living herbage perched on top of and rooting through a base of dead material and that large tussocks were readily upturned. It is perhaps through such disruption of the vegetation cover by natural disturbance that the typical weeds of this sub-community are able to attain a hold. Studies at Swyncombe Down in Oxfordshire, unsampled in this survey but evidently supporting similar vegetation (Smith *et al.* 1971), have also shown how, as downwind parts of the tussocks become frayed, they decompose, exposing the mineral soil to colonisation by just such coarse weeds.

Zonation and succession

Despite the widespread decline in pasturing over the lowland limestones, large stands of the community are rare because the bulk of the most suitable soils has passed into arable cultivation. Where it does occur, this vegetation usually forms part of zonations which are both related to edaphic variation but also a reflection of seral changes mediated by grazing.

The *Dactylis-Briza* sub-community is typically found as a narrow, patchy fringe towards the bottom of valley-side slopes or as small, isolated stands on local accumulations of colluvium or head. As the soil cover thins to more shallow, dry and oligotrophic rendzinas, it grades to more calcicolous swards. For the most part, these are various kinds of *Festuca-Avenula* grassland though, on the Durham Magnesian Limestone, this sub-community may pass to *Sesleria-Scabiosa* grassland.

The more extensive stands of the *Potentilla-Tragopogon* sub-community at Porton occur in analogous edaphically-related zonations, grading to *Festuca-Avenula* swards over shallower rendzinas and on the sorted soils of occupied ant-hills. Over very oligotrophic, flint-strewn soils, especially where there has been a history of rabbit-warrening, the sub-community gives way to the *Festuca-Hieracium-Thymus* grassland.

The most likely immediate seral precursors of the

Avenula pubescens grassland are perhaps the more meso-trophic *Festuca-Avenula* swards, like those in the *Holcus-Trifolium* and *Succisa-Leucanthemum* sub-communities. In more calcicolous vegetation, ant-hills may provide nuclei from which the ranker grasses can spread. Once established, the *Avenula pubescens* grassland can probably progress to scrub provided the sward is not so overwhelmingly dominated by a mattress-like or tus-socky cover of grasses as to leave no space for seedlings to gain a hold. The less rank vegetation of the *Dactylis-Briza* sub-community is more likely to be invaded and species like *Crataegus monogyna* and *Prunus spinosa* sometimes gain a vigorous hold in its deep, moist soils and from there slowly advance up valley sides.

Rabbits may play some part in keeping the vegetation of the *Potentilla-Tragopogon* sub-community in check but, even without any grazing, it seems doubtful whether the very rank swards here can be easily invaded by shrubs. They may therefore maintain themselves for very long periods with, perhaps, some measure of cyclical change as tussocks decay and the weeds come and go. At Swyncombe, Smith *et al.* (1971) noted that, if *Urtica dioica* gained a hold on bare patches, it precipitated a very rapid breakdown of grass litter which opened the way for colonisation by more extensive areas of weed vegetation.

Distribution
The *Dactylis-Briza* sub-community has been recorded from scattered localities over a variety of lowland limestones, the *Potentilla-Tragopogon* sub-community only from Porton Down and Swyncombe.

Affinities
Vegetation of this kind has been described but rarely sometimes being included within a 'Chalk grassland' community (e.g. Tansley 1939, Duffey *et al.* 1974, Ratcliffe 1977), sometimes being given separate status (e.g. Wells *et al.* 1976). Phytosociologically, it has affinities with both the Mesobromion (through the *Dactylis-Briza* sub-community) and ranker, mesotrophic grasslands of the Arrhenatheretalia (through the *Potentilla-Tragopogon* sub-community).

Floristic table CG6

	a	b	6
Festuca rubra	V (3–9)	IV (1–9)	V (1–9)
Lotus corniculatus	V (1–5)	IV (1–6)	IV (1–6)
Avenula pubescens	IV (1–5)	IV (1–9)	IV (1–9)
Avenula pratensis	IV (2–5)	IV (1–7)	IV (1–7)
Taraxacum officinale agg.	IV (1–3)	IV (1–3)	IV (1–3)
Pseudoscleropodium purum	III (1–4)	IV (1–8)	IV (1–8)
Leontodon hispidus	V (1–7)	III (1–8)	III (1–8)
Carex flacca	V (2–7)	III (1–5)	III (1–5)
Dactylis glomerata	V (2–6)	I (1–5)	II (1–6)
Linum catharticum	III (1–3)	I (1–2)	II (1–3)
Briza media	III (1–5)	I (1–5)	II (1–5)
Holcus lanatus	III (1–4)	I (1–6)	II (1–6)
Succisa pratensis	III (1–7)		I (1–7)
Danthonia decumbens	III (1–6)		I (1–6)
Viola hirta	II (1–3)	I (1–4)	I (1–4)
Plantago media	II (1–6)	I (1–4)	I (1–6)
Ranunculus bulbosus	II (1–3)	I (1–2)	I (1–3)
Helianthemum nummularium	II (1–6)	I (1–3)	I (1–6)
Cirsium palustre	II (1–3)		I (1–3)
Bellis perennis	II (1–3)		I (1–3)
Primula veris	II (1–3)		I (1–3)
Centaurea nigra	II (1–5)		I (1–5)
Listera ovata	II (1–3)		I (1–3)

Brachypodium sylvaticum	I (1–6)		I (1–6)
Fraxinus excelsior sapling	I (1–3)		I (1–3)
Leucanthemum vulgare	I (1–4)		I (1–4)
Cynosurus cristatus	I (1–5)		I (1–5)
Agrostis capillaris	I (1–3)		I (1–3)
Potentilla reptans	I (1)	IV (1–3)	III (1–3)
Tragopogon pratensis	II (1–3)	III (1–3)	II (1–3)
Vicia cracca	I (1–3)	III (1–3)	II (1–3)
Knautia arvensis	I (1)	II (1–3)	I (1–3)
Coeloglossum viride		I (1–3)	I (1–3)
Erigeron acer		I (1)	I (1)
Reseda luteola		I (1–3)	I (1–3)
Linaria vulgaris		I (1–3)	I (1–3)
Reseda lutea		I (1–3)	I (1–3)
Orobanche elatior		I (1–3)	I (1–3)
Phleum pratense pratense		I (1–3)	I (1–3)
Crepis capillaris		I (1–3)	I (1–3)
Phleum pratense bertolonii		I (1–3)	I (1–3)
Cynoglossum officinale		I (1)	I (1)
Cirsium arvense		I (3–6)	I (3–6)
Pastinaca sativa		I (1–2)	I (1–2)
Plantago lanceolata	IV (1–5)	III (1–7)	III (1–7)
Sanguisorba minor	IV (2–7)	III (1–7)	III (1–7)
Koeleria macrantha	III (1–4)	III (1–5)	III (1–5)
Cirsium acaule	III (1–5)	III (1–7)	III (1–7)
Galium verum	II (1–5)	II (1–4)	II (1–5)
Prunella vulgaris	II (1–4)	II (1–3)	II (1–4)
Festuca ovina	II (2–8)	II (1–9)	II (1–9)
Thymus praecox	I (1–3)	II (1–7)	II (1–7)
Agrimonia eupatoria	II (1–3)	I (1–3)	I (1–3)
Calliergon cuspidatum	II (1–4)	I (1–4)	I (1–4)
Centaurea scabiosa	II (1–5)	I (1–5)	I (1–5)
Carex caryophyllea	II (1–3)	I (1–4)	I (1–4)
Agrostis stolonifera	II (1–4)	I (1–3)	I (1–4)
Anthoxanthum odoratum	II (1–3)	I (1)	I (1–3)
Trifolium repens	II (1–4)	I (1–2)	I (1–4)
Trifolium pratense	II (1–3)	I (1)	I (1–3)
Campanula rotundifolia	II (1–2)	I (1–3)	I (1–3)
Scabiosa columbaria	II (1–4)	I (1–2)	I (1–4)
Achillea millefolium	I (2–4)	I (1–4)	I (1–4)
Asperula cynanchica	I (1–3)	I (1–3)	I (1–3)
Medicago lupulina	I (1–3)	I (1–5)	I (1–5)
Poa pratensis	I (1–3)	I (1–3)	I (1–3)
Hieracium pilosella	I (1–4)	I (1–4)	I (1–4)
Filipendula vulgaris	I (2–4)	I (1–6)	I (1–6)
Trisetum flavescens	I (1–4)	I (1–2)	I (1–4)
Fissidens cristatus	I (1–4)	I (1–2)	I (1–4)
Arrhenatherum elatius	I (1–4)	I (1–6)	I (1–6)
Anacamptis pyramidalis	I (1)	I (1–3)	I (1–3)

Floristic table CG6 (*cont.*)

	a	b	6
Cerastium fontanum	I (1–3)	I (1)	I (1–3)
Homalothecium lutescens	I (1)	I (1–4)	I (1–4)
Rumex acetosa	I (1)	I (1–3)	I (1–3)
Rhytidiadelphus triquetrus	I (1–4)	I (1–5)	I (1–5)
Ctenidium molluscum	I (1–6)	I (1–3)	I (1–6)
Fragaria vesca	I (1–3)	I (1–6)	I (1–6)
Cirsium vulgare	I (3)	I (1–3)	I (1–3)
Urtica dioica	I (1)	I (1–3)	I (1–3)
Hypnum cupressiforme	I (3)	I (1–8)	I (1–8)
Senecio jacobaea	I (1–4)	I (1–2)	I (1–4)
Veronica chamaedrys	I (1–2)	I (1–2)	I (1–2)
Pimpinella saxifraga	I (1–3)	I (1–3)	I (1–3)
Clinopodium vulgare	I (2–3)	I (1–3)	I (1–3)
Campylium chrysophyllum	I (2)	I (2–3)	I (2–3)
Gentianella amarella	I (2)	I (1–2)	I (1–2)
Polygala vulgaris	I (1–3)	I (1)	I (1–3)
Galium mollugo	I (1–3)	I (1)	I (1–3)
Rhytidiadelphus squarrosus	I (1–5)	I (1)	I (1–5)
Euphrasia officinalis agg.	I (1–4)	I (1)	I (1–4)
Number of samples	36	87	125
Number of species/sample	25 (8–40)	16 (7–29)	17 (7–40)

a   *Dactylis glomerata-Briza media* sub-community
b   *Potentilla reptans-Tragopogon pratensis* sub-community
6   *Avenula pubescens* grassland (total)

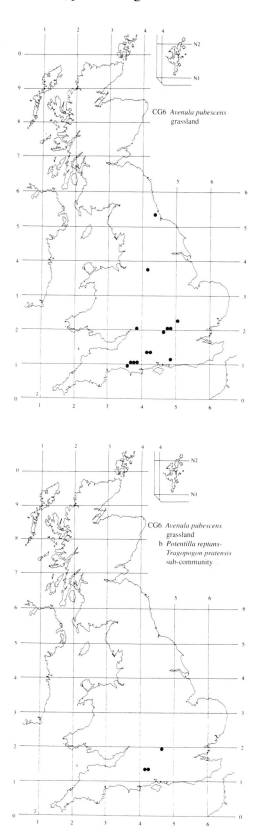

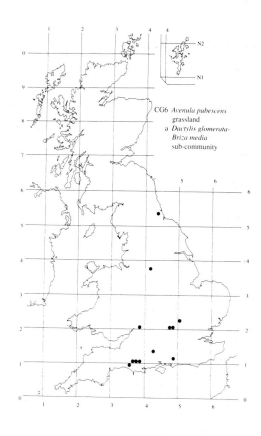

CG6 *Avenula pubescens* grassland

CG6 *Avenula pubescens* grassland
a *Dactylis glomerata-Briza media* sub-community

CG6 *Avenula pubescens* grassland
b *Potentilla reptans-Tragopogon pratensis* sub-community

CG7
Festuca ovina-Hieracium pilosella-Thymus praecox/ pulegioides grassland

Synonymy
Herbland   Tansley & Adamson 1925; Primitive Chalk
grassland   Tansley & Adamson 1925 *p.p.*; Rabbit-
grazed Chalk grassland   Tansley & Adamson 1925
p.p.; Breckland Grasslands A and B   Watt 1940; Old
ex-arable grassland   Cornish 1954; *Hornungia petraea*
stands   Ratcliffe 1959 *p.p.*; Annual community   Ratcliffe 1961 *p.p*; Pulpit Hill Field vegetation   Lloyd 1964;
Catapodium rigidum stands   Clark 1974 *p.p.*; *Festuca
ovina/Hieracium pilosella/Cladonia* spp. lichen-rich
grassland   Wells *et al.* 1976; Ant-hill vegetation   King
1977a *p.p.*; Breckland grasslands   Ratcliffe 1977 *p.p.*;
*Helictotrichon   pratense-Koeleria   cristata-Phleum
phleoides* grassland   Smith 1980; Lichen grasslands
Smith 1980 *p.p.*

Constant species
*Festuca ovina, Hieracium pilosella, Leontodon hispidus,
Thymus praecox/pulegioides, Hypnum cupressiforme.*

Rare species
*Artemisia campestris, Astragalus danicus, Carex erice-
torum, Galium parisiense, Himantoglossum hircinum,
Hornungia petraea, Medicago lupulina* ssp. *minima, M.
sativa* ssp. *falcata, M.* × *varia, Minuartia hybrida,
Phleum phleoides, Potentilla tabernaemontani, Silene
conica, S. otites, Thymus serpyllum, Veronica spicata, V.
verna, Pleurochaeta squarrosa, Bacidia muscorum, Buellia epigaea, Diploschistes scruposus* var. *bryophilus, Fulgensia fulgens, Lecidea decipiens, Squamaria lentigera,
Toximia caerulea* var. *nigricans, T. lobulata.*

Physiognomy
The swards of the *Festuca-Hieracium-Thymus* grassland
can be open or closed but, even where there is a
continuous cover of vegetation, this usually lacks the
dense, plush quality of the *Festuca-Avenula* swards.
Festuca ovina (only on rare occasions replaced by *F.
rubra*) is constant and frequently abundant, but it
generally occurs as small isolated tussocks or an open
network of somewhat attenuated, procumbent shoots.
Koeleria macrantha is the only other grass that is at all
common throughout, other important Mesobromion
species being markedly patchy or scarce. *Avenula pratensis*, for example, is frequent only in certain sub-
communities and, even then, may be sparse in heavily-
grazed swards. *Dactylis glomerata* and *Briza media*, two
other major *Festuca-Avenula* grasses, are here strikingly
uncommon.

In general, it is herbaceous dicotyledons, and
especially chamaephytes and therophytes, which give
the vegetation its distinctive character. Among the
former, *Hieracium pilosella* is usually the most frequent
and abundant: more than 1000 rosettes m^{-2} have been
recorded in this kind of vegetation (e.g. Bishop *et al.*
1978). *Thymus praecox* is also constant and it, too, can
be locally prominent. *T. pulegioides*, though inadequately distinguished from *T. praecox* in the bulk of the
data, is very characteristic of this community (Pigott
1955). In marked contrast to most kinds of *Festuca-
Avenula* grassland, *Helianthemum nummularium* is
scarce here, except in certain rather particular circumstances: it occurs, for example, in this vegetation on ant-
hills, though only where there are plants nearby in the
surrounding Mesobromion sward (King 1977a, b, c).
Sedum acre is occasional and it can be locally abundant.

Mesobromion rosette species are few in number;
indeed, hemicryptophytes in general are not very
numerous and often rather unevenly represented
between the different sub-communities. *Carex flacca*
and *Scabiosa columbaria*, for example, are here at most
occasional. More frequent are *Leontodon hispidus, Prunella vulgaris, Taraxacum officinale* agg. (often including
plants of the Erythrosperma section), *Cirsium acaule,
Sanguisorba minor* and *Lotus corniculatus*. This last,
together with the more uncommon and patchily distributed *Anthyllis vulneraria* and *Galium verum* (and
Sedum acre), can give stands a splash of yellow as their
flowers appear in mid-summer after the purple of the
thymes.

The open texture of this turf provides patches of bare ground which are exploited by coarse weeds and therophytes. Among the former, *Senecio jacobaea* (sometimes biennial) is the most frequent and characteristic of the whole community but there is occasionally some *Potentilla reptans* and, in some sub-communities, *Fragaria vesca* or *Rumex acetosa*. Though rarely abundant in individual stands, these species can give the swards a coarseness that is generally absent from the *Festuca-Avenula* grassland. Certain of the pauciennials, too, are large and conspicuous: *Crepis capillaris*, *Medicago lupulina*, *Erigeron acer* and *Blackstonia perfoliata*. Much more easily missed, especially when sampling is undertaken in summer, but nonetheless very typical of certain of these swards, are more diminutive therophytes, many of which behave as winter annuals, disappearing after flowering or persisting as shrivelled stems. *Linum catharticum* and *Gentianella amarella* occur occasionally throughout; less widely distributed are *Centaurium erythraea*, *Arenaria serpyllifolia*, *Aphanes arvensis*, *Myosotis ramosissima* and *Veronica arvensis* and, of much more restricted occurrence, a variety of national rarities (see below).

Bryophytes are very often an important component of the vegetation. *Hypnum cupressiforme* (usually var. *lacunosum* where different forms have been distinguished) is especially striking, being much more consistently frequent and abundant here than in the *Festuca-Avenula* grassland. *Pseudoscleropodium purum*, *Homalothecium lutescens* and *Fissidens cristatus* are also common with, rather more unevenly present in different sub-communities, *Dicranum scoparium* and *Weissia* cf. *microstoma*. Lichens, especially *Cladonia* spp., are particularly prominent in some kinds of sward.

Sub-communities

Koeleria macrantha sub-community: Primitive Chalk grassland Tansley & Adamson 1925 *p.p.*; *Hornungia petraea* stands Ratcliffe 1959 *p.p.*; Annual community Ratcliffe 1961 *p.p.*; *Catapodium rigidum* stands Clark 1974 *p.p.*; Ant-hill vegetation King 1977a. Mesobromion species, such as *Avenula pratensis* and *Sanguisorba minor* and, less markedly, *Carex flacca*, *Briza media* and *Scabiosa columbaria*, are a little more frequent here than in some other sub-communities and *Galium verum* and *Plantago lanceolata* also occur commonly. The basic floristics of the community are, however, preserved: *F. ovina*, *H. pilosella* and *Thymus* spp. make up the bulk of the vascular cover and within the usually rather open sward there are scattered hemicryptophytes, patches of pleurocarpous mosses and bare ground. On the latter, coarser weeds and pauciennials, apart from *Senecio jacobaea*, are rather infrequent but there are commonly some smaller therophytes. *Linum catharticum*, *Gentia-*

nella amarella and *Arenaria serpyllifolia* are the commonest species but mixtures of the following can also occur: *Myosotis ramosissima*, *Aphanes arvensis*, *Veronica arvensis*, *Erophila verna*, *Acinos arvensis*, *Desmazeria rigida*, *Cerastium semidecandrum*, *Saxifraga tridactylites*, *Cardamine hirsuta* and *Arabidopsis thaliana*. The national rarities *Hornungia petraea* and *Draba muralis* and the perennial *Potentilla tabernaemontani* have been recorded in this vegetation in Derbyshire (Ratcliffe 1959, 1961).

The prominence of these different components and their disposition within the sward vary somewhat with the physiography of the habitat. Over flat limestone surfaces and rounded knolls, for example, the vegetation occurs as a loose-textured turf with its elements closely intermingled. On fine talus, the vascular species are confined to crevices with the bryophytes spreading between. Around exposures of harder limestones, where this vegetation can occur as distinctive 'eye-brows', the vascular plants thin out over the shallower soil and the therophytes become more prominent on loose detritus. On ant-hills, some striking patterns can be encountered as this vegetation develops from those perennials in the disrupted sward which are able to respond to burial or the exposure of bare soil and therophytes which, though largely absent from the surrounding vegetation, seed in on to the heaps. Such patterns also change with the ageing of the hills and their abandonment (King 1977a, b, c: see below).

***Cladonia* spp. sub-community:** Breckland Grassland B Watt 1940; *Festuca ovina/Hieracium pilosella/Cladonia* spp. lichen-rich grassland Wells et al. 1976; Breckland grasslands Ratcliffe 1977 *p.p.*; *Helictotrichon pratense-Koeleria cristata-Phleum phleoides* grassland Smith 1980. In its classic localities in the Breckland of the Norfolk/Suffolk borders, this sub-community occurs as a more or less continuous, though often rather sparse and wiry, turf in which *F. ovina*, together with complementary proportions of *K. macrantha*, *H. pilosella* and *Astragalus danicus* (dfferential here), generally make up the bulk of the vascular component of the vegetation. At Porton Down, its only other known locality, *K. macrantha* is less frequent and prominent and *A. danicus* is absent though the general appearance of the vegetation remains the same. *Thymus praecox* and *T. pulegioides* (with very occasional *T. serpyllum*, essentially a plant of more calcifugous swards on sandier substrates) are now much less widespread than formerly, though they may be locally abundant and, with *A. danicus*, provide a flush of purple with their early summer flowers.

Most of the other associates are also decidedly patchy and the floristics and physiognomy of the vegetation are much less consistent than in the early descriptions of heavily-grazed swards (e.g. Watt 1940, 1957). Among the most frequent associates under those conditions, *Lotus corniculatus*, *Taraxacum officinale* agg. and *Senecio jacobaea* (typical of the whole community) and *Avenula pratensis* and *Galium verum* (characteristic, too, of the *Koeleria* sub-community) remain common, though rather uneven in their occurrence and abundance. Likewise, *Cerastium fontanum* and *Luzula campestris* (both good preferentials for this sub-community) are still frequent as scattered individuals. *Carex flacca*, *Leontodon hispidus*, *Cirsium acaule* and *Sanguisorba minor* remain distinctively sparse, the last three being much less frequent here (at least in Breckland) than in most other kinds of *Festuca-Hieracium-Thymus* grassland.

Apart from these species, however, different stands show some general and some peculiar variations which reflect the kinds of gains and losses which Watt (1957, 1974) noted in ungrazed stands. In the first place, the vegetation is now generally grassier and, in places, decidedly rank. *Avenula pratensis*, in particular, though not much more frequent than in the early accounts, is now often more abundant and *Holcus lanatus* and *Trisetum flavescens* can also be locally prominent. *Achillea millefolium*, *Plantago lanceolata* and *Trifolium repens* sometimes accentuate the mesophytic look of the sward. Coarse grasses (such as *Avenula pubescens* and *Bromus erectus*), though sometimes encountered by Watt, are, however, absent in these data and some other species which held their frequency, increased or appeared with a lack of grazing, are very sparse here. Among the former, for example, *Phleum phleoides* and *Silene otites* were only rarely encountered and, among Watt's newcomers, *Ononis repens* and *Anthyllis vulneraria* were uncommon in Breckland samples (though the latter can be conspicuous at Porton) and *Medicago sativa* ssp. *falcata* was absent.

Second, as with Watt, there are some striking absences from the data. Finer grasses, for example, like *Agrostis capillaris*, *A. canina* ssp. *montana* and *Anthoxanthum odoratum*, are no more than occasional and both *Carex caryophyllea* and *C. ericetorum* (the latter much more so than in Watt's ungrazed grasslands) are scarce. The present patchiness of *Thymus* spp. has already been noted and *Sedum acre*, another chamaephyte, is also now uncommon. But it is among the therophytes that the omissions are especially obvious. Although some caution is necessary before drawing too general conclusions from samples taken within a single season, especially a very dry one like 1976, all of the following, characteristic to varying degrees of grazed swards of this kind, are now apparently much rarer: *Aira praecox*, *Aphanes arvensis*, *Myosotis ramosissima*, *Rumex acetosella*, *Veronica arvensis* (also present in the more calcifugous grasslands), *Sagina apetala* spp. *apetala*, *Teesdalia nudicaulis* (more typical of the calcifugous swards but

occasionally present in grazed 'Grassland B'), *Aira caryophyllea*, *Centaurium erythraea*, *Crepis capillaris*, *Desmazeria rigida*, *Erophila verna*, *Sagina apetala* ssp. *erecta*, *S. nodosa*, *Trifolium dubium* and the rarities *Minuartia hybrida* and *Veronica verna* (particularly characteristic of the more calcicolous grazed Breckland grasslands). Even *Arenaria serpyllifolia*, *Linum catharticum*, *Arabis hirsuta* and *Medicago lupulina*, which held up somewhat without grazing in Watt's studies, were uncommon in our samples.

There are marked reductions, too, in the frequency and variety of bryophytes and lichens, though this sub-community still preserves some preferential features in these elements of the vegetation. Of the rich and chequered pattern of bryophytes typical of the grazed swards of 'Grassland B', *Hypnum cupressiforme*, *Pseudoscleropodium purum* and *Homalothecium lutescens* (typical of this whole community) and *Dicranum scoparium* (a good preferential species here) are the most frequent. Less common, though still distinctive of this sub-community are *Brachythecium albicans*, *Ceratodon purpureus*, *Bryum capillare* and, in its southern outpost in Britain, *Rhytidium rugosum*. Most of the distinctive acrocarpous mosses of the grazed Breckland grasslands (e.g. *Encalypta vulgaris*, *Tortula ruralis* ssp. *ruraliformis*, *Ditrichum flexicaule*, *Weissia* spp., *Fissidens cristatus*, *Plagiomnium rostratum* and *Rhodobryum roseum*) and even some of the more robust pleurocarps which persisted without grazing in Watt's samples (e.g. *Pleurozium schreberi*, *Hylocomium splendens*, *Rhytidiadelphus triquetrus*, *Thuidium abietinum*) seem, however, to have declined, together with the more delicate hepatics (*Ptilidium ciliare*, *Frullania tamarisci*, *Lophozia excisa*, *Lophocolea bidentata* s.l.).

Although, in grazed 'Grassland B', lichens typically occurred as individuals rather than as pure patches or extensive sheets (Watt 1940), they were, nonetheless, a striking element in the vegetation. They remain, as a group, a good diagnostic feature of this sub-community, but their frequency and abundance have declined such that now usually just two or three species (sometimes even fewer) are present together in a stand. The most frequent are *Cornicularia aculeata*, *Cladonia rangiformis*, *C. impexa*, *C. arbuscula*, *C. pyxidata* and *C. foliacea* and the thallose *Peltigera canina*. *P. rufescens* and *P. polydactyla*, which Watt noted as declining more markedly than *P. canina* without grazing, were not recorded in our Breckland samples, though they were common at Porton in the survey of Wells *et al.* (1976). Indeed, the Porton swards of this kind seem to be much more like grazed 'Grassland B' in their variety and abundance of lichens than do surviving Breckland stands. Some of the vegetation at Porton even attained the very open character of 'Grassland A' with its flint- and Chalk-strewn surface and encrusting lichens were, likewise, common there. On balance, however, even these swards seem to be better located here than in the next sub-community.

Ditrichum flexicaule-Diploschistes scruposus var. bryophilus sub-community: Breckland Grassland A Watt 1940; Breckland grasslands Ratcliffe 1977 *p.p.*; Lichen grasslands Smith 1980 *p.p.* Only three samples were taken of this vegetation which is of very restricted distribution (see below), but they are distinctive in relation to the other sub-communities and show some of the important features noted in the original account (Watt 1940) and in the more recent descriptions of later changes (Watt 1981*a*, *b*). First, the cover is here generally open with small raised tussocks of *F. ovina* occurring within a ground of *H. pilosella* and scattered patches of *T. praecox*. *T. pulegioides* and *T. serpyllum* were not recorded though they apparently occur here in small amounts. Second, other perennials, especially hemicryptophytes, are sparse: there are sometimes scattered plants of *Avenula pratensis*, *Koeleria macrantha*, *Astragalus danicus*, *Leontodon hispidus*, *Lotus corniculatus*, *Taraxacum officinale* agg., *Senecio jacobaea*, *Plantago lanceolata* and *Cerastium fontanum*. The fern *Botrychium lunaria*, the abundance of which was a very diagnostic feature of 'Grassland A', was rare but, as its shoots are avidly eaten by rabbits, it may remain commoner than our data imply. Third, pauciennials occur in the more open areas. The most frequent species were *Carlina vulgaris*, *Centaurium erythraea*, *Linum catharticum*, *Sagina nodosa* and *Arabis hirsuta* but, as in the previous sub-community, many once-common species are now apparently sparse, even those of wide distribution such as *Desmazeria rigida*, *Myosotis ramosissima* and *Veronica arvensis* as well as the rarities *Minuartia hybrida* and *Galium parisiense*. Fourth, bryophytes and lichens remain a prominent feature of the sub-community. Among the former, *Hypnum cupressiforme*, *Campylium chrysophyllum*, *Encalypta vulgaris* and, especially typical of this kind of grassland, *Ditrichum flexicaule*, were abundant in the samples. The most frequent lichens were *Cornicularia aculeata*, *Peltigera canina*, *Cladonia foliacea*, *C. papillaria* (restricted to this sub-community), *C. furcata*, *C. impexa* and *C. rangiformis*. Particularly distinctive here is a further suite of lichens encrusting the soil surface and the abundant exposed flints and Chalk rubble: *Diploschistes scruposus* var. *bryophilus*, *Squamarina lentigera* and *Buellia epigaea* were recorded in the samples but other species known from this vegetation are *Bacidia muscorum*, *Dermatocarpon hepaticum*, *Fulgensia fulgens*, *Lecidea decipiens*, *Toximia caerulea* var. *nigricans* and *T. lobulata* (James *et al.* 1977, Ratcliffe 1977).

Fragaria vesca-Erigeron acer sub-community: Herbland Tansley & Adamson 1925 *p.p.*; Old ex-arable grassland Cornish 1954 *p.p.*; Pulpit Hill Field vegetation Lloyd

& Pigott 1967. Although both *H. pilosella* and *Thymus* spp. remain constant here, they are generally less prominent in the swards and hemicryptophytes are more frequent and varied than in the preceeding sub-communities. *F. ovina* (sometimes totally replaced by *F. rubra*) remains very much the most frequent and abundant grass but, though *Koeleria macrantha* is only occasional, there is often some *Trisetum flavescens* in the sward, less commonly a little *Agrostis stolonifera* and, more locally, *Holcus lanatus*, *Avenula pubescens* and *Phleum pratense* ssp. *bertolonii*. Typically, however, this grass cover remains sparse and the vegetation is usually short and somewhat open.

Among the herbaceous dicotyledons, the most frequent perennials are *Leontodon hispidus*, *Taraxacum officinale* agg. *Senecio jacobaea*, *Prunella vulgaris*, *Cirsium acaule* and, especially distinctive here, *Fragaria vesca*. *Potentilla reptans*, *Clinopodium vulgare* and *Inula conyza* are much scarcer but can be locally prominent. Pauciennials are always an important component of the vegetation with coarser weed species, such as *Daucus carota*, *Erigeron acer* and *Crepis capillaris*, and less commonly *Pastinaca sativa* and *Arctium minus* agg., particularly noticeable. Smaller therophytes, such as *Linum catharticum*, *Gentianella amarella* and, preferential here, *Centaurium erythraea*, are also frequent. The rare species *Iberis amara*, *Teucrium botrys* and *Ajuga chamaepitys* have all been recorded in this vegetation. Occasional saplings of *Sambucus nigra* or *Cornus sanguinea* may add to the scruffy appearance of the vegetation but it never has a lush, rank character.

As usual in the community, bryophytes are frequent and sometimes abundant, though the species involved are few. *Hypnum cupressiforme*, *Homalothecium lutescens* and *Campylium chrysophyllum* form patches over the soil surface, but *Pseudoscleropodium purum* is noticeably scarce here. *Fissidens cristatus* and *Weissia* spp. are the most frequent acrocarps. Lichens are rare, though there may be occasional tufts of *Cladonia rangiformis* or *C. furcata*.

Medicago lupulina-Rumex acetosa sub-community. In its general features, the somewhat grassier character of the sward, the greater prominence of hemicryptophytes and the presence of a conspicuous weed element, this vegetation is similar to that of the previous sub-community, though some of the species involved are different. *F. ovina* tends to be more abundant here and it commonly makes up the bulk of the vegetation, often with small amounts of *Phleum pratense* ssp. *bertolonii* and *Koeleria macrantha* and, less commonly, with a little *Avenula pubescens*.

Scattered in this ground which, though a little more extensive than in the *Fragaria-Erigeron* sub-community, is again not tall or luxurious, are small patches of the chamaephytes *H. pilosella* and *Thymus* spp., and scattered rosettes of the same hemicryptophyte perennials with the addition, here, of *Sanguisorba minor*. Also preferential to this sub-community are *Rumex acetosa* and *Leontodon autumnalis* and, less commonly, *Ranunculus bulbosus* and *Viola hirta*. Ranker weeds are somewhat less conspicuous here (*Fragaria vesca* and *Potentilla reptans* occur occasionally) but, among the smaller pauciennials, *Medicago lupulina* and *Euphrasia officinalis* agg. are preferential and there are frequent records, too, for *Linum catharticum* and *Gentianella amarella*. *Iberis amara* and *Seseli libanotis* have been recorded in this vegetation.

The bryophytes are predominantly the more robust pleurocarpous mosses of the community with *Pseudoscleropodium purum* resuming its usually high frequency. Indeed, here it is the commonest species and often has extensive cover.

Habitat

The *Festuca-Hieracium-Thymus* grassland is characteristically a vegetation type of thin, stony, very free-draining and highly oligotrophic calcareous soils developed under more continental climatic conditions, with heavy rabbit-grazing and, sometimes, a history of past disturbance. It occurs, mainly over the Chalk, but also on some more southerly exposures of Carboniferous Limestone, on flat or gently-sloping sites, over knolls and around bedrock exposures, on some abandoned arable land and on spoil heaps. It can also be found where the typical edaphic, climatic and biotic factors coincide on the hills produced by mound-building ants.

The dry and impoverished nature of the soils is probably of prime importance in influencing both the floristics and the physiognomy of the community, perhaps of as much consequence as their generally calcareous and base-rich nature. In some habitats, the skeletal character of the soil cover is obvious: this vegetation can be found, for example, on those natural or man-made raw soils which develop in the crevices of certain rock exposures, between talus fragments or amongst disturbed rock waste where there is little more than an accumulation of wind-blown organic matter and mineral detritus (Tansley & Adamson 1925, Hope-Simpson 1940*b*, Ratcliffe 1959). More commonly, it occurs over more intact, extensive and somewhat deeper rendzinas but, wherever these have been analysed, they are strikingly poor in major nutrients. In the shallow, chalky soils at Lloyd's Pulpit Hill Field site in the Chilterns, for example (Lloyd 1964, Lloyd & Pigott 1967), and in the rendzinas under this kind of vegetation at Porton Down (Wells *et al*. 1976), nitrogen and phosphorus values were very low. In Breckland, the soils are of complex composition and provenance, being developed over Chalk with varying amounts of boulder clay and sand, disrupted by periglacial solifluction and

overlain to different degrees by aeolian material but here, too, the Newmarket brown rendzinas which typically underlie the community can be very oligotrophic (Watt 1936, 1940, Corbett 1973, Duffey 1976, Curtiss *et al*. 1976, *Soil Survey* 1983).

The effects of this can be seen throughout the vegetation in the generally low cover and poor vigour of the grasses and the prominence of herbaceous dicotyledons, especially those composites and labiates which can survive at low nutrient levels and those legumes which can obviate the need for soil nitrogen by having root-nodule bacteria. Though the large amounts of light reaching the sward surface enable low-growing species to persist, they are rarely able to produce lush growth and the consequent abundance of open ground leaves ample opportunity for the appearance of ephemerals and the spread of bryophytes and lichens. The importance of a poor nutrient supply in maintaining the balance between these components in this kind of vegetation was early confirmed by Farrow (1917), who applied farmyard manure to a Breckland sward and produced a more luxuriant herbage in which the balance of dominance was shifted towards the grasses. At Pulpit Hill, Lloyd (1964, Lloyd & Pigott 1967) demonstrated a similar response in *Festuca ovina*, *F. rubra* and *Koeleria macrantha*, suggesting that the normally poor availability of nitrogen in particular was a controlling factor in limiting their growth in the ungrazed swards there. *F. ovina* and *K. macrantha* were also the main species which responded to the addition of major nutrients in a series of fertiliser experiments on Breckland grasslands of this type carried out by Davy & Bishop (1984). An increase in their biomass and litter, especially marked in ungrazed plots, was accompanied by a catastrophic decline in *Hieracium pilosella* and a reduction in the cover of bryophytes and lichens. Much of the response of *H. pilosella* was attributed to the indirect effect of the more vigorous growth of its competitors but the addition of nutrients also greatly stimulated inflorescence production and accelerated the population decline in this species which, in Breckland, behaves as a monocarpic perennial (Bishop & Davy 1984). *Astragalus danicus*, a nationally rare and important species within the Breckland swards of this community, appeared to maintain itself in the fertilised plots, provided they were grazed.

Climatic conditions maintain and accentuate the harsh nature of the soil environment. The community is most characteristic of those eastern lowland limestones where the regional climate tends towards that of the Continental European mainland and outside this area generally occurs only where the local topoclimate brings conditions close to this, as on some south-facing slopes towards the upland margins (e.g. Ratcliffe 1961) or on the southern faces of ant-hills (e.g. King 1977*a*). In

community represents the most calcicolous extreme of a spectrum of oligotrophic communities, passing, over deeper and more acidic soils, to calcifugous grasslands and heaths. This is the kind of sequence first fully described by Watt (1936, 1940) from Breckland where both the *Ditrichum-Diploschistes* and *Cladonia* sub-communities occur as part of the range of his 'grass-heaths'. The former type of *Festuca-Hieracium-Thymus* grassland is very much confined to the most exposed and inhospitable, flinty rendzinas, being replaced over slightly deeper and less calcareous, though still base-rich, soils by the latter sub-community. This, in turn, gives way, over sandy brown earths, to *Festuca-Agros-tis-Rumex* grassland as calcicoles are eclipsed by calci-fuges and the grass cover thickens up somewhat but, in the harsh continental climate and with (at the time of Watt's survey) continued heavy rabbit-grazing and mole-disturbance, the general contribution made to the vegetation by therophytes, bryophytes and lichens remains important. As Watt stressed, the actual mani-festations of this sequence, and its continuation into very calcifuge vegetation over podzols, are complex and often incomplete at a single site. Moreover, even the most intact of these zonations are now much altered because of the reduction in rabbit-grazing.

However, it is clear that the various kinds of *Festuca-Hieracium-Thymus* grassland respond rather differently to the cessation of grazing. In the more extreme types of sward, such as those of the *Ditrichum-Diploschistes* sub-community, edaphic and climatic conditions are so harsh as to prevent the speedy development of a more intact and mesotrophic vegetation upon enclosure (Watt 1962, 1981a, b). Some stands of the *Koeleria* sub-community, such as those over very rocky cliff tops, are probably similarly uncongenial to the spread of grasses and the appearance of hemicryptophytes. In other cases, as with the *Fragaria-Erigeron* sub-community, artifici-ally enhanced oligotrophy seems to be so severe as to impede the formation of a more grassy sward (Lloyd 1964, Lloyd & Pigott 1967). With these vegetation types, the stabilisation of the soil surface and the accumulation of fertility is likely to be a very slow process (Watt 1981b).

The *Cladonia* sub-community, on the other hand, shows much more pronounced and rapid changes with enclosure, so that the vegetation comes to resemble something akin to a *Festuca-Avenula* sward (Watt 1957, 1962). At Porton, too, Wells *et al.* (1976) suggested that this might be the fate of ungrazed stands of the *Cladonia* sub-community. But such resemblances are likely to be temporary, because the *Festuca-Avenula* grassland is also dependent on grazing for its maintenance and, among herbaceous species, it is perhaps the ranker grasses which are the likelier candidates as eventual dominants after long enclosure or abandonment. Here,

too, though, there may be complexities: *Bromus erectus*, for example, in many ways eminently suited to these more oligotrophic, calcareous soils, seems strikingly slow to invade them, perhaps because of its poor disper-sal (Watt 1957, 1962, Lloyd 1964, Lloyd & Pigott 1967). *Avenula pratensis* may likewise be inhibited from spreading for a number of reasons not directly related to the poor nutrient content of the soils (Wells *et al.* 1976). Other possible colonisers, such as *Avenula pubescens* and *Festuca rubra*, though they can gain an initial hold on abandoned ant-hills, may need a considerable accumulation of fertility before they can spread exten-sively (Wells *et al.* 1976, King 1977a, c).

Even where changes in the herbaceous vegetation are fairly rapid and considerable, evidence suggests that, in some cases, they are likely to be overtaken by the invasion of those woody species which can establish themselves directly in the open and inhospitable con-ditions. At Lloyd's Pulpit Hill Field site, for example, *Cornus sanguinea* was noticeably common among the *Fragaria-Erigeron* vegetation and he suggested that a number of features gave it an advantage over *Crataegus monogyna*, a very abundant early invader of *Festuca-Avenula* swards: high percentage viability and quick germination of the seed, indeterminate shoot growth and the ability to sucker and spread vegetatively once established (Lloyd 1964, Lloyd & Pigott 1967). *C. sanguinea* also occurred at Porton though, here, *Juniper-us communis* ssp. *communis* was an important coloniser of swards of the *Cladonia* sub-community after myxo-matosis, forming open scrub (Wells *et al.* 1976). In Breckland, *Pinus sylvestris* is the most prominent woody species, seeding in from the extensive plantations nearby, and even the *Ditrichum-Diploschistes* vegetation would seem to be progressing to open pinewood (Watt 1957, 1962, 1974, 1981a).

Distribution

The community occurs in scattered localities over the Chalk of south-east England with outliers on south-facing slopes on the Yorkshire Wolds Chalk and the Carboniferous Limestone of Derbyshire and the Men-dips. Breckland still retains the greatest concentration and extent of this kind of vegetation but, though the *Ditrichum-Diploschistes* sub-community is wholly, and the *Cladonia* sub-community largely, confined there, their cover and composition have altered since the days of the early accounts. The former vegetation type has shrunk and changed less, though its extent in living memory was always very small: it still occurs as patches within a matrix of the latter sub-community over a few hectares of the most exposed ground at Lakenheath Warren, Suffolk. Several hectares of vegetation over flint-strewn soil at Deadman's Grave on the Icklingham Heaths, also in Suffolk, approach it in composition and physiognomy (Ratcliffe 1977). The *Cladonia* sub-community itself is, for the most part, much more restricted than formerly and surviving stands are often rank and encroached upon by scrub. Even where grazing continues, as at Eriswell Warrens near Lakenheath, cattle are sometimes pastured when the ground can become poached or strewn with fodder in winter (Ratcliffe 1977). Stands of the typically open and more species-rich vegetation of this sub-community remain at Lakenheath and, over the Norfolk border, at Weeting Heath and within the Stanford Practical Training Area where sheep are still pastured. The other locality for the *Cladonia* sub-community, at Porton Down, was not revisited for this survey.

Affinities

In its more extreme forms, the vegetation of the *Festuca-Hieracium-Thymus* grassland can be seen as the Continental counterpart of the *Festuca-Carlina* grassland and its possible relationships to the Xerobromion swards of the European mainland were noted by Watt (1962). With its poor representation of Mesobromion species, it clearly belongs among the more arid grasslands of the Festuco-Brometea and, like the *Festuca-Carlina* grassland, its open swards provide a congenial location in Britain for a number of species which are rare here but which, in Europe, are more common and widespread in dry, calcareous situations. In this community, however, the Oceanic Southern element of the *Festuca-Carlina* grassland is replaced by a more strictly Continental component and this seems to argue for a link with either the *Stipa*-dominated vegetation of the Festucion valesiacae or the grasslands of the Koelerio-Phleion phleoidis (e.g. Ellenberg 1978, Oberdorfer 1978). Certainly,

with its distinctive mixture of calcicolous perennials, therophytes, bryophytes and lichens, the *Festuca-Hieracium-Thymus* grassland represents one of the nearest approaches among British vegetation types to the steppe-grasslands of eastern Europe.

The prominent therophyte element in the community provides floristic links with a variety of other vegetation types in which these species, mostly characteristic of the Sedo-Scleranthetea, are able to exploit gaps in the sward produced by a variety of processes. Here, their prominence reflects edaphic and climatic features, such as the extent of sand-blow and the incidence of summer drought or winter frost-heave, and the activities of rabbits, ants and moles (Watt 1981*b*). Some calcareous maritime habitats, such as the tops of limestone cliffs and deposits of consolidated shell-sand, are characterised by a similar coincidence of factors and they may carry vegetation which is physiognomically and floristically close to the *Festuca-Hieracium-Thymus* grassland. For example, the *Arenaria serpyllifolia* sub-community of the *Armeria maritima-Cerastium diffusum* ssp. *diffusum* maritime therophyte vegetation, typical of the crumbling tops of limestone sea-cliffs, shares many species with the *Koeleria* sub-community here. Likewise, there is a considerable overlap between this sub-community and the assemblages of smaller annuals found on moderately stable dunes.

With other habitats, it is more a combination of biotic disturbance and calcareous soils which provides a link with the conditions here and, in this case, the floristic transitions involve the more weedy vegetation of the *Fragaria-Erigeron* and *Medicago-Rumex* sub-communities which show overlaps with more ephemeral communities of arable fields on limestone-derived soils.

Floristic table CG7

	a	b	d	e	7
Festuca ovina	V (2–9)	V (4–8)	V (1–8)	V (2–8)	V (1–9)
Hieracium pilosella	V (1–4)	IV (3–8)	V (1–8)	V (1–7)	V (1–8)
Hypnum cupressiforme	IV (1–6)	IV (2–6)	IV (1–8)	III (1–8)	IV (1–8)
Leontodon hispidus	IV (2–7)	II (2–5)	V (1–7)	V (2–8)	IV (1–8)
Thymus praecox/pulegioides	V (1–6)	II (2–5)	IV (1–6)	IV (1–7)	IV (1–6)
Koeleria macrantha	IV (1–7)	V (3–7)	II (1–4)	III (1–4)	III (1–7)
Avenula pratensis	III (1–4)	III (2–6)	I (1–2)	II (1–4)	II (1–6)
Galium verum	III (1–6)	III (2–5)	I (1–3)	II (1–4)	II (1–6)
Plantago lanceolata	III (1–4)	III (1–4)	I (1–4)	II (1–4)	II (1–4)
Sedum acre	II (1–3)	II (3)	I (1–3)	I (1–3)	I (1–3)
Arenaria serpyllifolia	II (1–3)	II (2–3)			I (1–3)
Myosotis ramosissima	I (1–2)	I (2–4)			I (1–4)
Aphanes arvensis	I (1–4)	I (3)			I (1–4)
Erophila verna	I (1–2)	I (2–4)			I (1–4)
Veronica arvensis	I (1–2)	I (1–2)			I (1–2)
Acinos arvensis	I (1–2)	I (1)			I (1–2)
Desmazeria rigida	I (1–2)	I (1)			I (1–2)
Carex flacca	II (1–5)	I (4)	I (1–4)	I (1–5)	I (1–5)
Briza media	II (1–6)	I (3–4)	I (2–4)		I (1–6)
Scabiosa columbaria	II (1–3)			I (1–2)	I (1–3)
Plantago media	I (1–4)	I (2–3)			I (1–4)
Dicranum scoparium	I (2–6)	III (3–6)	I (1–4)	I (1–3)	I (1–6)
Cerastium fontanum	I (1–3)	III (1–4)	I (1–3)	I (1)	I (1–4)
Astragalus danicus		III (2–5)			I (2–5)
Luzula campestris		III (2–5)			I (2–5)
Trifolium repens	I (1–3)	II (2–6)	I (2–3)	I (1–3)	I (1–6)
Holcus lanatus	I (3)	II (1–4)	I (1–3)	I (1–5)	I (1–5)
Achillea millefolium	I (1–3)	II (2–4)		I (1–2)	I (1–4)
Rumex acetosella		II (2–4)			I (2–4)
Agrostis capillaris		II (1–5)			I (1–5)
Anthoxanthum odoratum		II (3–4)			I (3–4)

Floristic table CG7 (*cont.*)

	a	b	d	e	7
Cladonia rangiformis		II (1–6)			I (1–6)
Cladonia impexa		II (2–6)			I (2–6)
Cornicularia aculeata		II (3–4)			I (3–4)
Cladonia arbuscula		II (2–8)			I (2–8)
Cladonia pyxidata		II (1–3)			I (1–3)
Cladonia foliacea		II (3)			I (3)
Ceratodon purpureus		II (2–4)			I (2–4)
Bryum capillare		II (1–4)			I (1–4)
Rhytidium rugosum		II (1–6)			I (1–6)
Brachythecium albicans		II (3–5)			I (3–5)
Erodium cicutarium		II (1–4)			I (1–4)
Peltigera canina		II (3–4)			I (3–4)
Galium saxatile		I (2–4)			I (2–4)
Cladonia fimbriata		I (2–3)			I (2–3)
Ptilidium ciliare		I (3–4)			I (3–4)
Cladonia gracilis		I (2–4)			I (2–4)
Sagina nodosa		I (3)			I (3)
Teesdalia nudicaulis		I (3–4)			I (3–4)
Cladonia tenuis		I (2–5)			I (2–5)
Polytrichum juniperinum		I (2–4)			I (2–4)
Cladonia squamosa		I (2–3)			I (2–3)
Barbula unguiculata		I (2–3)			I (2–3)
Sedum album		I (3)			I (3)
Carex ericetorum		I (2–3)			I (2–3)
Fragaria vesca	I (1–4)	I (1–3)	IV (1–4)	II (1–4)	II (1–4)
Erigeron acer	I (1–3)	I (1–3)	IV (1–3)		II (1–3)
Crepis capillaris	II (1–3)	I (1–2)	III (1–3)	I (1–2)	III (1–3)
Daucus carota	I (1–3)	I (1–2)	III (1–3)	I (2)	II (1–3)
Centaurium erythraea		I (1–3)	III (1–3)		I (1–3)
Agrostis stolonifera	I (1–2)	I (4)	II (1–4)	I (1)	I (1–4)
Campylium chrysophyllum	I (1–2)	I (1–3)	II (1–4)	I (1–2)	I (1–4)
Sambucus nigra sapling			I (1–3)		I (1–3)

CG8
Sesleria albicans-Scabiosa columbaria grassland

Synonymy
Magnesian Limestone Rough Pasture   Heslop-Harrison & Richardson 1953; *Seslerio-Helictotrichetum pratensis*   Shimwell 1968a.

Constant species
Avenula pratensis, Briza media, Carex flacca, Centaurea nigra, Festuca ovina, Galium verum, Helianthemum nummularium, Koeleria macrantha, Linum catharticum, Lotus corniculatus, Pimpinella saxifraga, Plantago lanceolata, Sanguisorba minor, Scabiosa columbaria, Sesleria albicans, Thymus praecox.

Rare species
Epipactis atrorubens, Linum perenne ssp. *anglicum, Primula farinosa, Sesleria albicans.*

Physiognomy
The *Sesleria albicans-Scabiosa columbaria* grassland comprises generally closed swards in which *S. albicans* is usually the most abundant grass, often dominating as vigorous tussocks, especially when the sward is ungrazed. *Festuca ovina* (occasionally with a little *F. rubra*, though rarely exceeded by it) and *Briza media* are constant, but somewhat variable in abundance. *Avenula pratensis* and *Koeleria macrantha* are very frequent, too, though usually in small amounts. There is occasionally a little *Agrostis capillaris, Cynosurus cristatus* and *Brachypodium sylvaticum* and, in one sub-community in particular, coarser species such as *Avenula pubescens, Dactylis glomerata* and *Arrhenatherum elatius* lend a rank appearance to the sward. *Bromus erectus* and *Brachypodium pinnatum* occur rarely, here very much towards their northern limits in Britain and the former prominent only on south-facing slopes. *Carex flacca* is very common and, with *F. ovina*, often makes up the bulk of the sward among the *S. albicans*; *C. caryophyllea* is also frequent, though usually not so abundant.

Although this vegetation has been very intensively sampled in relation to its extent and the amount of variation it shows, such that it appears somewhat over-solidly defined, the richness of the dicotyledonous element is a real feature of the community. As in the *Sesleria-Galium* grassland, *Thymus praecox* and *Helianthemum nummularium* are both constant and each can be locally prominent, and *Sanguisorba minor, Plantago lanceolata, Lotus corniculatus, Campanula rotundifolia, Linum catharticum* and *Euphrasia officinalis* agg. (mostly *E. nemorosa*) are very frequent as usually scattered individuals. Here, however, there is an additional enrichment from a group of species that are especially characteristic of lowland Mesobromion swards. Especially distinctive among these is *Scabiosa columbaria* but *Galium verum, Leontodon hispidus, Anthyllis vulneraria, Plantago media, Primula veris, Viola hirta, Centaurea scabiosa* and *Gentianella amarella* are also very common and, in two of the sub-communities, *Centaurea nigra* and *Pimpinella saxifraga* occur at high frequency.

Bryophytes vary considerably in their abundance, being generally sparse in the ranker swards, though more conspicuous and varied in damper sites and more open turf. The most frequent species throughout are *Ctenidium molluscum, Fissidens cristatus* and *Hypnum cupressiforme*.

Sub-communities

Hypericum pulchrum-Carlina vulgaris **sub-community:**
Seslerio-Helictotrichetum pratensis typicum and *caricetosum pulicariae* Shimwell 1968a. *S. albicans* is usually the dominant in the closed and sometimes tussocky swards here which are characterised by the preferential frequency of *Hypericum pulchrum, Carlina vulgaris, Polygala vulgaris*, and, less markedly, *Stachys betonica, Succisa pratensis, Achillea millefolium* and *Trisetum flavescens*. The rare orchid *Epipactis atrorubens* occurs in this vegetation at a few sites, sometimes in abundance, and with *Gymnadenia conopsea*, which is more widely distributed throughout the community, it can make a

splendid show when flowering in mid-summer. *Linum perenne* ssp. *anglicum* is also found in a very few stands and *Bromus erectus* and *Brachypodium pinnatum* are encountered very occasionally, locally rivalling *S. albicans* in prominence. On cooler and damper slopes, *Carex pulicaris*, which is, generally speaking, a rare plant in the community, may be present in considerable abundance, sometimes with much *Ctenidium molluscum* and *Calliergon cuspidatum* and some *Hylocomium splendens*. Here, too, there may be scattered plants of *Selaginella selaginoides* and *Pinguicula vulgaris*. It is in such vegetation, which could be characterised as a variant (cf. Shimwell 1968a), that the rare records in lowland Durham for *Antennaria dioica*, *Primula farinosa* and *Preissia quadrata* occur (Shimwell 1968a, Bellamy 1970).

Avenula pubescens sub-community: *Seslerio-Helictotrichetum pratensis helictotrichetosum pubescentis* Shimwell 1968a. Although *S. albicans* generally remains the most abundant species in this sub-community, there is greater variety here among the grasses, and the high frequency of *Avenula pubescens* and *Dactylis glomerata* and the more local prominence of *Arrhenatherum elatius* and *Bromus erectus* often gives the swards a coarse, tussocky appearance. *Trifolium repens* attains constancy here and the scattered occurrence of *Daucus carota* and, less frequently, *Senecio jacobaea*, *Rhinanthus minor* and *Cirsium vulgare* can increase the scruffy look of the vegetation. Seedling shrubs occur occasionally with *Crataegus monogyna*, *Ulex europaeus* and *Rosa* spp., including *R. pimpinellifolia*, *R. canina*, *R. eboracensis* and *R. afzeliana*. Where these begin to close up into scattered patches of scrub there is a consequent thinning of the grassland herbage.

Hieracium pilosella sub-community: *Seslerio-Helictotrichetum* Sub-association of *Encalypta vulgaris* and *Plantago maritima* Shimwell 1968a. In the more open vegetation here, the cover of *S. albicans* is usually reduced to small tussocks and the frequency of a number of generally common species falls, e.g. *Centaurea nigra*, *Pimpinella saxifraga*, *Briza media*, *Lotus corniculatus*, *Campanula rotundifolia*, *Primula veris* and *Leontodon hispidus*. By contrast, there is an increase in *Hieracium pilosella* and hemicryptophytes characteristic of open, disturbed or marginal habitats, e.g. *Hypochoeris radicata*, *Hypericum montanum*, *Plantago maritima*, *Medicago lupulina*, *Leontodon autumnalis* and Hieracia of the section Vulgata. Orchids are common with frequent records for *Gymnadenia conopsea* and, especially distinctive here, *Listera ovata* and *Coeloglossum viride*. Bryophyte cover is usually extensive among the turf with, in addition to *Ctenidium molluscum* and *Fissidens cristatus*, patches of small acrocarps such as *Encalypta vulgaris*, *E. streptocarpa*, *Bryum argenteum* and *Ceratodon purpureus*.

Habitat

The community is restricted to free-draining, calcareous soils over Magnesian Limestone in the cool, dry climate of lowland Durham. It is a plagioclimax vegetation maintained by the grazing of stock and rabbits and has been reduced in extent by changes in agricultural practice and myxomatosis as well as by quarrying of the bedrock. It is now largely confined to a few more intractable natural slopes and some artificial habitats such as abandoned quarries and road verges.

The climatic conditions in the region are intermediate between those of the lowland south and east and the north-west uplands. Even over the western Magnesian Limestone scarp, where most of the stands of the community occur and where rainfall is a little higher than over the plateau, the effect of the Pennine rainshadow is felt and the climate is as dry as or drier than much of the southern Chalk with an annual precipitation of only 700–750 mm and about 120 wet days yr^{-1} (Manley 1935, *Climatological Atlas* 1952, Ratcliffe 1968, Smith 1970). However, though the winters are relatively mild, the summers are cool with mean annual maximum temperatures around 26 °C (Conolly & Dahl 1970) and annual accumulated temperatures typical of northern lowland areas above a line from the Humber to the Severn. The climate is also considerably cloudier than to the south with 1–2 hours less bright summer sunshine and occasional inland penetration of the distinctive north-east coastal hahn markedly depressing air temperatures and insolation into June (Manley 1935, Smith 1970, Chandler & Gregory 1976). Compared with the Carboniferous Limestone uplands of western Durham, however, the growing season starts up to one month earlier and lasts up to 50 days longer (Smith 1976).

The general floristics of the community reflect these climatic features. Only on some north- and west-facing slopes, where winter frosts and snow-lie are a little more frequent and sunshine reduced, does the vegetation begin to approach the sub-montane character of the *Sesleria-Galium* grassland of the Pennine uplands with the appearance of such plants as *Carex pulicaris*, *Selaginella selaginoides*, *Antennaria dioica* and *Hylocomium splendens*. Conversely, it is only on warmer and sunnier, south-facing slopes that *Bromus erectus* makes any prominent contribution, and all other members of that Continental element which is so conspicuous in much of the south-eastern *Festuca-Avenula* grassland are quite absent. Rather, the bulk of the frequent members of the community are species which extend in Mesobromion grassland well beyond the Humber–Severn line but which do not penetrate far into the cooler, damper climate of the upland north-west. In many respects, this vegetation looks like the *Dicranum* sub-community of the *Festuca-Avenula* grassland with an overlying dominance of *S. albicans*.

Within the region, the *Sesleria-Scabiosa* grassland is largely restricted to those few areas of the Magnesian Limestone which are free of drift (Beaumont 1970, Stevens & Atkinson 1970). Although it is found at a few isolated sites over the east Durham plateau, it is most characteristic of the steeper slopes along the western scarp, usually between 45 and 160 m altitude. Here, the soils are kept permanently immature and the profiles are typical rendzinas, often very shallow, free-draining, rich in calcium and magnesium carbonates and with a pH generally above 7 (Frisby 1961, Shimwell 1968a, Stevens & Atkinson 1970). Such soils have been described from the area within the Cornforth and Middleham series (McKee 1965). Comparable lithomorphic profiles can be found over long-abandoned rock waste such as quarry spoil.

Both the *Carlina-Hypericum* and *Avenula* sub-communities occur over these rendzina soils on natural slopes and in older artificial habitats, their good representation of calcicoles and lack of Nardo-Galion species reflecting the maintenance of high base-status and lack of leaching in the dry climate. Where the profiles remain base-rich, the *Sesleria-Scabiosa* grassland will also extend some way on to deeper, and perhaps more mesotrophic, brown calcareous earths. Such soils develop where there is an accumulation of colluvium towards the bottom of slopes, over more marly or friable strata of limestone which weather more deeply and over more light-textured and calcareous till (Stevens & Atkinson 1970), and they have been mapped in Durham as part of the Aberford Series (*Soil Survey* 1983).

It is likely, though, that variations in grazing also play some part in the floristic differences between these sub-communities. The *Sesleria-Scabiosa* grassland is a plagioclimax vegetation, derived ultimately from the clearance of woodland which, though perhaps early on parts of the better-drained Magnesian Limestone (e.g. Bartley *et al.* 1976), seems not to have occurred extensively over much of eastern Durham until Romano-British times and to have experienced a number of regressions after that (Roberts *et al.* 1973, Donaldson & Turner 1977, Turner & Hodgson 1979). More recently, the typical pattern of agriculture over the Magnesian Limestone has been a mixed arable/pasture economy with deeper, drift-derived soils being ploughed for cereals, roots and leys and stands of *Sesleria-Scabiosa* grassland providing additional grazing for sheep, dairy cattle and stores (Warwick Percy 1970). There is no doubt that, over the gentler slopes around plateau knolls and above and below the western scarp, there has been some loss of the community with conversion of land to arable or sown grasslands and improvement with artificial fertilisers. However, judicious grazing has been an essential element in the maintenance of the short, species-rich swards of the community and the effects of

pastoral neglect are clearly visible in places with the encroachment of scrub. As over the southern Chalk, the demise of rabbits in the 1954/5 myxomatosis epidemic has played an important part here in the loss of the more close-cropped and varied grassland. It is possible that the *Avenula* sub-community represents, at least in part, a stage in the progression through ranker, tussocky swards to scrub with such relaxation in grazing pressure. Perhaps, too, differences in grazing stock have been involved because this kind of *Sesleria-Scabiosa* grassland shows structural and floristic features which are very characteristic of grazing by cattle as opposed to sheep. Some stands may also have been derived by the recolonisation of abandoned arable land.

As well as losses to agriculture, there has been some destruction of the habitat of the community with quarrying of the Magnesian Limestone. Prior to 1800, this was extracted locally as a building material but now the deposits are extensively exploited for high-grade dolomite, much prized as a flux in steel- and glass-making and for refractory products, and for aggregates. In addition to the irretrievable loss of stands of the *Sesleria-Scabiosa* grassland and the gross alteration of the scarp landscape of which it forms a part, the quarrying also deposits limestone dust over the surrounding vegetation. Abandoned quarry floors and faces and spoil heaps provide some compensation for these activities, creating habitats which are colonised, first by bryophytes and ephemerals, then by grassland, scrub and woodland. The *Hieracium* sub-community, typically found in such situations, clearly represents a stage in this process of recolonisation and, with its sometimes spectacular populations of orchids, may have much floristic interest. Stands of this kind of *Sesleria-Scabiosa* grassland and more long-established areas of the other sub-communities are, however, vulnerable to tipping of rubbish and industrial waste and to renewed mineral extraction as deposits elsewhere are worked out (Woodward 1970). The rocky verges of roads cut through the Magnesian Limestone, like sections of the A1(M), provide another man-made habitat for the development of the community.

Many of the often small stands of the *Sesleria-Scabiosa* grassland occur close to settlements and their use for recreation results in trampling, disturbance and eutrophication of the vegetation.

Zonation and succession

Since most remaining tracts of the community occur within enclosures sandwiched between plateau and vale agricultural land, natural zonations to other vegetation types developed in relation to soil sequences over the scarp are rare. In a few places, however, the *Carlina-Hypericum* sub-community can be seen grading through the *Avenula* sub-community to rank, mesotrophic swards over slope-foot colluvium.

More often, however, the community occurs in mosaics with open vegetation, rough grasslands and scrub which reflect seral progressions mediated by soil development and the amount of grazing. Colonisation of artificial habitats, such as quarry floors, ledges and spoil begins with the appearance of rank weeds such as *Epilobium angustifolium*, *Hypericum* spp. and *Erigeron acer* and, in damper places, patches of acrocarpous mosses (Shimwell 1968a, 1971b) and proceeds to the more stable vegetation of the *Hieracium* sub-community. Where grazing is absent, this can probably be quickly invaded by shrubs and trees but, where stock or rabbits have access and there is deeper accumulation of soil, it is possible that it develops into the *Carlina-Hypericum* sub-community. With light grazing, the tussocky swards of the *Avenula* sub-community may supervene and develop into patchy scrub with *Avenula pubescens* grassland or *Arrhenatheretum* between. The commonest scrub species in the area are *Crataegus monogyna*, *Prunus spinosa*, *Ulex europaeus* and *Rosa* spp. and the natural progression would be to the *Geranium* sub-community of the *Fraxinus-Acer-Mercurialis* woodland, stands of which occur in close association with *Sesleria-Scabiosa* grassland. In damper situations, *Salix* spp. may appear as early woody colonisers.

Distribution

The community occurs only in eastern Durham where the most extensive and varied tracts are found around Cassop Vale and Thrislington Plantation with smaller stands elsewhere along the western scarp of the Magnesian Limestone and at a few plateau localities. Abandoned quarries and road verges provide valuable secondary habitats.

Affinities

Essentially, the *Sesleria-Scabiosa* grassland represents a northern extension of the less Continental kind of Mesobromion grassland which is common in Derbyshire, the Yorkshire Wolds and the North York Moors into the peculiarly restricted British range of *S. albicans*. The affinities with the mainstream *Festuca-Avenula* grassland of the southern lowlands are much stronger than they are in the *Sesleria-Galium* grassland which is a more obvious transition to the montane vegetation of the Elyno-Seslerietea. Nonetheless, the most obvious position of the community is alongside the grassland in the Seslerio-Mesobromion sub-alliance (Shimwell 1968a, 1971b). Similar vegetation types have been described from France and Germany (Tüxen 1937, Schubert 1963, Stott 1970).

The core of the community is represented by the *Carlina-Hypericum* sub-community and the floristic trends seen within the other kinds of *Sesleria-Scabiosa* grassland reflect lines of variation seen among its more southerly equivalents. The *Hieracium* sub-community is a parallel vegetation type to the *Festuca-Hieracium-Thymus* grassland of very dry soils with heavy grazing and the *Avenula* sub-community equates with the more mesotrophic kinds of *Festuca-Avenula* grassland included in the *Holcus-Trifolium* sub-community.

Floristic table CG8

	a	b	c	8
Sesleria albicans	V (3–8)	V (1–8)	V (3–7)	V (1–8)
Carex flacca	V (1–7)	V (1–5)	V (1–5)	V (1–7)
Linum catharticum	V (1–4)	V (1–3)	V (1–3)	V (1–4)
Sanguisorba minor	V (1–5)	V (1–5)	V (1–3)	V (1–5)
Scabiosa columbaria	V (1–4)	IV (1–3)	IV (1–3)	IV (1–4)
Thymus praecox	IV (1–5)	V (1–5)	V (1–3)	IV (1–5)
Helianthemum nummularium	IV (1–5)	V (1–7)	IV (1–3)	IV (1–7)
Plantago lanceolata	IV (1–4)	V (1–3)	IV (1–3)	IV (1–4)
Avenula pratensis	IV (1–4)	IV (1–3)	V (1–5)	IV (1–5)
Lotus corniculatus	IV (1–4)	IV (1–3)	III (1–3)	IV (1–4)
Galium verum	IV (1–3)	III (1–3)	IV (1–3)	IV (1–3)
Koeleria macrantha	IV (1–3)	III (1–3)	V (1–3)	IV (1–3)
Festuca ovina	IV (1–7)	II (1–3)	IV (1–3)	IV (1–7)
Centaurea nigra	IV (1–3)	V (1–3)	I (1)	IV (1–3)
Briza media	IV (1–7)	IV (1–3)	II (1–3)	IV (1–7)
Pimpinella saxifraga	IV (1–2)	V (1–3)	II (1)	IV (1–3)

Stachys betonica	III (1–4)	II (1–3)		II (1–4)
Taraxacum officinale agg.	II (1)	III (1–3)		II (1–3)
Agrostis capillaris	II (1–3)	II (1–3)		II (1–3)
Cynosurus cristatus	II (1–4)	II (1–3)		II (1–4)
Rhytidiadelphus squarrosus	II (1)	I (1)		I (1)
Mnium hornum	I (1)	I (1)		I (1)
Linum perenne anglicum	I (1–6)	I (1)		I (1–6)
Viola riviniana	I (1)	I (1)		I (1)
Campylium chrysophyllum	I (1)	I (1)		I (1)
Hypericum pulchrum	III (1–3)	I (1)		II (1–3)
Carlina vulgaris	III (1–2)	I (1)		II (1–2)
Polygala vulgaris	III (1–2)			II (1–2)
Epipactis atrorubens	II (1–3)	I (1)		II (1–3)
Succisa pratensis	II (1–4)		I (1)	II (1–4)
Eurhynchium swartzii	II (1)	I (1)	I (1)	I (1)
Trisetum flavescens	II (1–4)			I (1–4)
Achillea millefolium	II (1)			I (1)
Pseudoscleropodium purum	II (1–3)			I (1–3)
Carex pulicaris	II (2–5)			I (2–5)
Danthonia decumbens	I (1–2)			I (1–2)
Bellis perennis	I (1–2)			I (1–2)
Calliergon cuspidatum	I (1–6)			I (1–6)
Trifolium medium	I (1–3)			I (1–3)
Plagiomnium undulatum	I (1–2)			I (1–2)
Vicia cracca	I (1–3)			I (1–3)
Knautia arvensis	I (1–2)			I (1–2)
Avenula pubescens	II (1–4)	V (1–5)	I (1)	III (1–5)
Daucus carota	I (1)	V (1)	I (1)	II (1)
Dactylis glomerata	II (1–4)	V (1–5)	II (1)	II (1–5)
Trifolium repens	II (1)	IV (1–3)	II (1)	II (1–3)
Senecio jacobaea	I (1)	II (1–3)	I (1)	I (1–3)
Ulex europaeus	I (3)	II (1)	I (1)	I (1–3)
Bromus erectus	I (4–6)	II (1–8)		I (1–8)
Arrhenatherum elatius		II (1–3)	I (1)	I (1–3)
Rhinanthus minor		I (1)		I (1)
Lathyrus pratensis		I (1)		I (1)
Cirsium vulgare		I (1)		I (1)
Hieracium pilosella	II (1–4)	II (1–3)	V (1–3)	II (1–4)
Hypochoeris radicata	II (1–2)		V (1–3)	II (1–3)
Hypericum montanum			V (1–3)	I (1–3)
Encalypta vulgaris			V (1–5)	I (1–5)
Prunella vulgaris		III (1–3)	IV (1–3)	II (1–3)
Medicago lupulina	II (1–3)	II (1–3)	IV (1–3)	II (1–3)
Plantago maritima	I (1–2)	I (1)	IV (1–3)	I (1–3)
Listera ovata	I (1–3)		IV (1)	I (1–3)
Hieracium sect. Vulgata	I (1–4)	I (1)	III (1)	I (1–4)
Coeloglossum viride	I (1)		III (1)	I (1)
Encalypta streptocarpa	I (1)		II (1–5)	I (1–5)
Bryum argenteum			II (1)	I (1)

Floristic table CG8 (*cont.*)

	a	b	c	8
Leontodon autumnalis			II (1)	I (1)
Ceratodon purpureus			II (1–3)	I (1–3)
Campanula rotundifolia	IV (1–3)	III (1)	II (1–3)	III (1–3)
Primula veris	IV (1–3)	III (1–3)	II (1)	III (1–3)
Leontodon hispidus	IV (1–4)	III (1)	II (1–5)	III (1–5)
Carex caryophyllea	III (1–3)	IV (1–3)	V (1–3)	III (1–3)
Viola hirta	III (1–3)	IV (1)	V (1–3)	III (1–3)
Anthyllis vulneraria	III (1–5)	IV (1–5)	IV (1–3)	III (1–5)
Centaurea scabiosa	III (1–3)	IV (1–3)	IV (1–3)	III (1–3)
Plantago media	III (1–5)	III (1–3)	III (1–3)	III (1–5).
Gentianella amarella	III (1–3)	III (1–3)	V (1–3)	III (1–3)
Ctenidium molluscum	III (1–4)	I (1)	IV (1–3)	II (1–4)
Gymnadenia conopsea	III (1–3)	I (1)	V (1–3)	II (1–3)
Euphrasia officinalis agg.	III (1–3)	III (1–3)	II (1–3)	III (1–3)
Brachypodium sylvaticum	II (1–4)	III (1–3)	II (1–3)	II (1–4)
Fissidens cristatus	III (1–3)	I (1)	III (1)	II (1–3)
Trifolium pratense	I (1–2)	II (1)	III (1)	II (1–2)
Cerastium fontanum	II (1–3)	II (1)	II (1)	II (1–3)
Festuca rubra	II (1–6)	II (1–3)	I (1)	II (1–6)
Hypnum cupressiforme	II (1–3)	I (1)	II (1–3)	II (1–3)
Number of samples	35	12	8	55
Number of species/sample	32 (25–40)	29 (26–32)	32 (27–38)	31 (25–40

a   *Hypericum pulchrum-Carlina vulgaris* sub-community
b   *Avenula pubescens* sub-community
c   *Hieracium pilosella* sub-community
8   *Sesleria albicans-Scabiosa columbaria* grassland (total)

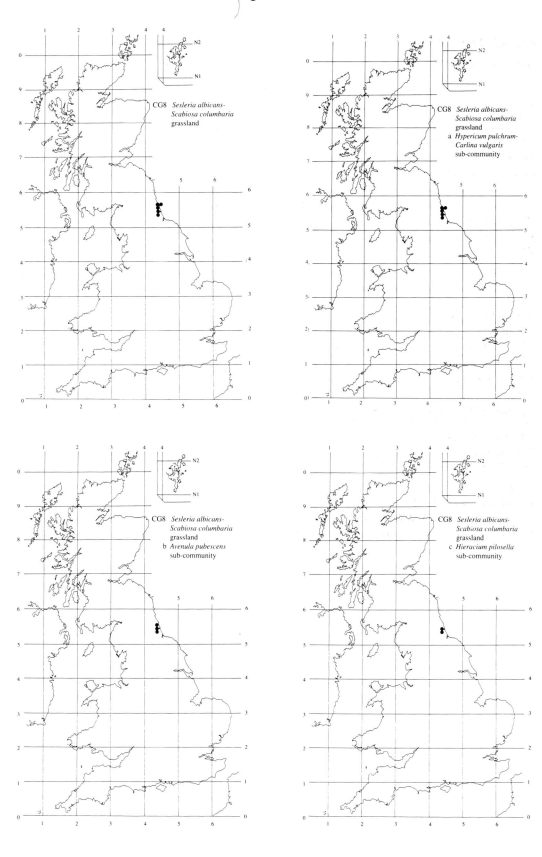

CG9
Sesleria albicans-Galium sterneri grassland

Synonymy
Limestone hill pasture   Smith & Rankin 1903; *Festucetum ovinae* Moss 1911 *p.p.*; Limestone grassland Tansley 1939 *p.p.*; Carboniferous Limestone grassland Pigott 1956, Ratcliffe 1977 *p.p.*; Sugar Limestone grassland Pigott 1956; Limestone Grassland E Sinker 1960; *Potentilla fruticosa* localities   Elkington & Woodell 1963 *p.p.*; *Seslerieto-Festucetum*   Ratcliffe 1965; 'Ordinary' Limestone grassland   Ratcliffe 1965; *Seslerio-Caricetum pulicariae*   Shimwell 1968a; *Asperulo-Seslerietum typicum, Helianthemum* variant   Shimwell 1968a; *Potentilletum fruticosae*   Shimwell 1968a *p.p.*; *Agrosto-Festucetum, Sesleria* facies   Eddy *et al.* 1969; *Dryas octopetala* localities   Elkington 1971 *p.p.*; *Festuca ovina-Sesleria albicans* vegetation   Jeffrey & Pigott 1973; *Seslerio-Caricetum pulicariae*   Shimwell 1968a emend. Jones 1973; *Thymo-Festucetum sesleriosum* Evans *et al.* 1977; *Thymo-Agrosto-Festucetum*   Evans *et al.* 1977 *p.p.*; *Sesleria albicans* grassland   Ratcliffe 1977; Dry sugar-limestone grassland   Ratcliffe 1978.

Constant species
Briza media, Campanula rotundifolia, Carex flacca, Festuca ovina, Galium sterneri, Helianthemum nummularium, Koeleria macrantha, Linum catharticum, Sesleria albicans, Thymus praecox, Viola riviniana, Ctenidium molluscum.

Rare species
Alchemilla glaucescens, A. minima, Aster linosyris, Carex capillaris, C. ericetorum, Cypripedium calceolus, Draba incana, Dryas octopetala, Galium sterneri, Gentiana verna, Helianthemum canum, Hypochoeris maculata, Kobresia simpliciuscula, Minuartia verna, Myosotis alpestris, Polygala amara, Potentilla crantzii, P. fruticosa, P. tabernaemontani, Primula farinosa, Sesleria albicans, Thalictrum alpinum, Veronica spicata, Viola rupestris, Tortella densa.

Physiognomy
The *Sesleria albicans-Galium sterneri* grassland can occur as closed or open swards or, in very rocky situations, as on stabilised talus and small rock ledges or over pavement clints, be reduced to fragmentary assemblages, in which one or more of the major species of the community attain local prominence. The vegetation can be short, but a few centimetres high, with an even close-cropped appearance when grazed, or be taller and decidedly tussocky. *Sesleria albicans* is the most frequent grass and, though it can vary in abundance from dominant to sparse, it is frequently conspicuous with its tufts of rather stiff, glaucous leaves, choked below by persistent sheaths. When growing vigorously and ungrazed, plants can attain a considerable size, spreading laterally by long creeping rhizomes. Then the species behaves more like a winter-green chamaephyte, though its slowly-growing leaves are often masked by dead material which can give a brown tinge to the sward in winter (Dixon 1982).

Koeleria macrantha and *Festuca ovina* are also constant components of the vegetation and, in some sub-communities, they can be abundant, occasionally attaining co-dominance with *S. albicans*. *F. rubra* is generally uncommon here and rarely prominent, a marked contrast to the *Festuca-Agrostis-Thymus* grassland. *Briza media* and *Avenula pratensis* are also frequent but somewhat unevenly represented throughout and generally not abundant. There is occasionally a little *Danthonia decumbens* and, in some sub-communities, *Agrostis capillaris* and *Deschampsia cespitosa* are common though rarely prominent. There are almost always some sedges in the sward. *Carex flacca* and *C. caryophyllea* are most frequent overall and the former especially can be abundant. In certain kinds of *Sesleria-Galium* grassland, *C. pulicaris, C. panicea* and (much less commonly) *C. capillaris* and another cyperaceous species, *Kobresia simpliciuscula*, also attain prominence.

The woody chamaephytes *Thymus praecox* and *Helianthemum nummularium* are both constant and, though the latter is somewhat patchy in its occurrence, each can be locally abundant as prostrate or low sub-shrubs. In contrast to many lowland calcicolous grasslands, however, *Hieracium pilosella* is generally infrequent. Another mat-former and a very characteris-

withdrawn. Succession has never been followed in detail but the most likely invaders would seem to be *Crataegus monogyna*, *Prunus spinosa*, *Corylus avellana*, *Sorbus aucuparia*, *Fraxinus excelsior* and *Taxus baccata* with, on drier sites, *Betula pendula* and, over more mesotrophic soils, *Acer pseudoplatanus* and *Ulmus glabra*. Mixtures of these species commonly occur as dense thickets on scree slopes which are ungrazed and they can expand out from grikes on inaccessible stretches of pavement. On drier and warmer slopes, the succession can be to *Fraxinus-Acer-Mercurialis* woodland but *Acer campestre* itself, and two common invaders of grasslands on Carboniferous Limestone further south, *Cornus sanguinea* and *Rhamnus catharticus*, are very much towards their northern limit in this area and are only sparsely seen in developing or mature woodlands. On moister and cooler slopes, the *Fraxinus-Sorbus-Mercurialis* woodland, the climax community on base-rich soils in the north and west of Britain, may be a more likely development.

Distribution

The community is confined to the Carboniferous Limestone of the Morecambe Bay area, to the Craven district of North Yorkshire and to the borders of Cumbria, Durham and North Yorkshire around Upper Teesdale. The sub-communities show a fairly marked distribution pattern within this range. The *Helianthemum-Asperula* sub-community is confined to a few cliffs in the western lowlands on Scout, Underbarrow and Whitbarrow Scars, on Humphrey Head and around Silverdale, though stands approaching it in composition occur on south-facing slopes in Craven, as for example above Grassington. The Typical sub-community is the most widely distributed and extensive type and it also occurs on the Morecambe Bay exposures though it is especially characteristic of the Craven district, where large stands are common between 250 and 500 m on the Great Scar Limestone of the southern dales. The *Carex pulicaris-Carex panicea* sub-community is more restricted: it occurs around the Malham area in Craven, on exposures around the headwaters of the Lancashire Lune and in Upper Teesdale. The *Carex-Kobresia* sub-community is confined to the sugar-limestone in the last area, where some stands have been lost with the flooding of the Cow Green reservoir. The *Saxifraga-Cochlearia* community also occurs in Upper Teesdale and at scattered localities over the fells to the immediate west.

Affinities

British grasslands in which *S. albicans* plays a prominent role occupy a floristic and geographical position which is intermediate between the lowland southern Mesobromion swards and those upland communities of the north and west where montane calcicoles replace those with Continental affinities, and Nardo-Galion species become a constant accompaniment to calcicoles of broader ecological amplitude. Just as the *Sesleria-Scabiosa* grassland can be seen as a close relative of the *Festuca-Avenula* grassland, so the *Sesleria-Galium* grassland can be seen as having strong affinities with the *Festuca-Agrostis-Thymus* grassland, though its particular floristic features amply justify its treatment as a distinct vegetation type.

In this country, therefore, *S. albicans* has its major locus in communities which are transitional to the high mountain vegetation generally placed in the Elyno-Seslerietea (Braun-Blanquet 1948, Ellenberg 1978) or some modification of this class (like the Seslerietea of Oberdorfer 1978). Intermediate swards of this kind have sometimes been accommodated within a special subdivision of the Mesobromion, such as the Seslerio-Mesobromion (Oberdorfer 1957, Shimwell 1968a, 1971a). On such a view, the core of this community is represented by the Typical sub-community and similar vegetation to this has been described from both Germany and France (e.g. Tüxen 1937, Schubert 1963, Stott 1970).

From such a core, the community shows a number of floristic trends within the other sub-communities. That of *Helianthemum-Asperula* has close affinities with drier *S. albicans* grasslands of common occurrence in the Continental lowlands of northern France and sometimes placed in a division of the Xerobromion, the Seslerio-Xerobromion (Allorge 1921–2, Oberdorfer 1957, Braun-Blanquet & Braun-Blanquet 1971). It is also similar to some less xeric vegetation of the Burren Carboniferous Limestone (e.g. Braun-Blanquet & Tüxen 1952, Ivimey-Cook & Proctor 1966a) and was treated as a sub-association of an Irish community, the *Asperulo-Seslerietum*, seen as having a foothold in the more equable parts of the Pennines, by Shimwell (1968a).

Floristic transitions to the mires of the Caricion davallianae, such as are seen in the *Carex pulicaris-Carex panicea* sub-community, have also been described from the Continent (e.g. Bresinsky 1965), though much more common is the montane and sub-alpine *S. albicans* vegetation, to which the *Carex-Kobresia* and *Saxifraga-Cochlearia* sub-communities are the nearest approach in Britain (e.g. Braun-Blanquet & Tüxen 1952, Pignatti & Pignatti 1975). However, the particular mixtures of Northern Montane and Arctic-Alpine species found at Upper Teesdale in the former sub-community, appear to be unique.

Floristic table CG9

	a	b	c	d	e	9
Sesleria albicans	V (1–5)	V (3–9)	V (1–9)	V (3–7)	V (1–7)	V (1–9)
Thymus praecox	V (1–5)	V (2–7)	V (1–5)	V (1–4)	V (1–5)	V (1–7)
Koeleria macrantha	IV (1–5)	IV (1–5)	V (1–5)	IV (1–5)	IV (1–3)	IV (1–5)
Galium sterneri	IV (1–3)	IV (1–4)	IV (1–3)	IV (1–3)	V (1–3)	IV (1–4)
Campanula rotundifolia	IV (1–3)	V (1–5)	IV (1–3)	V (1–3)	III (1–3)	IV (1–5)
Linum catharticum	IV (1–3)	V (1–4)	IV (1–3)	V (1–3)	III (1)	IV (1–4)
Festuca ovina	IV (1–3)	V (1–9)	IV (1–7)	V (1–7)	V (3–7)	IV (1–9)
Ctenidium molluscum	V (1–3)	III (1–3)	IV (1–5)	IV (1–3)	III (1–3)	IV (1–5)
Viola riviniana	III (1)	III (1–4)	IV (1–3)	IV (1–4)	V (1–3)	IV (1–4)
Briza media	I (2)	IV (1–4)	IV (1–3)	V (2–5)	III (1–3)	IV (1–5)
Carex flacca	IV (1–5)	V (2–6)	IV (1–5)	III (1–3)	III (1–3)	IV (1–6)
Helianthemum nummularium	IV (1–7)	II (1–5)	V (1–3)	IV (1–4)	III (1–3)	IV (1–7)
Leontodon hispidus	IV (1–3)	I (1–4)	II (1–5)			I (1–5)
Homalothecium lutescens	IV (1–3)	II (1–3)	I (1)			I (1–3)
Encalypta streptocarpa	IV (1–3)		I (1)	I (3)		I (1–3)
Asperula cynanchica	IV (3–5)	I (1)				I (1–5)
Helianthemum canum	IV (1–3)			I (3)		I (1–3)
Hippocrepis comosa	III (1–3)	I (1)	I (1)	I (2)		I (1–3)
Scabiosa columbaria	III (1–3)		I (1)			I (1–3)
Carlina vulgaris	III (1–3)	I (1–3)	I (1)		I (1)	I (1–3)
Weissia sp.	III (1)					I (1)
Viola hirta	II (1–3)	II (1–3)		II (1)		I (1–3)
Anthyllis vulneraria	I (1)				I (1)	I (1)
Sedum acre	I (1–2)					I (1–2)
Potentilla erecta	II (1)	II (1–3)	IV (1–5)	IV (1–4)	V (1–3)	III (1–5)
Cornicularia aculeata			IV (1–3)	V (1–3)	IV (1–3)	III (1–3)
Carex pulicaris			V (1–5)	II (1–3)	V (1–5)	III (1–5)
Carex panicea			V (1–3)	V (1–5)	II (1–3)	II (1–5)
Hylocomium splendens			IV (1–5)	III (1–2)	III (1–3)	II (1–5)
Frullania tamarisci	I (1)		II (1)	III (1–3)	III (1)	II (1–3)
Polygala vulgaris			III (1–3)	I (1)	II (1)	I (1–3)
		I (1–3)	II (1–3)		I (3)	I (1–3)

Species					
Parnassia palustris		II (1–3)		I (1)	I (1–3)
Primula farinosa		II (1–3)	I (1)		I (1–3)
Dryas octopetala		II (7–9)	I (3)		I (3–9)
Molinia caerulea		II (1–3)			I (1–3)
Pinguicula vulgaris		II (1–3)			I (1–3)
Carex hostiana		II (1–5)			I (1–5)
Pimpinella saxifraga		II (1–3)			I (1–3)
Calluna vulgaris		I (3–5)			I (3–5)
Empetrum nigrum		I (1–5)			I (1–5)
Rhodobryum roseum		I (1)			I (1)
Thalictrum minus		I (3)			I (3)
Racomitrium lanuginosum		II (1–3)	V (1–5)	II (3–5)	II (1–5)
Gentiana verna		I (1)	IV (1–3)	I (1)	I (1–3)
Carex capillaris		I (1–3)	IV (2–5)		I (1–5)
Kobresia simpliciuscula			IV (1–8)		I (1–8)
Cetraria islandica			IV (1–2)		I (1–2)
Viola rupestris		I (1–3)	III (1–4)		I (1–4)
Selaginella selaginoides		I (1–3)	III (1–4)		I (1–4)
Polygonum viviparum		I (3)	III (1–3)	I (1)	I (1–3)
Antennaria dioica			III (1–3)		I (1–3)
Plantago maritima	I (1)	I (1)	II (1–2)		I (1–2)
Saxifraga hypnoides	IV (1–4)	I (1)		IV (1–5)	I (1–5)
Cochlearia alpina	III (1–4)			IV (1–3)	I (1–3)
Euphrasia confusa	III (1–3)	II (1–3)	I (1)	IV (1–3)	I (1–3)
Draba incana				III (1–3)	I (1–3)
Plagiochila asplenoides		I (1)		III (1–3)	I (1–3)
Tritomaria quinquedentata		I (1)		III (1)	I (1)
Myosotis alpestris				III (1–3)	I (1–3)
Veronica officinalis				III (1)	I (1)
Polytrichum juniperinum				III (1)	I (1)
Barbilophozia barbata		I (1)	I (1)	II (1)	I (1)
Rhytidiadelphus triquetrus			I (1)	II (1)	I (1)
Blepharostoma trichophyllum			I (1)	II (1)	I (1)
Carex caryophyllea	II (1)	III (1–3)	III (1–3)	V (1–5)	III (1–5)
Avenula pratensis	III (1–3)	IV (1–5)	II (1–4)	IV (1–3)	III (1–5)
Lotus corniculatus	III (1–3)	IV (1–3)	IV (1–3)	III (1–3)	III (1–3)

Floristic table CG9 (*cont.*)

	a	b	c	d	e	9
Tortella tortuosa	IV (1-3)	III (1-4)	IV (1-5)	III (1-5)	III (1-3)	III (1-5)
Hypnum cupressiforme	II (1)	IV (1-5)	II (1-3)	IV (1-3)	V (1-2)	III (1-5)
Euphrasia officinalis agg.	III (1-3)	III (1-3)	IV (1-3)	IV (1-3)	II (1)	III (1-3)
Ditrichum flexicaule	IV (1-3)	I (1)	I (1-3)	IV (1-3)	II (1-3)	II (1-3)
Dicranum scoparium	IV (1-3)	IV (1-5)	II (1-3)	I (1)	II (1)	II (1-5)
Plantago lanceolata	III (1-3)	II (1-4)	III (1-3)	II (1-3)	IV (1-3)	II (1-4)
Pseudoscleropodium purum	III (1)	I (1-4)	IV (1-3)	I (1-3)	II (1)	II (1-4)
Fissidens cristatus	III (1)	II (1-3)	III (1-3)	II (1)	I (1)	II (1-3)
Danthonia decumbens	II (1-3)	II (1-6)	I (1-5)	II (1-3)	I (1)	II (1-6)
Sanguisorba minor	II (1-3)	II (1-4)	III (1-3)			II (1-4)
Hieracium pilosella	I (1-2)	III (1-4)	I (1-3)	II (1-3)		II (1-4)
Racomitrium canescens		II (1-3)	I (1)	III (1-4)		II (1-3)
Prunella vulgaris		II (1-3)	II (1-3)		III (1)	I (1-3)
Cladonia rangiformis		I (1-3)	I (1)	I (3)	III (1-3)	I (1-3)
Gentianella amarella	II (1)	I (1)	II (1)	II (1-3)	III (1)	I (1-3)
Trifolium repens	I (1)	II (1-4)	II (1-3)	I (1)	III (1-3)	I (1-4)
Agrostis capillaris		I (1-3)	III (1-3)	I (1)	III (1-3)	I (1-3)
Festuca rubra		II (1-9)			III (1-3)	I (1-9)
Deschampsia cespitosa			II (1-3)		III (1-3)	I (1-3)
Succisa pratensis	III (1-3)		II (1-3)	II (1-2)	II (1)	I (1-3)
Cladonia pocillum		II (1-5)	I (1-3)	II (1-3)	II (1-3)	I (1-5)
Achillea millefolium	II (1)	I (3-4)		I (1)	III (1)	I (1-4)
Scapania aspera	II (1)			II (1-3)	II (1)	I (1-3)
Minuartia verna		I (1-3)	II (1-3)	II (1-3)	II (1-3)	I (1-3)
Neckera crispa		I (1-2)	I (1)		II (1)	I (1-2)
Bellis perennis		I (3-4)	II (1)		I (1)	I (1-4)
Cerastium fontanum		I (1-3)	I (1-3)	I (1-3)	II (1-3)	I (1-3)
Rhytidiadelphus squarrosus		II (1-4)	II (1-7)		II (7)	I (1-7)
Lophocolea bidentata s.l.			II (1)	I (1)	I (1)	I (1)
Taraxacum officinale agg.	I (1)	II (1-3)	I (1)		I (1)	I (1-3)
Galium verum	I (3)	I (1-4)	II (1-3)			I (1-4)
Thuidium tamariscinum			I (1-3)	I (1)	II (1)	I (1-3)

	a	b	c	d	e	9
Epipactis atrorubens	I (1)		I (1)			I (1)
Hypericum pulchrum	I (1)		I (1)			I (1)
Dactylis glomerata		I (3)	I (1)			I (1-3)
Carex ericetorum	I (1-3)		I (1)			I (1-3)
Cynosurus cristatus		I (2)	I (1-3)			I (1-3)
Potentilla tabernaemontani			I (1)		I (1)	I (1)
Brachypodium sylvaticum		I (1-3)	I (3)			I (1-3)
Rhytidium rugosum			I (1-5)	I (3)		I (1-5)
Cirsium vulgare			I (1)		II (1)	I (1)
Number of samples	14	54	28	19	9	124
Number of species/sample	28 (24-32)	19 (13-31)	32 (22-48)	31 (20-42)	37 (34-42)	26 (13-48)

a *Helianthemum canum-Asperula cynanchica* sub-community

b Typical sub-community

c *Carex pulicaris-Carex panicea* sub-community

d *Carex capillaris-Kobresia simpliciuscula* sub-community

e *Saxifraga hypnoides-Cochlearia alpina* sub-community

9 *Sesleria albicans-Galium sterneri* grassland (total)

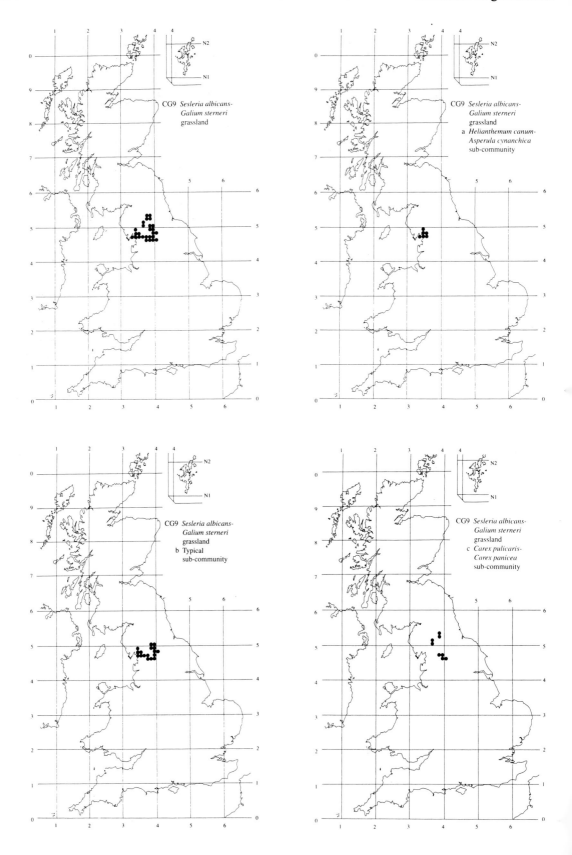

CG9 *Sesleria albicans-Galium sterneri* grassland

CG9 *Sesleria albicans-Galium sterneri* grassland
a *Helianthemum canum-Asperula cynanchica* sub-community

CG9 *Sesleria albicans-Galium sterneri* grassland
b Typical sub-community

CG9 *Sesleria albicans-Galium sterneri* grassland
c *Carex pulicaris-Carex panicea* sub-community

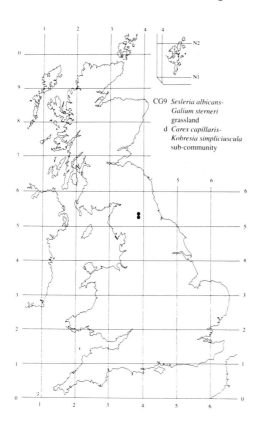

CG9 *Sesleria albicans-Galium sterneri* grassland
d *Carex capillaris-Kobresia simpliciuscula* sub-community

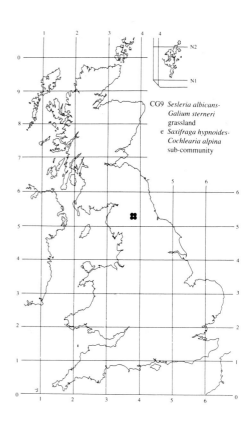

CG9 *Sesleria albicans-Galium sterneri* grassland
e *Saxifraga hypnoides-Cochlearia alpina* sub-community

CG10
Festuca ovina-Agrostis capillaris-Thymus praecox grassland

Synonymy
Agrostis-Festuca grassland *auct. angl. p.p.*; Basic grassland Ratcliffe 1959; Species-rich *Agrosto-Festucetum* McVean & Ratcliffe 1962; *Saxifrageto-Agrosto-Festucetum* McVean & Ratcliffe 1962; *Festuca ovina/Agrostis tenuis* Type D grassland King 1962; *Festuca-Agrostis* Type 9 grassland King & Nicholson 1964; *Festuco-Poetum* Shimwell 1968a *p.p.*; Herb-rich *Agrostis-Festuca* grassland Ward *et al.* 1972; *Festuco-Nardetum* Jones 1973 *p.p.*; *Achilleo-Festucetum tenuifoliae* Birse & Robertson 1976; *Thymo-Agrosto-Festucetum* Evans *et al.* 1977; *Trifolio-Agrosto-Festucetum* Evans *et al.* 1977 *p.p.*; *Thymo-Festucetum boreale* Evans *et al.* 1977 *p.p.*; *Viola-Festuca-Agrostis* nodum Huntley 1979 *p.p.*; *Galium sterneri-Helictotrichon pratense* community Birse 1980; *Galium verum-Koeleria cristata* community Birse 1980.

Constant species
Agrostis capillaris, Campanula rotundifolia, Festuca ovina, F. rubra, Plantago lanceolata, Potentilla erecta, Prunella vulgaris, Thymus praecox, Viola riviniana, Hylocomium splendens.

Rare species
Alchemilla filicaulis ssp. *filicaulis, A. wichurae, Carex capillaris, C. montana, C. rupestris, Draba incana, Galium sterneri, Minuartia verna, Myosotis alpestris, Omalotheca supina, Sagina saginoides, Salix herbacea, Sibbaldia procumbens, Tofieldia pusilla.*

Physiognomy
The *Festuca-Agrostis-Thymus* grassland occurs as generally closed swards, often close-cropped by heavy grazing, in which the most obvious distinguishing feature, compared with the *Sesleria albicans* grasslands and lowland calcicolous communities, is the consistent prominence of various Nardo-Galion species. Among the grasses, for example, *Agrostis capillaris* is as frequent

and abudant as *Festuca ovina* (sometimes recorded here as *F. tenuifolia* but rarely as *F. vivipara*) and these two species together often comprise the basis of the turf. *Festuca rubra* is also very frequent, though it is usually not so abundant as *F. ovina*. Common, too, but somewhat unevenly represented in the different sub-communities, are *Nardus stricta, Anthoxanthum odoratum* and *Danthonia decumbens* and each of these can also be locally prominent. Much more patchy and rarely at high cover are *Poa pratensis/subcaerulea* (inadequately distinguished in some data), in drier stands *Agrostis canina* ssp. *montana* and *Deschampsia flexuosa* and in damper sites *A. canina spp. canina* and *D. cespitosa*.

Some sedges are frequently present but, again, there is a shift in emphasis among these species as compared with lowland swards. *Carex flacca*, for example, is never more than occasional and *C. caryophyllea*, though frequent in one sub-community, is scarce in the others. As in other upland calcicolous grasslands, certain kinds of *Festuca-Agrostis-Thymus* sward are characterised by the presence, together and at high frequency, of *C. pulicaris* and *C. panicea*. *C. pilulifera* is occasional in drier stands, *C. binervis* in wetter and the Arctic-Alpine rarity *C. capillaris* is very occasionally found.

Thymus praecox is the commonest dicotyledonous associate and it is often abundant, but that other important chamaephyte of lowland calcicolous swards, *Helianthemum nummularium*, though it typically occurs in this vegetation at many of its upland, northern stations, is scarce in the community as a whole. Also constant are *Viola riviniana, Plantago lanceolata, Campanula rotundifolia* and *Prunella vulgaris* with, less frequently, *Ranunculus acris, Lotus corniculatus* and *Euphrasia officinalis* agg. (including *E. nemorosa, E. confusa, E. scottica* and, very rarely, the Arctic rarity, *E. frigida* where taxa have been distinguished). Among this component, too, Nardo-Galion species can be important. *Potentilla erecta* is a constant throughout and *Galium saxatile* and *Luzula campestris* (rarely *L. multiflora*) are frequent in

Selaginella selaginoides	I (1–2)	III (1–3)	V (1–3)	III (1–3)
Ctenidium molluscum	I (1–6)	III (1–4)	V (1–4)	III (1–6)
Saxifraga aizoides	I (1–3)	II (1–4)	IV (1–4)	II (1–4)
Bellis perennis	I (1–4)	II (1–4)	IV (1–4)	II (1–4)
Ditrichum flexicaule	I (1–3)	I (1)	IV (1–4)	II (1–4)
Tortella tortuosa	I (1–3)	I (1–3)	IV (1–3)	I (1–3)
Briza media	I (1–4)	II (1–4)	III (1–4)	I (1–4)
Polygonum viviparum	I (1–4)	II (1–4)	III (1–4)	I (1–4)
Saxifraga oppositifolia		II (1–3)	III (1–4)	I (1–4)
Pinguicula vulgaris		I (1)	III (1–3)	I (1–3)
Plagiochila asplenoides	I (1)	I (1)	II (1–3)	I (1–3)
Leontodon autumnalis	I (1–4)	I (1–3)	II (1–3)	I (1–4)
Racomitrium canescens	I (1–3)	I (1–3)	II (1–3)	I (1–3)
Antennaria dioica	I (1–3)	I (1–4)	II (1–3)	I (1–4)
Avenula pratensis	I (1–4)	I (1–4)	II (1–3)	I (1–4)
Carex demissa	I (1–3)	I (1–3)	II (1–4)	I (1–4)
Thalictrum alpinum		I (1–4)	II (1–6)	I (1–6)
Silene acaulis		I (1–4)	II (1–4)	I (1–4)
Nardus stricta	III (1–4)	III (1–7)	III (1–5)	III (1–7)
Ranunculus acris	III (1–4)	III (1–4)	III (1–3)	III (1–4)
Calluna vulgaris	III (1–4)	III (1–4)	II (1–3)	III (1–4)
Pseudoscleropodium purum	III (1–4)	III (1–3)	II (1–3)	III (1–4)
Lotus corniculatus	III (1–6)	III (1–5)	II (1–3)	III (1–6)
Danthonia decumbens	III (1–4)	II (1–6)	III (1–4)	III (1–6)
Rhytidiadelphus triquetrus	II (1–4)	II (1–4)	III (1–4)	II (1–4)
Euphrasia officinalis agg.	II (1–4)	II (1–3)	II (1–3)	II (1–4)
Deschampsia cespitosa	II (1–5)	I (1–8)	III (1–3)	II (1–8)
Hypnum cupressiforme	II (1–4)	II (1–3)	I (1–3)	II (1–4)
Hieracium spp.	I (1–3)	II (1)	II (1)	I (1–3)
Frullania tamarisci	I (1–2)	II (1–3)	II (1–3)	I (1–3)
Thuidium tamariscinum	II (1–6)	I (1–4)	I (1)	I (1–6)
Racomitrium lanuginosum	I (1)	II (1–4)	I (4)	I (1–4)
Thuidium delicatulum	I (1–2)	II (1–3)	I (1)	I (1–3)
Breutelia chrysocoma	I (1–4)	I (1–4)	I (1–3)	I (1–4)
Luzula multiflora	I (1–3)	I (1)	I (1–3)	I (1–3)
Vaccinium myrtillus	I (1–4)	I (1–4)	I (1–3)	I (1–4)
Rhytidiadelphus loreus	I (1–4)	I (1)	I (1)	I (1–4)
Vaccinium vitis-idaea	I (1)	I (1–4)	I (1)	I (1–4)
Ptilidium ciliare	I (1)	I (1)	I (1)	I (1)
Blechnum spicant	I (1)	I (1)	I (1)	I (1)
Deschampsia flexuosa	I (1–4)	I (1–3)	I (1–3)	I (1–4)
Alchemilla filicaulis vestita	I (1–3)	I (1–4)	I (5)	I (1–5)
Peltigera canina	I (1–4)	I (1–3)	I (1–3)	I (1–4)
Drepanocladus uncinatus	I (1–4)	I (1–3)	I (1–3)	I (1–4)
Galium boreale	I (1)	I (1–3)	I (1–4)	I (1–4)
Fissidens osmundoides	I (1–3)	I (1–3)	I (1–3)	I (1–3)
Alchemilla filicaulis filicaulis	I (1–3)	I (1–3)	I (1–3)	I (1–3)
Helianthemum nummularium	I (1–6)	I (4–7)	I (1–4)	I (1–7)
Botrychium lunaria	I (1–2)	I (1–3)	I (1)	I (1–3)
Fissidens adianthoides	I (1–3)	I (1–3)	I (1–3)	I (1–3)
Calliergon cuspidatum	I (1–3)	I (1–3)	I (1–3)	I (1–3)
Hypochoeris radicata	I (4)	I (1–3)	I (1–3)	I (1–4)

Floristic table CG10 (*cont.*)

	a	b	c	10
Pteridium aquilinum	I (1–4)	I (1–3)	I (1–3)	I (1–4)
Euphrasia scottica	I (1–3)	I (1–3)	I (1–3)	I (1–3)
Carex nigra	I (1)	I (1–4)	I (1–3)	I (1–4)
Bryum pseudotriquetrum	I (1–3)	I (1–3)	I (1–3)	I (1–3)
Gentianella campestris	I (1–3)	I (1–3)	I (1–3)	I (1–3)
Oxalis acetosella	I (1–3)	I (1–3)		I (1–3)
Luzula sylvatica	I (1–3)	I (1–3)		I (1–3)
Molinia caerulea	I (1–4)	I (1–4)		I (1–4)
Rhizomnium punctatum	I (4)	I (1–3)		I (1–4)
Galium sterneri	I (1–3)	I (1–3)		I (1–3)
Erica cinerea	I (1–3)	I (1–5)		I (1–5)
Scapania undulata	I (1–3)	I (1–3)		I (1–3)
Koeleria macrantha	I (1–4)	I (1)		I (1–4)
Poa pratensis	I (1–4)	I (1–3)		I (1–4)
Cynosurus cristatus	I (1–3)	I (1–2)		I (1–3)
Carex binervis	I (1–4)	I (1)		I (1–4)
Lathyrus montanus	I (1–3)	I (1–3)		I (1–3)
Potentilla sterilis	I (1–4)	I (1)		I (1–4)
Aira praecox	I (1–3)	I (1–3)		I (1–3)
Alchemilla xanthochlora	I (1–3)	I (1–3)		I (1–3)
Empetrum nigrum nigrum	I (1–3)	I (1–3)		I (1–3)
Mnium hornum	I (1–3)	I (1–4)		I (1–4)
Ranunculus repens	I (1–3)	I (1–3)		I (1–3)
Bryum pallens	I (1–3)	I (1–3)		I (1–3)
Cardamine pratensis	I (1–3)	I (1)		I (1–3)
Euphrasia confusa	I (1–3)	I (1–3)		I (1–3)
Juncus squarrosus	I (1–7)	I (1–3)		I (1–7)
Senecio jacobaea	I (1–3)		I (1–3)	I (1–3)
Empetrum nigrum hermaphroditum		I (1–3)	I (1–3)	I (1–3)
Potentilla crantzii		I (1–3)	I (1–6)	I (1–6)
Cetraria islandica		I (1–3)	I (1–3)	I (1–3)
Solidago virgaurea		I (1–3)	I (1–3)	I (1–3)
Angelica sylvestris		I (1–3)	I (1–3)	I (1–3)
Rubus idaeus		I (1–3)	I (1)	I (1–3)
Number of samples	78	34	12	124
Number of species/sample	30 (14–47)	36 (12–48)	40 (23–64)	32 (12–64)

a　*Trifolium repens-Luzula campestris* sub-community
b　*Carex pulicaris-Carex panicea* sub-community
c　*Saxifraga aizoides-Ditrichum flexicaule* sub-community
10　*Festuca ovina-Agrostis capillaris-Thymus praecox* grassland (total)

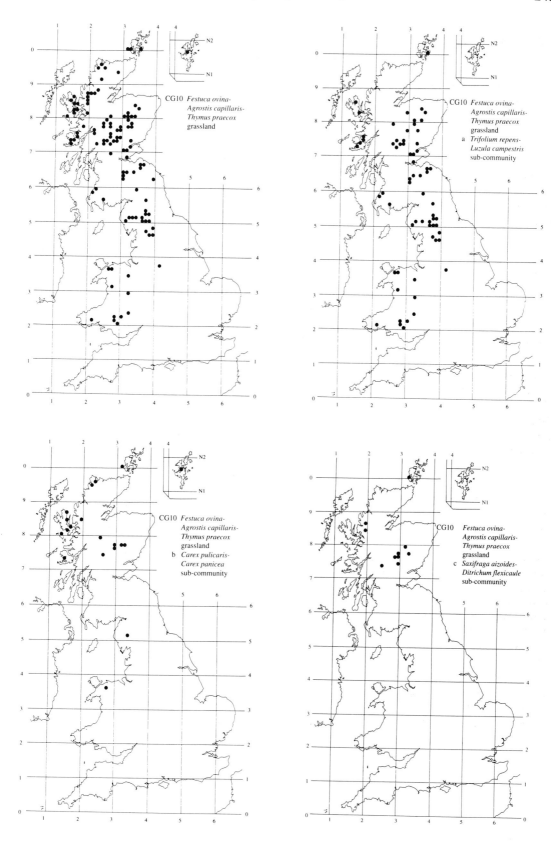

CG10 *Festuca ovina-*
Agrostis capillaris-
Thymus praecox
grassland

CG10 *Festuca ovina-*
Agrostis capillaris-
Thymus praecox
grassland
a *Trifolium repens-*
Luzula campestris
sub-community

CG10 *Festuca ovina-*
Agrostis capillaris-
Thymus praecox
grassland
b *Carex pulicaris-*
Carex panicea
sub-community

CG10 *Festuca ovina-*
Agrostis capillaris-
Thymus praecox
grassland
c *Saxifraga aizoides-*
Ditrichum flexicaule
sub-community

CG11
Festuca ovina-Agrostis capillaris-Alchemilla alpina grass-heath

Synonymy

Alpine pasture Smith 1900; Arctic-Alpine grassland Smith 1911 *p.p.*; *Alchemilleto-Agrosto-Festucetum* McVean & Ratcliffe 1962; Species-rich *Agrosto-Festucetum* McVean & Ratcliffe 1962 *p.p.*; *Achilleo-Festucetum tenuifoliae* Birse & Robertson 1976 *p.p.*; Montane *Festuca vivipara-Agrostis canina* grassland Ratcliffe 1977 *p.p.*; *Alchemilla alpina-Festuca-Vaccinium* nodum Huntley 1979 *p.p.*;

Constant species

Agrostis capillaris, Alchemilla alpina, Anthoxanthum odoratum, Festuca ovina/vivipara, Galium saxatile, Nardus stricta, Potentilla erecta, Thymus praecox, Vaccinium myrtillus, Hylocomium splendens.

Rare species

Alchemilla filicaulis spp. *filicaulis, A. wichurae, Cerastium alpinum, Diphasium alpinum, Juncus trifidus, Luzula spicata, Minuartia sedoides, Omalotheca supina, Polystichum lonchitis, Potentilla crantzii, Sibbaldia procumbens, Barbilophozia lycopodioides, Herbertus stramineus, Lophozia obtusa, Mastigophora woodsii.*

Physiognomy

Generally speaking, it is the prominence of *Alchemilla alpina* which gives the *Festuca-Agrostis-Alchemilla* swards their distinctive stamp. It is often the most abundant species present and the low cover of leaves produced from its creeping branches makes the epithet 'grassland' somewhat inappropriate for the vegetation. A number of grasses are, however, frequent and abundant and they often comprise the bulk of the sward between and among the *A. alpina* plants. *Agrostis capillaris* and *Festuca ovina* (*F. tenuifolia* in Birse & Robertson 1976 and Birse 1980) are the commonest species throughout, though *F. vivipara* here becomes almost as frequent as the latter and sometimes as abundant. *F. rubra*, however, is less common than in the closely-related *Festuca-Agrostis-Thymus* grassland. *Anthoxan-*

thum odoratum and *Nardus stricta* are also very frequent, though their abundance varies between the two sub-communities. Other grasses occasionally recorded are *Deschampsia cespitosa, D. flexuosa* and *Agrostis canina.*

The floristic shift among the sedge component visible in moving from the lowland calcicolous swards through the *Sesleria-Galium* grassland into the *Festuca-Agrostis-Thymus* grassland continues here. *Carex flacca* and *C. caryophyllea* are now no more than occasional and, though *C. pulicaris* and *C. panicea* are frequent and especially distinctive of one of the sub-communities, the commonest species overall are *C. pilulifera* and, more occasionally, *C. binervis. C. bigelowii* is scarce but sometimes locally prominent. In general, however, sedges do not make a major contribution to the vegetation cover, occurring usually as scattered shoots within the sward.

Apart from *Thymus praecox*, which is constant and often quite abundant, the commonest dicotyledonous associates are the Nardo-Galion species *Potentilla erecta* and *Galium saxatile. Vaccinium myrtillus* is very common, too, though it typically occurs as scattered sprigs, only very rarely assuming local prominence in clumps. *Calluna vulgaris* is a little patchier in its occurrence and similarly sparse. Other frequent species are *Viola riviniana, Ranunculus acris, Campanula rotundifolia* and *Selaginella selaginoides*, but all these tend to be better represented in one or other of the sub-communities. *Euphrasia officinalis* agg. (including, on Skye at least (Birks 1973), *E. nemorosa, E. confusa, E. brevipila* and *E. micrantha*) and *Prunella vulgaris* are both less common here than in the *Festuca-Agrostis-Thymus* grassland. *Polygonum viviparum, Thalictrum alpinum,* montane Hieracia and *Oxalis acetosella* are occasionally encountered and some stands have such species as *Succisa pratensis, Trollius europaeus, Geum rivale* and *Geranium sylvaticum* which, in ungrazed swards, can assume prominence. Rarer montane species are not common but *Potentilla crantzii* and *Silene acaulis* have

Carex binervis	II (3)	III (1–4)	II (1–4)
Deschampsia cespitosa	II (1–5)	III (1–4)	II (1–5)
Polygonum viviparum	II (2–4)	III (1–4)	II (1–4)
Rhytidiadelphus triquetrus	II (1–4)	III (1–4)	II (1–4)
Thuidium tamariscinum	II (1–4)	II (1–4)	II (1–4)
Euphrasia officinalis agg.	II (2–3)	II (1–4)	II (1–4)
Oxalis acetosella	II (1–3)	II (1–3)	II (1–3)
Deschampsia flexuosa	II (1–3)	II (1–3)	II (1–3)
Festuca rubra	II (1–3)	II (1–4)	II (1–4)
Hieracium spp.	II (1–3)	II (1–3)	II (1–3)
Hypnum cupressiforme	II (1–3)	II (1–6)	II (1–6)
Prunella vulgaris	II (1–3)	II (1–4)	II (1–4)
Agrostis canina	II (2–5)	II (1–3)	II (1–5)
Thalictrum alpinum	II (2–3)	II (1–4)	II (1–4)
Luzula multiflora	II (1–3)	II (1–3)	II (1–3)
Geum rivale	II (1–3)	II (1–3)	II (1–3)
Veronica officinalis	II (1–2)	I (1–3)	II (1–3)
Luzula campestris	II (2–3)	I (1–2)	II (1–3)
Breutelia chrysocoma	II (2–4)	I (1)	I (1–4)
Polytrichum alpinum	II (1–5)	I (1–3)	I (1–5)
Succisa pratensis	II (1–2)	I (1–3)	I (1–3)
Ctendium molluscum	II (3–4)	I (1–3)	I (1–4)
Luzula sylvatica	II (1)	I (1–3)	I (1–3)
Empetrum nigrum hermaphroditum	II (1–4)	I (1)	I (1–4)
Alchemilla filicaulis vestita	II (3–4)	I (1–3)	I (1–4)
Trifolium repens	I (1–3)	II (1–4)	I (1–4)
Danthonia decumbens	I (2)	II (1–3)	I (1–3)
Linum catharticum	I (1–2)	II (1–3)	I (1–3)
Filipendula ulmaria	I (1)	II (1–4)	I (1–4)
Taraxacum officinale agg.	I (1–2)	II (1–3)	I (1–3)
Leontodon autumnalis	I (1)	II (1–3)	I (1–3)
Cerastium fontanum	I (2)	II (1–3)	I (1–3)
Plagiomnium undulatum	I (2)	I (1)	I (1–2)
Carex caryophyllea	I (3)	I (1)	I (1–3)
Coeloglossum viride	I (1)	I (1)	I (1)
Antennaria dioica	I (1–2)	I (1–4)	I (1–4)
Carex bigelowii	I (1–4)	I (1)	I (1–4)
Huperzia selago	I (1)	I (1)	I (1)
Galium boreale	I (2)	I (1–3)	I (1–3)
Frullania tamarisci	I (1)	I (1)	I (1)
Diplophyllum albicans	I (1)	I (1)	I (1)
Achillea millefolium	I (1–4)	I (1–3)	I (1–4)
Drepanocladus uncinatus	I (1–2)	I (1)	I (1–2)
Botrychium lunaria	I (2–3)	I (1–3)	I (1–3)
Cladonia pyxidata	I (1–2)	I (1)	I (1–2)
Plantago maritima	I (1)	I (1)	I (1)
Tritomaria quinquedentata	I (1)	I (1)	I (1)
Saxifraga aizoides	I (1–2)	I (1–3)	I (1–3)
Barbilophozia lycopodioides	I (1)	I (1)	I (1)

Floristic table CG11 (*cont.*)

	a	b	11
Silene acaulis	I (1)	I (1)	I (1)
Saxifraga oppositifolia	I (3–5)	I (1)	I (1–5)
Potentilla crantzii	I (2–3)	I (1)	I (1–3)
Number of samples	43	20	63
Number of species/sample	30 (13–58)	35 (19–51)	32 (13–58)

a   Typical sub-community

b   *Carex pulicaris-Carex panicea* sub-community

11 *Festuca ovina-Agrostis capillaris-Alchemilla alpina* grass-heath (total)

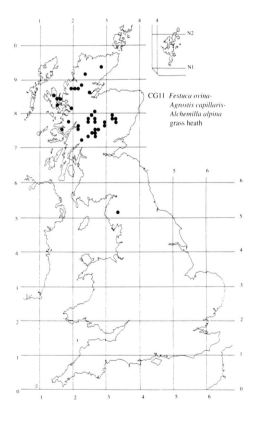

CG11 *Festuca ovina-Agrostis capillaris-Alchemilla alpina* grass heath

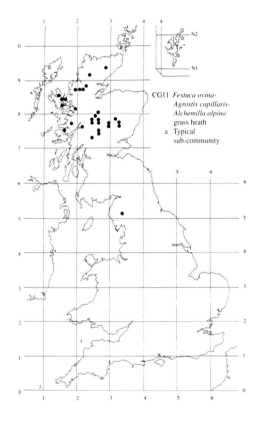

CG11 *Festuca ovina-Agrostis capillaris-Alchemilla alpina* grass heath
a Typical sub-community

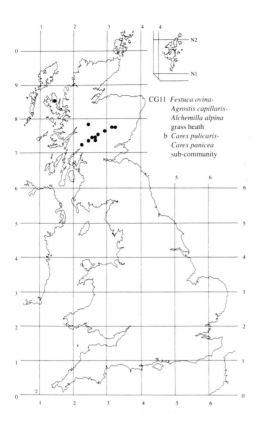

CG11 *Festuca ovina-*
Agrostis capillaris-
Alchemilla alpina
grass heath
b *Carex pulicaris-*
Carex panicea
sub-community

CG12

Festuca ovina-Alchemilla alpina-Silene acaulis dwarf-herb community

Synonymy

Arctic-Alpine grassland Smith 1911 *p.p.*; *Sibbaldia procumbens* nodum Poore 1955b *p.p.*; Dwarf Herb nodum McVean & Ratcliffe 1962; *Myosotis alpestris* localities Elkington 1964 *p.p.*

Constant species

Agrostis capillaris, Alchemilla alpina, Deschampsia cespitosa, Festuca ovina/vivipara, Luzula spicata, Selaginella selaginoides, Sibbaldia procumbens, Silene acaulis, Thymus praecox.

Rare species

Alchemilla filicaulis ssp. *filicaulis, Cerastium alpinum, Draba norvegica, Euphrasia frigida, Loiseleuria procumbens, Minuartia sedoides, Myosotis alpestris, Omalotheca supina, Poa alpina, Potentilla crantzii, Sagina saginoides, Saxifraga stellaris, Sibbaldia procumbens, Veronica alpina, Aulacomnium turgidum, Barbilophozia hatcheri, B. lycopodioides, Lescuraea incurvata.*

Physiognomy

The *Festuca-Alchemilla-Silene* community comprises rich mixtures of grasses and herbaceous dicotyledons which, with bryophytes, form a very distinctive kind of short (usually less than 10 cm) and sometimes rather open sward. Floristically, the vegetation has much in common with the *Festuca-Agrostis-Alchemilla* grass-heath. *Festuca ovina, F. vivipara, Agrostis capillaris* (with some *A. canina*) and *Nardus stricta* all remain frequent here and, with *Deschampsia cespitosa*, which rises to constancy in this community, they can make up a substantial proportion of the sward. *Alchemilla alpina, Thymus praecox* and *Selaginella selaginoides* also remain very common and *A. alpina* especially can be locally abundant. Usually, however, it is the cushion-forming *Silene acaulis* and, less frequently, *Minuartia sedoides* or the tufted *Sibbaldia procumbens* which give the vegetation its unmistakeable stamp, forming on occasion a discontinuous velvety carpet in which the other species are rooted. Although each of these occurs

in other montane vegetation types, among the more calcicolous communities it is here that they make their strongest contribution.

Moreover, they are accompanied, with varying degrees of frequency, by a wide variety of montane plants, mostly Arctic-Alpines, some with very restricted distributions in Britain, which make this one of the most renowned of our vegetation types. Most frequent among these are *Luzula spicata, Saxifraga oppositifolia* and *Polygonum viviparum* with, less commonly, *Antennaria dioica, Salix herbacea, Omalotheca supina, Thalictrum alpinum, Juncus trifidus, Euphrasia frigida, Epilobium anagallidifolium, Alchemilla filicaulis* ssp. *filicaulis* and especially distinctive of this community, *Sagina saginoides* and *Cerastium alpinum* (this last sometimes hybridising with *C. fontanum* which is occasional in the community: Stace 1975). More rarely, there are records for *Myosotis alpestris* (confined to this vegetation type and to calcicolous swards in higher reaches of the northern Pennines where the plants are distinctly smaller: Pigott 1956a, Elkington 1964), *Saxifraga hypnoides, S. aizoides, S. stellaris, S. nivalis, Diphasium alpinum, Loiseleuria procumbens, Veronica alpina, Poa alpina, Draba norvegica, Potentilla crantzii* and *Juncus triglumis.*

The Nardo-Galion and mesophytic elements prominent in the *Festuca-Agrostis-Thymus* and *Festuca-Agrostis-Alchemilla* grasslands are here much less obvious: *Potentilla erecta, Galium saxatile, Anthoxanthum odoratum, Viola riviniana* and *Ranunculus acris* are, at most, occasional and *Vaccinium myrtillus, Calluna vulgaris* and *Carex pilulifera* rare. Interestingly, sedges as a group are noticeably scarce here: apart from *C. pilulifera*, there are only very occasional records for *C. pulicaris, C. panicea, C. demissa* and *C. bigelowii; C. flacca* and *C. caryophyllea* are absent and they are not replaced by the Arctic-Alpine calcicoles *C. capillaris, C. atrata* and *C. rupestris*, species which are most characteristic of the high-altitude *Dryas-Silene* ledge vegetation.

Other vascular species of note in the community are

CG13
Dryas octopetala-Carex flacca heath

Synonymy

Dryas heaths Poore & McVean 1957, Gimingham 1972 *p.p.*; *Dryas-Carex flacca* nodum McVean & Ratcliffe 1962; *Dryas-Carex rupestris* nodum McVean & Ratcliffe 1962 *p.p.*; *Plantagino-Dryadetum* Shimwell 1968a; *Dryas octopetala* localities Elkington 1971 *p.p.*; Invernarver *Calluna-Arctostaphylos* heath Ward 1971a; *Dryas octopetala* heath Ratcliffe 1977 *p.p.*

Constant species

Bellis perennis, Carex flacca, Dryas octopetala, Festuca ovina, Linum catharticum, Lotus corniculatus, Plantago lanceolata, P. maritima, Thymus praecox, Viola riviniana, Ditrichum flexicaule.

Rare species

Agropyron donianum, Alchemilla glaucescens, Arctostaphylos uva-ursi, Arenaria norvegica ssp. *norvegica, Carex capillaris, C. rupestris, Draba incana, Dryas octopetala, Epipactis atrorubens, Oxytropis halleri, Amblystegium compactum, Brachythecium erythrorrhizon, Schistidium apocarpum* var. *homodictyon, S. trichodon, Seligeria trifaria, Tortella densa.*

Physiognomy

The *Dryas octopetala-Carex flacca* heath has a low patchy cover of sub-shrubs over what is essentially a sub-montane grassland sward. *D. octopetala* is usually the most abundant species, though its appearance (and the look of the vegetation as a whole) varies with the amount of grazing: in close-cropped swards, it has a prostrate much-nibbled habit but where there is no grazing it can grow more bushy and floriferous. Even in ungrazed stands, however, the plants often have small leaves, which suggests some genotypic distinction from the forms typical of the montane *Dryas-Silene* community (Elkington 1971). Intermixed with the *D. octopetala* there is commonly, in one sub-community, some *Calluna vulgaris* and, in the other, a little *Salix repens* and

Empetrum nigrum ssp. *nigrum* (less often *E. nigrum* ssp. *hermaphroditum*). Very occasionally, *Arctostaphylos uva-ursi* is found and, in some stands, it replaces *D. octopetala* in vegetation which is otherwise unchanged (e.g. Ward 1971a, b, Birks 1973). *Thymus praecox* is constant and locally abundant.

Grasses and sedges show considerable diversity in one sub-community but the species common throughout are few. *Festuca ovina* and *Carex flacca* are both constant and they often make up the bulk of the sward between the sub-shrubs but other species are only occasional and rarely abundant: *Koeleria macrantha, Agrostis capillaris, Festuca rubra* and *Carex panicea. Festuca vivipara* has been recorded in only one of the sub-communities.

Although some montane species occur in this vegetation (including certain Arctic-Alpines which attain their lowest altitudinal limit in the community), most of the other common herbaceous associates are species well represented in the more calcicolous sub-montane grasslands. Thus, there are frequently scattered plants of *Viola riviniana, Plantago lanceolata, Linum catharticum, Lotus corniculatus, Bellis perennis, Antennaria dioica, Selaginella selaginoides* and, more occasionally, *Cerastium fontanum* and *Luzula campestris*. In contrast to communities like the *Festuca-Agrostis-Thymus* grassland and the *Festuca-Agrostis-Alchemilla* grass-heath, there is, though, no consistently strong Nardo-Galion element present. One other distinction from these swards is the constancy here of *Plantago maritima* but, though the *Dryas-Carex* heath can occur in very close contact with coastal grasslands and heaths, other maritime species, such as *Armeria maritima* and *Scilla verna*, are characteristically scarce in the community.

The bryophyte component of the vegetation is frequently rich and varied but the species common throughout the community are few : *Ditrichum flexicaule* and *Hylocomium splendens* are most frequent with *Fissidens cristatus, Hypnum cupressiforme* and *Homalothecium lutescens* occasional. There is sometimes a little *Cladonia rangiformis*.

Sub-communities

Hieracium pilosella-Ctenidium molluscum **sub-community:** *Dryas-Carex rupestris* nodum   Poore & McVean 1957, McVean & Ratcliffe 1962 *p.p.*; *Dryas-Carex flacca* nodum   McVean & Ratcliffe 1962 *p.p.*; *Plantagino-Dryadetum typicum* Shimwell 1968*a*; *Dryas octopetala-Carex flacca* Association   Birks 1973. Although *D. octopetala* is frequently joined here by *Calluna vulgaris*, the most obvious feature of the vegetation is a richness and diversity among the herbaceous and bryophyte elements. *Carex flacca* and *C. panicea* are often accompanied by *C. pulicaris* and, as well as *Festuca ovina*, there are frequent records for *Danthonia decumbens* and *Anthoxanthum odoratum*. Then, among the dicotyledons, *Hieracium pilosella*, *Potentilla erecta*, *Prunella vulgaris*, *Hypericum pulchrum* and *Succisa pratensis* occur commonly with, less often, *Hypochoeris radicata* and *Euphrasia officinalis* agg. (including *E. confusa* and *E. scottica*). *Rubus saxatilis* and the rare *Epipactis atrorubens* are scarce but distinctive. Bryophytes are often abundant in crevices within or over the surfaces of the exposures among which the vegetation is disposed: *Ctenidium molluscum* and *Tortella tortuosa* are constant and *Thuidium tamariscinum*, *Racomitrium lanuginosum*, *Dicranum scoparium*, *Breutelia chrysocoma*, *Neckera crispa* and *Hylocomium brevirostre* occasional. *Rhytidium rugosum* and *Tortella densa* occur rarely.

Within this general definition, the vegetation shows a certain amount of physiognomic and floristic variation which reflects, on a smaller scale, that seen within communities like the *Sesleria-Galium* and *Festuca-Agrostis-Thymus* grasslands. Over the fractured tops of low-altitude exposures, and especially where these face south or west and catch more of the sun, the sward is often open and fragmentary and the vegetation comes to resemble the more xeric swards of the English and Welsh coastal limestones. Where the exposures are cut by grikes, as on southern Skye (Birks 1973) or in parts of the mainland Durness Limestone (Ratcliffe 1977), the inaccessibility of the vegetation to grazing may allow a more luxuriant growth and there can be an additional enrichment from tall herbs such as *Filipendula ulmaria*, *Centaurea nigra*, *Solidago virgaurea* and *Daucus carota* and ferns like *Polystichum aculeatum*, *P. lonchitis*, *Asplenium viride* and *A. marinum* growing from sheltered crevices. In other places, there is flushing from springs and seepage lines and the sward can have an unusual abundance of *Selaginella selaginoides* and *Carex panicea*, and species like *Pinguicula vulgaris*, *Saxifraga aizoides*, *Molinia caerulea* and *Schoenus nigricans* may make an appearance (e.g. the flushed facies of Birks 1973). Conversely, the vegetation may show signs of a floristic transition to a more calcifugous sward with a local prominence of *Calluna vulgaris*, *Danthonia decumbens* and *Anthoxanthum odoratum* (e.g. Birks 1973 leached facies). Finally, where this sub-community reaches its upper altitudinal limit on ungrazed ledges above Inchnadamph, it begins to approach the *Dryas-Silene* community in its structure and composition with occasional records for Arctic-Alpine rarities such as *Carex rupestris*, *C. capillaris* and *Draba incana* and some of the very few British occurrences of the Arctic-Subarctic *Agropyron donianum* and *Arenaria norvegica* ssp. *norvegica* (Raven & Walters 1956, Ratcliffe 1977). Here, too, the bryophyte component may be enriched with such rare sub-montane calcicoles as *Amblystegium compactum*, *Schistidium apocarpum* var. *homodictyon*, *S. trichodon*, *Seligeria trifaria* and *Tortula princeps* (Ratcliffe 1977).

Salix repens-Empetrum nigrum ssp. *nigrum* **sub-community:** *Dryas-Carex flacca* nodum   Poore & McVean 1957, McVean & Ratcliffe 1962 *p.p.*; *Dryas-Carex rupestris* nodum   McVean & Ratcliffe 1962 *p.p.*; *Plantagino-Dryadetum* maritime Sub-association Shimwell 1968*a p.p. Dryas octopetala* remains frequent and abundant here but the sub-shrub canopy is more varied and extensive with some *Salix repens* and *Empetrum nigrum* ssp. *nigrum* and, more occasionally, *E. nigrum* ssp. *hermaphroditum* and *Juniperus communis* (approaching ssp. *nana* in its procumbent habit). Among the herbaceous species, *Galium verum*, *Campanula rotundifolia* and *Luzula campestris* are frequent preferentials, and *Polygonum viviparum*, *Orchis mascula* and *Listera ovata* distinctive occasionals. As in the former sub-community, various Arctic-Alpines can be found in some stands, here descending almost to sea-level, e.g. *Saxifraga aizoides*, *S. oppositifolia*, *Draba incana*, *Silene acaulis*, and this vegetation also provides a major locus for the lovely Alpine rarity *Oxytropis halleri*. Preferential bryophytes are *Pseudoscleropodium purum*, *Rhytidiadelphus triquetrus*, *Homalothecium lutescens* and *Scapania undulata* with, less often, *Thuidium delicatulum* and *Pleurozium schreberi*. *Brachythecium erythrorrhizon* occurs at its only known British locality in this sub-community (Barkman 1955).

Again, there is some structural and floristic variation in the vegetation but here this is due not only to the configuration of bedrock exposures (which, as before, can be variously broken into small cliffs and pavement with flushing or drift cover) but also to the amount and disposition of wind-blown shell-sand over which this sub-community commonly occurs. Where this becomes fixed among the low branches of the sub-shrubs, there can be extensive stands of more stable machair-like vegetation with an even or hummocky surface. Elsewhere, localised erosion produces a more open and fragmentary cover and there may be occasional occurrences of therophytes typical of unstable maritime habitats (as in some samples of the maritime Sub-association of the *Plantagino-Dryadetum* of Shimwell 1968a). In other places, a thin layer of shell-sand may be blown over adjacent peat deposits so that the vegetation takes on some characteristics of blanket-mire edge vegetation with a more extensive cover of *Empetrum nigrum*.

Habitat

The *Dryas-Carex* heath is restricted to calcareous lithomorphic soils in the cool, oceanic lowlands of north-west Scotland. Although the community extends up to about 140 m on the hills above Inchnadamph, most stands occur below 70 m and, along the north Sutherland coast, they can be found down to about 15 m above sea-level. The regional climate is cool with a mean annual maximum temperature generally less than 23 °C (Conolly & Dahl 1970), but the annual accumulated temperature is of the same order as that over much of lowland Scotland and the winters especially are distinctly milder than on higher ground to the east with only 20–60 frost days and February minima often a degree or so above freezing (*Climatological Atlas* 1952). Although annual precipitation reaches over 1600 mm along the north-western seaboard, with up to 200 wet days yr^{-1} (Ratcliffe 1968), there is little or no snow-lie. The summers are humid but there is rather less daytime cloudiness than on higher ground inland.

Within this area of regional climate, the community occurs only where the influence of calcareous soil parent material is sufficiently great to offset the effect of leaching in the high rainfall. The major calcareous bedrocks in the region are the Durness Limestone, which is exposed on southern Skye and on the mainland along the Moine Thrust from Inverpolly through Inchnadamph to Durness itself on the north coast, and the Jurassic limestones on Skye and Raasay (Birks 1973, Ratcliffe 1977, Whittow 1979). Over these exposures, the *Dryas-Carex* heath is largely confined to the drift-free surfaces of cliffs, talus and, on the Durness Limestone, pavements, which carry shallow rendzinas. These are sometimes little more than skeletal accumulations of rock fragments and organic detritus but, even where the soil cover is more substantial, the profiles are consistently base-rich with pH values generally between 6.5 and 7.5 (Poore & McVean 1957, McVean & Ratcliffe 1962, Shimwell 1968a, Elkington 1971, Birks 1973). There is usually much free calcium carbonate, although over dolomitised sections of the Durness Limestone (as around Durness itself), magnesium may be the major cation present (Elkington 1971). Sometimes deeper profiles with a reddish B horizon are encountered but the community does not extend far on to drift-contaminated soils except where these are flushed with calcareous waters. Fragmentary stands of *Dryas-Carex* heath

also occur over similar rendziniform soils on the more calcareous parts of the Skye Tertiary basalts, as for example where these have tumbled in the complex of rotational landslips beneath The Storr and The Quiraing but the *Dryas* vegetation of more high-altitude ledges on the Trotternish ridge is best considered as fragments of the *Dryas-Silene* community.

Along the north coast of Scotland, fine calcareous shell-sand, blown from the shore and deposited over the surfaces of projecting headlands (not always themselves made up of calcareous bedrocks), provides an important alternative soil parent material. Here, the soils are calcareous sand pararendzinas, sometimes shallow smears over rock or peat, in other cases, where topographic variation or the vegetation itself has encouraged accumulation, of more substantial depth, but always, as before, base-rich. Very fine and extensive stands of the community occur on this substrate around Durness and, further east, at Invernarver (Ratcliffe 1977). Along this coast, the *Dryas-Carex* heath is found quite close to sea-level but, though the climate is considerably drier here than along the north-west seaboard, rainfall seems to be sufficient to ameliorate any deposition of salt-spray.

The milder character of the climate is reflected in the vegetation in its generally sub-montane nature. Only where the winters become sharper on higher-altitude ledges above Inchnadamph and along the Sutherland coast, do other Arctic-Alpines apart from *D. octopetala* make an appearance in the vegetation and, even here, they do not make a consistent contribution. Rather, there is something of a resurgence in the community of Mesobromion calcicoles, such as *Carex flacca* and *Linum catharticum*, which gives the vegetation a similar feel to many lowland calcicolous swards. Indeed, where the cover becomes fragmented over the fractured tops of cliffs close to sea-level, the community looks like a northern oceanic equivalent of the Xerobromion swards of the south-west. Such similarities are accentuated by the highly calcareous nature of the soils which, as a rule, prevents the development of any marked Nardo-Galion element such as is characteristic of other sub-montane swards in the north-west.

What part climate and soil variations play in influencing the floristic differences between the sub-communities is unknown. Strictly speaking, the *Salix-Empetrum* sub-community is no more maritime than the *Hieracium-Ctenidium* sub-community, even though it occurs more frequently in close association with maritime grasslands and heaths near to the sea (cf. Shimwell 1968a). Part of the answer may be found in the coincidental occurrence of the different substrates characteristic of the community in areas of slightly different climate. The *Hieracium-Ctenidium* sub-community, with its abundance of calcicolous bryophytes, is more common over the hard limestones along the wetter and milder north-west coast, whereas the *Salix-Empetrum* sub-community is more abundant along the drier, cooler north coast where its greater diversity of sub-shrubs is perhaps easily able to survive the shifting of shell-sand.

Although *Empetrum nigrum* itself is unpalatable to sheep (Bell & Tallis 1973), it is also possible that differences in grazing intensity play some part in the composition and distribution of the sub-communities. There is no doubt that the community as a whole attains its greatest luxuriance when it is ungrazed and, at some sites, like Durnesss (Ratcliffe 1977), the boundaries of stands apparently coincide with grazing limits but, on Skye at least (Birks 1973), this vegetation is quite heavily grazed by sheep and, in parts, rabbits, and nibbling perhaps contributes to the generally low and varied swards of the *Hieracium-Ctenidium* sub-community which is most abundant there. The *Salix-Empetrum* sub-community, on the other hand, may be grazed less, though no data are available to confirm this.

Zonation and succession

Most often, zonations involving the *Dryas-Carex* heath are related to edaphic variation which is dependent upon the topography and geology. The exposures of calcareous bedrocks in the north-west of Scotland, of which the *Hieracium-Ctenidium* sub-community is especially characteristic, are usually of limited extent and surrounded by more acid rocks. Stands of this vegetation are thus often small and they make up part of the patchworks of calcicolous communities that provide such a startling intrusion of fresh green into the dull brown cover that characterises much of this area. The outcrops are also frequently broken into cliffs, and dissected by crevices and grikes. Typically, therefore, this sub-community occurs in complex mosaics with fern-domination in crevice protorendzinas and woodland fragments in sheltered clefts and ravines. Such patterns are very well seen in the Suardal area of Skye (Birks 1973) and, to a more limited extent, around Durness (Ratcliffe 1977). At higher altitudes, on the hills above Inchnadamph, comparable mosaics involve montane chasmophytic vegetation, tall-herb communities and willow scrub (Poore & McVean 1957, McVean & Ratcliffe 1962, Ratcliffe 1977).

Very often, too, the exposures, especially on flatter land at lower altitudes, are smeared with varying amounts of drift which introduces a further element of complexity. Then the *Dryas-Carex* heath may pass to more calcifugous heath, perhaps through an intermediate zone in which *Calluna vulgaris* and Nardo-Galion grasses show an increase in cover. Flushing can complicate the pattern still further. Where this is intermittent, the community may persist with scattered Caricion davallianae species in the sward but, where it is more pronounced, there are transitions to sub-montane soligenous mires. Where base-poor waters flow through drift-contaminated soils adjacent to *Dryas-Carex* heath, the community may show a sharp transition to some kind of calcifuge mire.

Comparable patterns to these can be seen within and around areas of shell-sand along the north Scottish coast, where the *Salix-Empetrum* sub-community is more common, though here the zonations can have more diffuse boundaries because of the continuously variable thicknesses of sand that are laid down. The open chasmophytic communities of bare rock are also replaced by vegetation typical of raw sand and, with increasing proximity to the sea, there is the additional feature of maritime influence. Thus, around Invernarver, *Dryas-Carex* heath occurs behind dune communities and among maritime heath, notably the *Empetrum* sub-community of the *Calluna-Scilla* heath. Disturbed areas within stands of the community may carry small patches of the *Armeria-Cerastium* maritime therophyte community. Further away from the sea, where the influence of deposited sand peters out, there can be gradations to various kinds of inland heath over acidic mineral soils and peat. As before, varying degrees of flushing with waters of differing base-status result in the scattered occurrence of small-sedge mires of various kinds (Ratcliffe 1977).

Both sub-communities are also frequently found in close association with small patches of sub-montane swards, such as the *Festuca-Agrostis-Thymus* grassland or, along the north Scottish coast, the *Festuca-Plantago* maritime grassland. At somewhat higher altitudes, as along the slopes of the Trotternish ridge in Skye or above Inchnadamph, it may grade to the *Festuca-Agrostis-Alchemilla* grass-heath (Birks 1973, Ratcliffe 1977). The presumption is that such zonations are mediated primarily by grazing which, when very prolonged or heavy, is supposed to wipe out the *D. octopetala* (McVean & Ratcliffe 1962). However, the survival of the community under grazing on Skye (Birks 1973, see above), like the persistence of *D. octopetala* in *Sesleria-Galium* grassland on the Pennines, suggests that this relationship may be quite complex (Elkington 1971). Enclosure experiments at Inchnadamph may help reveal some of the answers to this problem (Ratcliffe 1977).

Distribution

The community is confined to the lowlands of north-west Scotland. The *Hieracium-Ctenidium* sub-community is the more widespread, the *Salix-Empetrum* sub-community having been recorded only from Raasay and the Sutherland coast.

Affinities

The *Dryas-Carex* heath, though it grades floristically to the *Dryas-Silene* community at higher altitudes, is quite

CG14
Dryas octopetala-Silene acaulis ledge community

Synonymy

Glen Lochay *Dryas* heath Poore & McVean 1957; *Dryas-Salix reticulata* nodum McVean & Ratcliffe 1962; *Dryas-Carex rupestris* nodum McVean & Ratcliffe 1962 *p.p.*; *Salico-Dryadetum* Shimwell 1968*a*; Cairnwell Limestone vegetation Coker 1969 *p.p.*; *Dryas octopetala* localities Elkington 1971 *p.p.*; *Dryas octopetala* heath Ratcliffe 1977 *p.p.*; Cliff ledge communities Jermy *et al.* 1978 *p.p.*; *Dryas octopetala-Salix reticulata* nodum Huntley 1979; *Viola-Festuca-Agrostis* nodum Huntley 1979 *p.p.*; *Alchemilla glabra-Sedum rosea* nodum Huntley 1979 *p.p.*

Constant species

Alchemilla alpina, Campanula rotundifolia, Carex capillaris, C. pulicaris, Dryas octopetala, Festuca ovina/vivipara, Hieracium spp., *Polygonum viviparum, Saxifraga aizoides, S. oppositifolia, Selaginella selaginoides, Silene acaulis, Thalictrum alpinum, Thymus praecox, Viola riviniana, Ctenidium molluscum, Ditrichum flexicaule, Hylocomium splendens, Rhytidiadelphus triquetrus, Tortella tortuosa.*

Rare species

Alchemilla filicaulis ssp. *filicaulis, Astragalus alpinus, Bartsia alpina, Carex atrata, C. capillaris, C. rupestris, C. vaginata, Cerastium alpinum, C. arcticum, Draba incana, Dryas octopetala, Euphrasia frigida, Minuartia sedoides, Orthilia secunda, Oxytropis halleri, Polystichum lonchitis, Potentilla crantzii, Pyrola rotundifolia, Salix arbuscula, S. lapponum, S. myrsinites, S. reticulata, Sesleria albicans, Tofieldia pusilla, Veronica fruticans.*

Physiognomy

The *Dryas-Silene* community comprises rich, varied and luxuriant mixtures of dwarf shrubs, tall herbs, sedges and grasses among a carpet of cushion herbs and bryophytes. Stands are frequently small and fragmentary, the vegetation being usually disposed over the ledges of rock outcrops. *Dryas octopetala* is the most frequent and often the most abundant component of the open and uneven cover of dwarf shrubs and it occurs here as taller and larger-leaved plants than those found in the *Dryas-Carex* heath and the *Sesleria-Galium* grassland, variation which reflects the lack of grazing in this community but which also seems to have some genetic basis (Elkington 1971). Then, there is commonly one (very occasionally more than one) of a number of rare Arctic-Alpine willows. *Salix reticulata* is the most frequently encountered and the cover of its low creeping branches can rival or exceed that of *D. octopetala*. In some stands, it is joined or replaced by *S. arbuscula* with its more robust bushy growth and there are rare records, too, for *S. myrsinites* and *S. lapponum*. Hybrids between these species are sometimes encountered though their parentage is often difficult to determine (Meikle 1975): there are records in the data for *S.* × *boydii* (which is probably *S. lapponum* × *reticulata*) and for *S.* × *phaeophylla* (originally thought to be *S. lapponum* × *myrsinites*, but now re-determined as *S. lapponum* × *herbacea*). Other more widely distributed willows which make a rare contribution to the dwarf-shrub component are *S. aurita* and *S. repens*.

Intermixed with the *D. octopetala* and willows are usually small clumps of ericoids. *Vaccinium vitis-idaea* is the most common of these with, less frequently, *V. uliginosum, V. myrtillus* and *Calluna vulgaris. Empetrum nigrum* ssp. *hermaphroditum* is occasional, too, and it can attain local prominence. There is very frequently a little *Alchemilla alpina* putting up its leaves among the shrubs and, more occasionally, some *A. glabra, A. filicaulis* ssp. *filicaulis* and *A. filicaulis* ssp. *vestita. Thymus praecox* is constant with its sprawling mats typically anchored by single thick primary roots penetrating crevices (Pigott 1955). *Rubus saxatilis* occurs very occasionally.

From among these plants, there protrude the shoots of a diversity of tall herbs, the variously-coloured flowers of which contribute greatly to the attractiveness of this vegetation. Among the commonest of these are *Succisa pratensis, Angelica sylvestris, Geranium sylvaticum* and numerous montane Hieracia, particularly the very handsome taxa of the sections Alpine, Subalpina and Cerinthoidea which frequently occur in distinctive assortments on the ledges of different mountain ranges (e.g. Sell & West 1955, 1965, 1968, Raven & Walters 1956, Kenneth & Stirling 1970). More occasionally, *Geum rivale, Filipendula ulmaria, Solidago virgaurea* and the Arctic-Alpine *Rhodiola rosea* and *Saussurea alpina* can be found and clumps of *Luzula sylvatica* are sometimes conspicuous.

Then, there is an equally rich and diverse lower tier of grasses, sedges and smaller herbaceous species. Among the grasses, *Festuca ovina* and *F. vivipara* are both frequent and the former especially can be patchily abundant. There is also commonly some *Deschampsia cespitosa, Anthoxanthum odoratum* and *Avenula pratensis* and *Nardus stricta, Festuca rubra* and *Agrostis capillaris* also occur occasionally. Most of the high-altitude Scottish localities of *Sesleria albicans* are in this vegetation and it can be locally abundant. The commonest sedges are *Carex pulicaris* and the Arctic-Alpine *C. capillaris* and both, especially the former, can have high cover. The community also provides one of the major loci for the much rarer Arctic-Alpine sedges, *C. rupestris* (easily confused with *C. pulicaris*, especially when vegetative) and the more robust *C. atrata* and *C. vaginata. C. flacca* is occasional too.

Among the herbaceous associates with the grasses and sedges the most frequent species are *Polygonum viviparum, Thalictrum alpinum, Campanula rotundifolia, Viola riviniana, Coeloglossum viride, Galium boreale, Pinguicula vulgaris, Linum catharticum* and *Asplenium viride* with, less commonly, *Luzula multiflora, Antennaria dioica, Parnassia palustris, Tofieldia pusilla, Huperzia selago* and the two rare wintergreens, *Pyrola rotundifolia* ssp. *rotundifolia* and *Orthilia secunda*, these sometimes in local colonies of numerous individuals. Other montane rarities occurring at very low frequencies here are *Minuartia sedoides* and *Oxytropis halleri* (Alpine), *Cerastium arcticum* (Arctic-Subarctic) and (all Arctic-Alpine) *Cerastium alpinum, Potentilla crantzii, Draba incana, Euphrasia frigida, Polystichum lonchitis, Veronica fruticans, Bartsia alpina* and the very rare *Astragalus alpinus*.

The lowest component of the vegetation comprises lush cushions of herbs and bryophytes. *Silene acualis* is a constant of the community and usually the most abundant species of this layer but there are very frequently scattered sprawls of *Selaginella selaginoides, Saxifraga oppositifolia* and *S. aizoides*. Among the bryophytes, bulky mosses usually make up most of the cover with *Ctenidium molluscum, Tortella tortuosa, Rhytidiadelphus triquetrus, Ditrichum flexicaule* and *Hylocomium splendens* especially frequent and abundant and, occurring more occasionally, *Racomitrium lanuginosum, Distichium capillaceum, Campylium stellatum* var. *proten-*

sum, Rhytidiadelphus loreus, Thuidium delicatulum, Dicranum scoparium, Breutelia chrysocoma, Hypnum cupressiforme and *Pseudoscleropodium purum*. Species of more restricted distribution which can be found in this vegetation are *Rhytidium rugosum* (particularly distinctive of *Dryas* vegetation in Scotland), *Entodon concinnus, Plagiopus oederi, Orthothecium rufescens* and *Anoectangium aestivum*. Hepatics are, generally speaking, much less frequent and conspicuous than mosses but *Frullania tamarisci, Aneura pinguis* and *Plagiochila asplenoides* are occasionally found and there are sparse records for *Tritomaria quinquedentata, Metzgeria leptoneura* and *Herbertus stramineus*. Lichens are not abundant but, among the cushions, there are sometimes scattered plants of *Cetraria islandica, Cladonia pyxidata, C. gracilis, C. arbuscula, Peltigera canina* and *P. aphthosa* and James (1965; see also James *et al.* 1977) has described a distinctive suite of saxicolous and terricolous species from rock faces, talus and mica-schist soils occurring in association with this community.

Habitat

The *Dryas-Silene* community is strictly confined to ungrazed crags and ledges of calcareous bedrocks, mostly in the montane regions of Scotland, with often lithomorphic, base-rich soils.

The altitudinal range of this vegetation and the regional climate it experiences are not quite so narrow and extreme as those typical of the *Festuca-Alchemilla-Silene* dwarf-herb community. Although stands tend to attain their greatest richness and luxuriance at higher levels, being found at up to about 900 m, and their mean altitude (just over 700 m) is well above that of the submontane calcicolous grasslands, the community can occur down to about 300 m, provided suitable topographies occur. Like many of the vegetation types of the Scottish Highlands, it descends lower in moving northwest as the climate becomes increasingly severe at equivalent altitudes, but it extends somewhat further into the more equable oceanic fringes than does the *Festuca-Alchemilla-Silene* community, though its stands become more fragmentary and isolated towards this margin of its range. Thus, outlying occurrences of *D. octopetala* with various sparse assortments of other Arctic-Alpines, such as are found on high-altitude ledges in Mull (Jermy & Crabbe 1978), Rhum (Ratcliffe 1977), Skye (Birks 1973) and Hoy (Prentice & Prentice 1975) and, further afield and yet more fragmentarily, on the Glyders (Ratcliffe 1977) and Helvellyn (Ratcliffe 1960, 1977), are a somewhat pale reflection of the full richness of the Highland stands (see also Elkington 1971). Furthermore, along the north-west coastal fringes of Scotland, the vegetation tends to show some floristic transitions to the sub-montane *Dryas-Carex* heath which is common there.

This slightly wider range than that typical of the *Festuca-Alchemilla-Silene* dwarf-herb community takes in areas with a climate that is a little warmer. The mean annual maximum temperature is 22 °C (Conolly & Dahl 1970), 23 °C if the Lakeland occurrences are included, 24 °C with those in north Wales, and the winters especially tend to be milder, with February minima sometimes above freezing (*Climatological Atlas* 1952). Although, in the north-west Highlands, annual precipitation is extremely heavy, with over 2400 mm and more than 220 wet days yr^{-1} (*Climatological Atlas* 1952, Ratcliffe 1968), the community extends somewhat further into the drier east than does the *Festuca-Alchemilla-Silene* community, both to the north in Sutherland and to the south around Caenlochan and Clova where the annual precipitation is about 1000 mm. Despite these features, however, the general character of the climate is still cold, wet and windy and the topographic conditions characteristic of the community considerably exacerbate certain elements of it.

The *Dryas-Silene* vegetation is always found on more calcareous bedrocks and, though these are quite varied because of the accidents of geological deposition, the climatic range of the community largely coincides with the distribution of the Dalradian metasediments of the central and southern Highlands. Here, the glories of the rich schist and limestone ledges of Ben Lawers, Beinn Laoigh, Creag Mhor, Meall Ghaordie, Meall na Samhna, Carn Gorm and Caenlochan-Clova have been long extolled. Further north, good stands also occur on the more sparsely distributed calcareous rocks of the Moine Assemblage around the Glen Feshie area of the Cairngorms, Beinn Dearg and Ben Hope. Outside this area, suitable rocks are not so widespread and often occur towards the limit of the climatic range of the community so that stands are often fragmentary, but rich examples can be seen on the Devonian Old Red Sandstone of Ben Griam More (providing a far-flung eastern locality), on the Durness Limestone above Inchnadamph and on Glas Cnoc, and on Tertiary basalt on Beinn Iadain. More fragmentary stands occur on Tertiary basalt in Mull, Skye and Rhum and on Lewisian gneiss around Letterewe. On Helvellyn, the very small amounts of Arctic-Alpine *Dryas* vegetation are found on more calcareous parts of the Borrowdale Volcanics and in Snowdonia on pumice.

Over these rocks, the community is confined to fractured cliff tops, ledges and inaccessible tumbles of talus. Softer deposits, like the Dalradian mica-schists, weather very readily and provide an abundance of suitable sites with broken rock faces and many small ledges. Harder rocks often crop out more dramatically and have fewer but larger ledges on which the vegetation is perched in spectacular 'hanging gardens'. But, whatever the particular physiognomy of the vegetation, there are two features of this crag environment of particular importance to the community.

The first is edaphic. Despite the generally high rainfall and the great potential for leaching, especially in the wetter north-west, the soils are maintained in a skeletal and base-rich condition by constant renewal of raw mineral material and the usually limited areas of bare rock over which they can develop. Freeze-thaw weathering may be of especial importance here: the climate is generally cold in winter and, though snow-falls can be frequent and appreciable, the exposure of the crags to wind means that they are generally kept free of its insulating cover. Even where more substantial amounts of rock fragments and organic detritus accumulate on larger ledges, horizon differentiation will be hindered by the churning of cryoturbation and solifluction over inclined surfaces. The soils are thus probably consistently rich in free calcium carbonate and their surface pH is always high, frequently more than 7 (McVean & Ratcliffe 1962, Coker 1969, Elkington 1971). Often, too, there is enrichment by the percolation or dripping of calcareous waters through and over the soils but the profiles are characteristically free-draining. Indeed, on fractured cliff tops with a southerly aspect which catch more of the small amounts of summer sun, the soils can become quite dry. In general, therefore, the profiles are much more like typical rendzinas than are those found under the *Festuca-Alchemilla-Silene* community (McVean & Ratcliffe 1962, Coker 1969, Elkington 1971, Huntley 1979), though more extensive and deeper soils may approach brown calcareous earths with a more integrated mull structure and slight differentiation of a B horizon.

The occurrence of such soils in areas of generally harsh montane climate exerts a major influence on the floristics of the community. It is in this vegetation that the richest mixtures of stricter Arctic-Alpine calcicoles occur and the combination of these features in less extreme situations has provided a relict habitat for the survival of various members of its flora in more outlying geographical localities (e.g. Raven & Walters 1956, Ratcliffe 1959a, b, 1960, McVean & Ratcliffe 1962, Coker 1969, Conolly & Dahl 1970, Elkington 1971). Here, too, there is a slight resurgence in some calcicoles which are, generally speaking, more widespread in lowland swards than they are throughout sub-montane calcicolous communities, e.g. *Carex flacca, Avenula pratensis, Linum catharticum*. Moreover, the contribution made by Nardo-Galion species is very slim and the somewhat chionophilous feel of the *Festuca-Alchemilla-Silene* community is not noticeable in this vegetation.

The second characteristic of the habitat of the community is that it is inaccessible to sheep and deer so that the vegetation is ungrazed. The effect of this is clearly seen in the prominence of dwarf shrubs and tall herbs and the general luxuriance of the herbage (McVean & Ratcliffe 1962, Elkington 1971, Huntley 1979), features which are abruptly lost around the margins of stands which grazing animals can reach. No member of these components of the vegetation is confined to this particular community: the willows, for example, can form more shrubby stands of scrub proper, the ericoids are widespread in other kinds of ledge vegetation as well as in heaths, the tall herbs are characteristic of various ungrazed habitats (not all of them montane or even sub-montane) and *D. octopetala* also occurs in the *Dryas-Carex* heath. This particular combination of grazing-sensitive species is, however, unique to this vegetation and its particular attractiveness rendered all the more valuable by the likelihood that it represents but a fragment of what was previously a more widespread cover over ungrazed areas of base-rich soils throughout the Scottish Highlands.

Zonation and succession

Most often, zonations between the community and other vegetation types are a direct reflection of the extent of grazing, though edaphic transitions related to bedrock and soil type are also found.

Very commonly, the *Dryas-Silene* community passes, where ledges become more accessible, to grazed calcicolous vegetation over the more intact soils of the surrounding smoother topography. The communities involved in such transitions vary with altitude. Around higher crags, there is typically a zonation to the *Festuca-Alchemilla-Silene* community in which the representation of those Arctic-Alpines tolerant of grazing is maintained. Such mosaics are well seen towards the upper slopes of Ben Lawers, on Beinn Dearg and Ben Alder (Ratcliffe 1977) and, on Helvellyn, the very small fragments of the community survive among more extensive areas of dwarf-herb vegetation (Ratcliffe 1960, 1977). Towards lower altitudes, the community gives way to the more sub-montane vegetation of the *Festuca-Agrostis-Alchemilla* grass-heath and the *Festuca-Agrostis-Thymus* grassland as on the lower slopes of Ben Lawers, on Beinn Laoigh, in the Caenlochan–Clova area and on Meall Ghaordie and Beinn Iadain. It is very likely that each of these three communities has, in part, been derived by grazing from the *Dryas-Silene* vegetation which has been progressively restricted to the more inaccessible crags but it should be remembered that, around the margins of these exposures, there is frequently a coincidental edaphic shift to more intact and sometimes deeper soils, even though the underlying bedrock remains calcareous.

It may be presumed that the frequent instability of the crag environment with its exposure to erosion by frost and wind, and the often precarious hold of the vegetation on ledges, help maintain the characteristic patchwork of plants of differing stature. For the most part,

stands are at altitudes which are too high for colonisation by trees and much of this vegetation represents a climax dependent on the extreme climatic and edaphic conditions. Sometimes, however, the willows form a more intact cover, shading out many of the species of the community, and transitions to such low scrub can be seen on Meall na Samhna, Carn Gorm and Beinn Dearg (Ratcliffe 1977).

Grazing-related zonations such as these can be complicated where there are edaphic variations attributable to differences in the underlying bedrock and/or the character of percolating waters. Then, other less calcicolous vegetation types may occur in close association with the *Dryas-Silene* community. For example, where isolated calcareous crags intrude into areas of acid bedrocks, a common feature among the Moine and Lewisian Assemblages, *Nardus-Galium*, *Juncus-Festuca* or *Deschampsia-Galium* grasslands may surround stands of the community. In other cases, complexes of ledges run across geological boundaries or receive irrigation with waters of varying calcium carbonate content. Where flushing with less calcareous waters occurs, the *Dryas-Silene* vegetation can pass into other kinds of ungrazed tall-herb communities in which calcicoles are much more poorly represented. Such transitions are a prominent feature of parts of Ben Lawers, Beinn Laoigh, the Caenlochan–Clova area, Meall Ghaordie and Beinn Dearg (Ratcliffe 1977).

Distribution

The centre of distribution of the *Dryas-Silene* community is in the central and southern Scottish Highlands with outlying localities in the north-west Highlands and fragmentary stands in North Wales, Cumbria, Mull, Rhum, Skye and Orkney.

Affinities

Although much British vegetation containing *D. octopetala* is varied and sometimes fragmentary, this community emerges as clearly distinct from the *Dryas-Carex* heath of low altitudes in north-west Scotland, although transitional stands do occur on ledges above Inchnadamph. It unites samples from two of the fragmentary noda recognised by McVean & Ratcliffe (1962) and is essentially similar to the *Salico-Dryadetum* of Shimwell (1968a). Among British calcicolous communities, it represents the nearest approach to the montane dwarf-shrub heaths of the Elyno-Seslerietea (recast by Oberdorfer (1978) as the Seslerietea variae). Within this class, the *Dryas-Silene* community is closest to the kinds of Scandinavian vegetation included by Nordhagen (1928, 1936, 1955) in the Kobresio-Dryadion and it shares a number of species with his rather compendious *Kobresieto-Dryadetum*: *D. octopetala*, *Salix reticulata*, *Carex capillaris*, *C. rupestris*, *C. atrata*, *Astragalus alpinus* and *Bartsia alpina*.

Floristic table CG14

Dryas octopetala	V (2–9)	*Aneura pinguis*	II (1–3)
Polygonum viviparum	V (1–4)	*Distichium capillaceum*	II (1–4)
Selaginella selagionides	V (1–3)	*Tofieldia pusilla*	II (1–3)
Hylocomium splendens	V (1–4)	*Parnassia palustris*	II (1–3)
Ctenidium molluscum	V (1–4)	*Nardus stricta*	II (2–4)
Thymus praecox	IV (1–5)	*Rhodiola rosea*	II (1–3)
Silene acaulis	IV (1–6)	*Saussurea alpina*	II (1–4)
Carex capillaris	IV (1–4)	*Rhinanthus minor*	II (1–2)
Saxifraga aizoides	IV (1–4)	*Mnium hornum*	II (1–4)
Alchemilla alpina	IV (1–5)	*Geum rivale*	II (1–4)
Hieracium spp.	IV (1–3)	*Rhytidiadelphus loreus*	II (1–3)
Saxifraga oppositifolia	IV (2–5)	*Luzula multiflora*	II (2–3)
Thalictrum alpinum	IV (1–3)	*Plagiochila asplenoides*	II (1–3)
Campanula rotundifolia	IV (1–3)	*Filipendula ulmaria*	II (1–2)
Rhytidiadelphus triquetrus	IV (1–5)	*Ranunculus acris*	II (1–3)
Carex pulicaris	IV (1–4)	*Thuidium delicatulum*	II (1–4)
Tortella tortuosa	IV (1–4)	*Solidago virgaurea*	II (1–3)
Festuca ovina	IV (3–9)	*Festuca rubra*	II (2–6)
Ditrichum flexicaule	IV (1–6)	*Agrostis capillaris*	II (1–3)
Viola riviniana	IV (1–4)	*Dicranum scoparium*	II (1–3)
Racomitrium lanuginosum	III (1–4)	*Carex rupestris*	II (2–4)
Succisa pratensis	III (1–4)	*Breutelia chrysocoma*	II (1–5)
Salix reticulata	III (1–8)	*Hypnum cupressiforme*	II (1–3)
Angelica sylvestris	III (1–3)	*Pseudoscleropodium purum*	II (1–3)
Coeloglossum viride	III (1–3)	*Lotus corniculatus*	II (1–4)
Geranium sylvaticum	III (1–3)	*Calluna vulgaris*	II (1–5)
Asplenium viride	III (1–3)	*Potentilla erecta*	II (1–3)
Vaccinium vitis-idaea	III (1–5)	*Carex atrata*	II (1–4)
Galium boreale	III (1–4)	*Thuidium abietinum*	I (1–3)
Deschampsia cespitosa	III (2–5)	*Salix arbuscula*	I (1–5)
Avenula pratensis	III (1–4)	*Alchemilla filicaulis vestita*	I (1–3)
Pinguicula vulgaris	III (1–3)	*Cetraria islandica*	I (1–3)
Anthoxanthum odoratum	III (2–4)	*Rhytidium rugosum*	I (1–3)
Festuca vivipara	III (1–4)	*Minuartia sedoides*	I (1–3)
Linum catharticum	III (1–3)	*Drepanocladus uncinatus*	I (1–4)
Antennaria dioica	II (1–4)	*Cerastium alpinum*	I (1–3)
Pyrola rotundifolia	II (1–4)	*Ptilium crista-castrensis*	I (1–3)
Alchemilla glabra	II (1–4)	*Tritomaria quinquedentata*	I (1–3)
Vaccinium uliginosum	II (2–5)	*Racomitrium canescens*	I (1–3)
Luzula sylvatica	II (1–4)	*Rhytidiadelphus squarrosus*	I (1–3)
Vaccinium myrtillus	II (1–3)	*Galium saxatile*	I (1–3)
Empetrum nigrum hermaphroditum	II (1–5)	*Thuidium tamariscinum*	I (1–3)
Euphrasia officinalis agg.	II (1–3)	*Hieracium pilosella*	I (1–3)
Frullania tamarisci	II (1–3)	*Hypericum pulchrum*	I (1–3)
Carex flacca	II (1–3)	*Prunella vulgaris*	I (1)
Campylium stellatum var. *protensum*	II (1–3)	*Bellis perennis*	I (1)
Huperzia selago	II (1–4)	*Luzula campestris*	I (1)
Alchemilla filicaulis filicaulis	II (1–2)	*Cerastium fontanum*	I (1–3)
		Scapania aspera	I (1–3)

Floristic table CG14 (*cont.*)

Entodon concinnus	I (1)		
Peltigera canina	I (1–3)	*Draba incana*	I (1)
Saxifraga hypnoides	I (1–3)	*Barbilophozia barbata*	I (1)
Anemone nemorosa	I (1)	*Bryum pseudotriquetrum*	I (1)
Armeria maritima	I (1–4)	*Orthilia secunda*	I (1)
Crepis paludosa	I (1–2)	*Sphagnum capillaceum*	I (1)
Cladonia pyxidata	I (1)	*Polystichum lonchitis*	I (1)
Cladonia gracilis	I (1–2)	*Peltigera aphthosa*	I (1)
Arabis hirsuta	I (1)	*Heracleum sphondylium*	I (1)
Gentianella campestris	I (1–2)	*Salix aurita*	I (1)
Veronica fruticans	I (1)	*Bartsia alpina*	I (1)
Botrychium lunaria	I (1–3)	*Ptilidium ciliare*	I (1–2)
Fissidens adianthoides	I (1–2)	*Nardia scalaris*	I (1)
Potentilla crantzii	I (1–3)	*Euphrasia frigida*	I (1)
Orthothecium rufescens	I (1)	*Deschampsia flexuosa*	I (2)
Fissidens osmundoides	I (1–2)	*Taraxacum officinale* agg.	I (1–2)
Preissia quadrata	I (1)	*Encalypta streptocarpa*	I (1)
Plagiobryum zierii	I (1)	*Solorina saccata*	I (1)
Metzgeria leptoneura	I (1)	*Rubus saxatilis*	I (2–4)
Plagiopus oederi	I (1–4)	*Leontodon autumnalis*	I (1–2)
Blindia acuta	I (1–4)	*Plantago lanceolata*	I (1–2)
Carex vaginata	I (1–2)	*Plantago maritima*	I (1–2)
Trollius europaeus	I (1–4)	*Lathyrus montanus*	I (1)
Salix myrsinites	I (2–3)	*Danthonia decumbens*	I (1–2)
Sesleria albicans	I (1–7)	*Cerastium arcticum*	I (1)
Cladonia arbuscula	I (1–3)	*Salix × phaeophylla*	I (5)
Salix herbacea	I (1–3)	*Salix × boydii*	I (3)
Luzula spicata	I (1)		
Salix lapponum	I (1–3)	Number of samples	33
Equisetum variegatum	I (1)	Number of species/sample	42 (23–62)

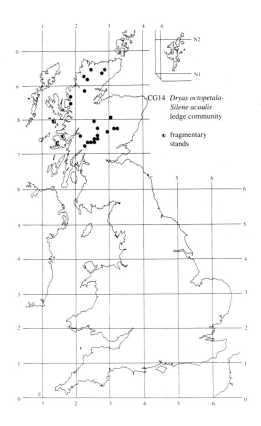

CG14 *Dryas octopetala-Silene acaulis* ledge community

o fragmentary stands

CALCIFUGOUS GRASSLANDS AND MONTANE COMMUNITIES

INTRODUCTION TO CALCIFUGOUS GRASSLANDS AND MONTANE COMMUNITIES

The sampling of calcifugous grasslands and montane vegetation

Compared with their calcicolous counterparts, swards with a prevailingly calcifuge flora, long known in Britain as acid or acidic grasslands, have commanded somewhat sporadic interest. In the earliest descriptive accounts, for example, fine-leaved grasslands in the uplands received much less attention than communities dominated by *Molinia caerulea* and *Nardus stricta* (Tansley 1911) and, only with the agriculturalists' concern to recognise different grades of hill pasture, was more precise classification of the better-quality swards undertaken. Then, though it quickly became apparent that such grasslands in the north and west were generally similar to grazed calcifugous swards and grass-heaths persisting locally in the south-east, no concerted effort was made to integrate these vegetation types into a single framework (Tansley 1939). Also, subsequent attempts to develop an overall perspective on these grasslands have usually relied on rather compendious definitions of a '*Festuca-Agrostis* complex', within which somewhat disparate kinds of sward have been treated together. For the present survey, then, we were very concerned to begin afresh, benefiting from this wealth of experience, but sampling the full range of calcifugous grasslands from throughout Britain and, within the context of the broader survey, defining communities free of preconceived notions about how they ought to relate to one another, or to more calcicolous and mesotrophic swards and floristically similar heaths and mires.

We wished to do full justice to the distinctive elements among our montane vegetation. These were among the first plant communities in Britain to be described in a phytosociological fashion (Poore 1955c, Poore & McVean 1957, McVean & Ratcliffe 1962), and we were very conscious of our indebtedness to the insights of these early accounts. At the same time, we tried to avoid any over-representation of rare or more species-rich assemblages among the data, and many samples were

collected from commonplace and run-down grasslands, from among derelict heaths and mires, and in the improved and afforested landscapes that are now so widespread through the upland fringes. As always, samples were located strictly according to the floristic and structural homogeneity of the vegetation and, though more intricately patterned swards and chaotically varied stands of tall-herb and fern vegetation sometimes posed problems, uniformity was usually not hard to detect.

Where more obvious repetitive patterning occurred within an assemblage, with associates organised around some tussocky dominant, for example, or where plants were disposed over patterned ground, such variation was treated as an integral part of the vegetation and sampled in its entirety. Where distinctly different assemblages were closely juxtaposed in a mosaic, these were sampled separately and notes made on the relationship of the vegetation types one to another. Usually quadrats of 2×2 m were adequate to provide a representative sample of the vegetation, with 4×4 m sometimes necessary in more coarsely structured assemblages. Oddly-shaped stands, such as were found on narrow ledges and banks, on snow-bed cornices and talus fans, were sampled using differently shaped quadrats of the appropriate area, while small stands were sampled in their entirety. Some vegetation included here, like the tall-herb communities, showed complex vertical structuring but it was always sufficient to sample this in its entirety rather than distinguishing layers for species recording.

In the usual way, all vascular plants, bryophytes and macrolichens were recorded from each sample plot and Domin scores assigned to them. Particular care was taken with the identification of grasses, although it was sometimes very difficult, in close-grazed swards, to be certain of the diagnosis of cropped vegetative shoots. *Festuca ovina* and *F. rubra* could be especially awkward in this respect and there were also many occasions on which it seemed best to conflate *F. vivipara* with the former. The subspecies of *Agrostis canina* could also be difficult to identify and *Deschampsia cespitosa* was usually recorded to the species, though the likely presence of ssp. *alpina* is noted in the text. *Bromus hordeaceus* ssp. *hordeaceus* was generally distinguishable although, by summer, the remains of this and other diminutive ephemerals could be shrivelled beyond recognition in more drought-prone swards. *Alchemilla filicaulis* ssp. *filicaulis* and ssp. *vestita* were recorded separately, but *Taraxacum officinale* and *Rubus fruticosus* were scored as the aggregates, and *Lophocolea bidentata s.l.* could include some infertile *L. cuspidata*. *Euphrasia officinalis* was usually recorded as the aggregate, although it was sometimes possible to recognise *E. frigida*. Apart from *Hieracium pilosella*, Hieracia were recorded to the genus, although the prominence of particular sections among the vegetation sampled is remarked upon in the community descriptions.

The quantitative floristic information was often supplemented by notes on the structure of the vegetation providing details of such things as the dominance of particular species or groups of plants, the growth form and vigour of different elements and any patterning among the associates. Indications of phenological relationships were recorded and the influence of accumulating litter or standing dead material. The contribution of the stand to the overall landscape was outlined, with descriptions of zonations or abrupt transitions to neighbouring vegetation and any suggestion of successional processes (Figure 24).

Slope, aspect and altitude were recorded in the usual fashion for each sample, and details of geology and soil provided where possible. In addition to noting the nature of the bedrock and any superficials, it was often helpful to have information on the physiography and microrelief, and the character of any rock exposures – the disposition of cliff ledges and faces, and the size, shape and stability of rock debris in screes and boulder fields. The pattern of drainage was also noted, particularly where there was flushing or parching, and a record made of any effects of wind, snow or frost on the substrate or the vegetation. Where a soil pit was dug and the kind of profile recorded, special care was taken to observe any signs of podzolisation, gleying, frost-heave or solifluction. Biotic impacts on the herbage or the habitat were catalogued, especially the influence of grazing by stock or wild herbivores through cropping of the vegetation, trampling or manuring, of the burning of heathy grasslands, of draining and other kinds of agricultural improvement, and of the pressure of recreational activities. Many of these data were of an informal qualitative character, reinforcing once again just how much we are still in need of systematic data on such things as soil fertility and reaction, snow-lie, and the carrying capacity and grazing efficiency of different kinds of sward.

As so often in our work, we were the beneficiaries of many data and a great deal of experience which greatly enhanced our coverage and understanding of these vegetation types. We were especially fortunate not just to have access to published and unpublished material from early studies of our montane vegetation (Poore 1955c, Poore & McVean 1957, McVean & Ratcliffe 1962), but also to have the expertise of Dr Derek Ratcliffe at the service of the research team. Continuing work at the Macaulay Institute (Birse & Robertson 1976, Birse 1980, 184) added greatly to our coverage of the Scottish uplands outside the Highlands and, along with published studies of particular sites or mountain massifs, like those of Ratcliffe (1959), Welch (1967), Eddy *et al.* (1969), Edgell (1969), Birks (1973), Prentice & Prentice (1975) and Huntley (1979), we were able to make use of unpublished reports from the NCC and the Institute of Terrestrial Ecology, such as those by Meek (1976), Evans *et al.* (1977), Hill & Evans (1978) and Ferreira (1978). In our own sampling, we were especially concerned to spread our interest across the full range of calcifuge swards and upland vegetation types, so as to set our understanding of the more distinctive and unusual communities in the broader context of a national survey. Over 1900 samples were available for analysis with geographical coverage as shown in Figures 25 and 26.

Data analysis and the description of calcifugous grassland and montane communities

As usual, data analysis used just the floristic records from the samples for producing the classification of these vegetation types, with environmental information being employed as a test of the ecological meaning of the emergent groups. Quantitative records for all the vascular plants, bryophytes and macrolichens were used, without any special weighting of rarities or supposed indicators of particular kinds of vegetation or habitat conditions. For this phase of the work, we were greatly indebted to Dr Hilary Birks who, in the space of a brief sub-contract to the project, produced a comprehensive preliminary classification of these vegetation types, together with related heaths and mires. Mr Steve Ridgill, then of the NCC Chief Scientist Directorate Uplands Team, was kindly freed to assist with summarising the results of these computer analyses.

In all, we have recognised 21 communities in this section of the work though, as always among the various volumes, readers may be disconcerted about exactly what has been included here. Figuring prominently are calcifuge swards dominated by such grasses as *Festuca ovina*, *F. rubra*, *Agrostis capillaris*, *A. curtisii*, *Anthoxanthum odoratum*, *Deschampsia flexuosa* and *Nardus*. These include the bulk of the Nardo-Galion grasslands in Britain and five communities have been distinguished

loped from lime-poor rocks through upland Britain, the altogether more calcifuge *Luzula sylvatica-Vaccinium myrtillus* community (U16) is characteristic. There, on humic rankers kept permanently moist by the humid climate, protection from grazing permits no chance of sustained succession to scrub, but encourages mixtures of *L. sylvatica* and *V. myrtillus* along with *Deschampsia flexuosa*, *Galium saxatile*, *Oxalis acetosella* and *Blechnum spicant*, and bryophytes like *Rhytidiadelphus loreus*, *R. squarrosus* and *Dicranum scoparium*. Taller dicotyledons are generally scarce, but the rare *Cicerbita alpina* usually belongs in this kind of vegetation and, where shelter and seepage bring increased humidity and some nutrient enrichment, ferns like *Dryopteris dilatata*, *Gymnocarpium dryopteris*, *Thelypteris phegopteris* and *T. limbosperma* can become very abundant, providing a clear floristic link with the Quercion forests of base-poor soils. By contrast, where stock have access, an increase in calcifuge grasses mediates transitions to close-cropped Nardo-Galion swards. To emphasise its floristic distinction from the *Luzula-Geum* community, it seems best to locate the *Luzula-Vaccinium* vegetation in the Dryoptero-Calamagrostidion, an alliance of less basiphilous tall-herb assemblages.

Calcifuge fern communities

The prevalence of grazing through much of upland Britain means that it is often difficult to follow lengthy successions on the base-poor soils that are so widespread in the cooler and wetter parts of the country. Early stages in colonisation, however, are often to be seen on fresh exposures of acid rocks in screes and boulder fields. Here, mosses like *Andreaea rupestris* and *Racomitrium fasciculare* are among the first invaders of rock surfaces with *R. heterostichum* and *Dicranella heteromalla* appearing on slowly-accumulating mineral detritus, and *Diplophyllum albicans* in sheltered crevices. At lower altitudes in the north and west, and especially in more oceanic regions where humid conditions combine with equable temperatures, the fern *Cryptogramma crispa* is an early colonist among such assemblages on the raw acid soils developing between coarser and more stable rock debris. The structural organisation of the vegetation is strongly controlled by the disposition of more sheltered spaces among the talus and enrichment of the associated flora depends much on the accumulation of mor from decaying fern fronds. Typically, though, the *Cryptogramma-Deschampsia flexuosa* community (U21) that develops in such situations has an open cover of the fern, together with small tussock grasses like *D. flexuosa*, *Festuca ovina*, *Nardus* and *Anthoxanthum*, and chamaephyte or rhizomatous herbs such as *Galium saxatile*, *Huperzia selago* and *Diphasium alpinum*. Some bryophytes can persist from earlier stages, but others like *Campylopus paradoxus*, *Poly-*

trichum formosum and *Dicranum scoparium* also appear in the shelter of the rocks and herbage. In phytosociological schemes, *Cryptogramma* is regarded as characteristic of the order Androsacetalia, but central European vegetation with this fern is more obviously montane than our own.

Cryptogramma does persist at higher altitudes but, on upper mountain slopes, and especially in drier regions, its survival is dependent on long snow-lie for the provision of a habitat protected from desiccation and frost. Above 600 m, though, and particularly on north- and east-facing slopes where low temperatures are combined with sheltered and humid conditions, the Arctic-Alpine fern *Athyrium distentifolium* also becomes common on the humic rankers that develop among blocky acid rock debris. In the *Cryptogramma-Athyrium* community (U18) of such snow-beds, clumps of the two ferns are often abundant, the late spring melt triggering a flush of frond growth and keeping the soils moist through the brief summer. The associated flora includes other chionophilous herbs and bryophytes like *Saxifraga stellaris*, *Omalotheca supina*, *Kiaeria starkei* and *Dicranum fuscescens*, but there are also fragments of assemblages that, over unbroken stretches of humic soils irrigated by frigid melt-waters, would thicken up into a damp calcifuge sward. *Deschampsia flexuosa* and *D. cespitosa* are often common, along with *Alchemilla alpina*, *Viola palustris*, *Rumex acetosa*, *Rhytidiadelphus loreus*, *Polytrichum alpinum* and *Hylocomium splendens*. Vegetation of this kind has generally been included among the more acidophilous tall-herb communities of the Dryoptero-Calamagrostidion.

Another fern that is able to colonise fairly bare mineral soils among acid rocks is *Blechnum spicant* but, with *Thelypteris limbosperma*, this is more commonly found among fern-dominated vegetation on maturer base-poor soils with a humose top in the cooler and wetter parts of Britain. The *Thelypteris limbosperma-Blechnum* community (U19) is most characteristic of brown podzolic soils developed on sheltered hill slopes and ravine sides in the mountainous west of the country. There, high rainfall keeps the climate consistently humid and there is local relief from desiccating winds, but the ground is sufficiently steep always to maintain free drainage. *T. limbosperma* is generally abundant with scattered *Blechnum* and an associated flora typical of moist, base-poor mineral soils: *F. ovina*, *A. capillaris*, *Vaccinium myrtillus*, *Galium saxatile*, *Potentilla erecta*, *Oxalis acetosella*, *Viola riviniana*, *Dicranum scoparium*, *Hylocomium splendens* and *Pleurozium schreberi*. Grazing mediates some variation in the community and there are often continuous gradations to grassy swards with relict fern clumps or, in more inaccessible situations, to Quercion or Dicrano-Pinion woodland of a mesophytic character. Indeed, the *Thelypteris-Blechnum* vegetation could be considered as an immature assemblage in either of these alliances, although it is probably more sensible to place it among the Dryoptero-Calamagrostidion communities.

Much more widespread and extensive than any of these kinds of fern vegetation are bracken-dominated stands. *Pteridium aquilinum* is tolerant of quite a wide range of climatic and edaphic conditions, although it becomes overwhelmingly abundant within rather clearly defined environmental limits and in particular circumstances. For vigorous growth, it prefers deeper, drought-free, lime-poor soils and is very well adapted to capitalise on injudicious forest clearance and pasturing. Very often, the stands are an obvious replacement for woodland, where bracken has been able to expand vigorously when released from the check of canopy shade. From the more fertile soils in lowland Britain, more mesophytic bracken vegetation has been included in a *Pteridium-Rubus fruticosus* underscrub (W25) which is described in this scheme among the woodlands and scrub of Volume 1. On moist, but well-aerated, base-poor soils of lower fertility, such as occur locally

among the heaths and commons of our lowlands, but are found much more extensively through the sub-montane north and west, bracken dominates in a more calcifuge *Pteridium-Galium saxatile* community (U20). Associates are usually few and sparse because, by mid-summer, the shade cast by the annual crop of fronds is often extremely dense, and the accumulation of litter each autumn so bulky and smothering. Scattered and puny plants of *F. ovina*, *D. flexuosa*, *A. capillaris*, *Anthoxanthum*, *G. saxatile* and *P. erecta* are the usual companions, together with occasional *Dicranum scoparium*, *Pleurozium schreberi*, *Rhytidiadelphus squarrosus*, *Hypnum cupressiforme* and *Pseudoscleropodium purum*. Most of the variation within this kind of vegetation relates to the prominence of a grassy or heathy element, and the community is often found in mosaics and zonations that are under the control of grazing and burning and part of long and complex processes of woodland clearance or degeneration. In phytosociological terms, the *Pteridium-Galium* vegetation is hard to place, though ecologically it can be considered as a replacement for Quercion forest.

KEY TO CALCIFUGOUS GRASSLANDS AND MONTANE COMMUNITIES

With something as complex and variable as vegetation, no key can pretend to offer an infallible short cut to diagnosis. The following should therefore be seen as a crude guide to identifying the types of calcifugous grassland and montane vegetation in the scheme and must always be used in conjunction with the data tables and community accounts. It relies on floristic (and, to a lesser extent, physiognomic) features of the vegetation and demands a knowledge of the British vascular flora and some bryophytes and lichens. It does not make primary use of any habitat features, although these can provide a valuable confirmation of a diagnosis.

Because the major distinctions between the vegetation types in the classification are based on inter-stand frequency, the key works best when sufficient samples of similar composition are available to construct a constancy table. It is the frequency values in this (and, in some cases, the ranges of abundance) which are then subject to interrogation with the key. Most of the questions are dichotomous and notes are provided at particularly difficult choices or where confusing mosaics and zonations are likely to be encountered.

Samples should always be taken from homogeneous stands and be of 2 × 2 m or 4 × 4 m according to the scale of the vegetation or, where complex patterns occur, of identical size but irregular shape. Very small stands can be sampled in their entirety.

---

1   Open or closed grassy swards dominated by one or more of *Festuca ovina, Agrostis capillaris, A. curtisii* or *Deschampsia flexuosa* with *Galium saxatile* and *Potentilla erecta* often common; *Nardus stricta, Deschampsia cespitosa, Juncus squarrosus* and *Carex bigelowii* usually occasional at most but, even if frequent, then not present in any abundance   2

F. ovina, A. capillaris, A. curtisii and *D. flexuosa* can be frequent but typically subordinate in cover   15

2   Generally closed swards dominated by either *D. flexuosa* or *A. curtisii*, often with a little *Calluna vulgaris*; *F. ovina* and *A. capillaris* can be frequent but always subordinate in cover   3

D. flexuosa and *A. curtisii* can occur but never as overwhelming dominants   5

3   *A. curtisii* constant and often very abundant with frequent *Danthonia decumbens, Ulex gallii, Polygala serpyllifolia* and *Carex pilulifera*

      U3 *Agrostis curtisii* grassland

Local prominence of *Calluna, U. gallii* and *Erica cinerea*, together with *Molinia caerulea* or of *Vaccinium myrtillus* and *Nardus*, can mark a shift to various kinds of damp heath, for which this grassland can be a temporary post-burn replacement.

D. flexuosa dominant with *A. curtisii, Danthonia, U. gallii, P. serpyllifolia* and *C. pilulifera* rare or absent

      U2 *Deschampsia flexuosa* grassland   4

4   *V. myrtillus* and *Empetrum nigrum* ssp. *nigrum* very common in small amounts with *Molinia, Juncus effusus, J. squarrosus* or *Eriophorum vaginatum* occasionally quite prominent; *Pleurozium schreberi* frequent, sometimes with *Polytrichum commune*

      U2 *Deschampsia flexuosa* grassland
      Vaccinium myrtillus sub-community

F. ovina and *A. capillaris* frequent with little or no *V. myrtillus* or *E. nigrum* ssp. *nigrum* or other listed associates; *Galium saxatile* and *Potentilla erecta* common with occasional to frequent *Pteridium aquilinum, Rumex acetosa* and *R. acetosella; Epilobium angustifolium* and birch and oak saplings sometimes quite prominent

> **U2** *Deschampsia flexuosa* grassland
> *Festuca ovina-Agrostis capillaris* sub-community

5   Usually open swards dominated by small tussocks of *F. ovina* and, more sporadically, *A. capillaris*, sometimes with *Koeleria macrantha; R. acetosella* very common with occasional to frequent *Hieracium pilosella* and *Senecio jacobaea* but *Anthoxanthum odoratum, G. saxatile* and *P. erecta* of restricted occurrence; small ephemeral dicotyledons or lichens sometimes very prominent

> **U1** *Festuca ovina-Agrostis capillaris-Rumex acetosella* grassland   6

Generally closed swards dominated by mixtures of *F. ovina, F. rubra, Agrostis capillaris* and *Anthoxanthum* with frequent *G. saxatile, P. erecta, Luzula campestris* and *Viola riviniana,* but *R. acetosella, Senecio jacobaea* and small ephemerals scarce; *Hypnum cupressiforme, Rhytidiadelphus squarrosus, Pseudoscleropodium purum* and *Dicranum scoparium* common but lichens usually sparse

> **U4** *Festuca ovina-Agrostis capillaris-Galium saxatile* grassland   11

6   *F. ovina* and *R. acetosella* very much the commonest vascular plants in often very open tussocky swards with *Dicranum scoparium, Polytrichum piliferum, P. juniperinum* and *Ceratodon purpureus* frequent and patchily abundant and lichens varied and extensive, commonly including *Cornicularia aculeata, Cladonia arbuscula, C. impexa, C. tenuis, C. foliacea* and *C. uncialis*

> **U1** *Festuca ovina-Agrostis capillaris-Rumex acetosella* grassland
> *Cornicularia aculeata-Cladonia arbuscula* sub-community

Above-listed bryophytes and lichens usually no more than occasional and rather sparse elements in the sward
7

7   Mixtures of *F. ovina, A. capillaris* and *K. macrantha* generally form the bulk of the sward with frequent *R. acetosella, S. jacobaea, H. pilosella* and rich and varied contingents of small ephemerals including *Aira praecox, Erodium cicutarium, Teesdalia nudicaulis, Aphanes arvensis, Myosotis ramossisima, Erophila verna* and *Veronica arvensis; Astragalus danicus* can be locally prominent

> **U1** *Festuca ovina-Agrostis capillaris-Rumex acetosella* grassland
> *Erodium cicutarium-Teesdalia nudicaulis* sub-community

Above-listed ephemerals no more than a very occasional element in the sward   8

8   *Anthoxanthum* frequent and *Holcus lanatus* occasional in an often more extensive grassy cover with *Lotus corniculatus, Galium verum* and *Achillea millefolium* common

> **U1** *Festuca ovina-Agrostis capillaris-Rumex acetosella* grassland
> *Anthoxanthum odoratum-Lotus corniculatus* sub-community

Above-listed grasses and dicotyledons usually no more than occasional   9

9   *D. flexuosa* common with *G. saxatile* and *P. erecta*

> **U1** *Festuca ovina-Agrostis capillaris-Rumex acetosella* grassland
> *Galium saxatile-Potentilla erecta* sub-community

Above-listed species usually no more than occasional
10

10   *F. rubra* often partly replacing *F. ovina* with *Hypochoeris radicata* constant, *Centaurium erythraea* and *Leontodon taraxacoides* occasional; *Sedum anglicum* and *Umbilicaria rupestris* sometimes prominent

> **U1** *Festuca ovina-Agrostis capillaris-Rumex acetosella* grassland
> *Hypochoeris radicata* sub-community

Above-listed species all rare

> **U1** *Festuca ovina-Agrostis capillaris-Rumex acetosella* grassland
> Typical sub-community

11   *F. rubra* often co-dominant with *A. capillaris* and *Anthoxanthum; Holcus lanatus* common and sometimes moderately abundant with occasional to frequent *Poa pratensis, Dactylis glomerata* and *Cynosurus cristatus; Galium saxatile* and *Potentilla erecta* rather less common than usual but *Trifolium repens, Achillea millefolium* and *Cerastium fontanum* common and *Prunella vulgaris, Veronica chamaedrys* and *Bellis perennis* occasional

> **U4** *Festuca ovina-Agrostis capillaris-Galium saxatile* grassland
> *Holcus lanatus-Trifolium repens* sub-community

U8 *Carex bigelowii-Polytrichum alpinum* sedge-heath
Polytrichum alpinum-Ptilidium ciliare sub-community

C. bigelowii often co-dominant with or subordinate to *D. fuscescens* with frequent *P. alpinum* and *R. lanuginosum*; *V. myrtillus*, *F. ovina/vivipara*, *Agrostis capillaris* and *Galium saxatile* all common

U8 *Carex bigelowii-Polytrichum alpinum* sedge-heath
Dicranum fuscescens-Racomitrium lanuginosum sub-community

Around the margins of snow-beds through the central Scottish Highlands or over slopes with patchworks of more and less exposed ground, it can be difficult to distinguish these vegetation types from *Nardus-Carex* grass-heath, but there *Nardus* and, in more free-draining situations, *V. myrtillus* become much more prominent while *P. alpinum* and *D. fuscescens* are reduced in frequency and cover.

35   Vegetation often rather grassy or heathy with mixtures of *C. bigelowii*, *F. ovina/vivipara*, *D. flexuosa* and *V. myrtillus* quite often co-dominant with *R. lanuginosum*; *Galium saxatile*, *Potentilla erecta* and *Carex pilulifera* frequent

U10 *Carex bigelowii-Racomitrium lanuginosum* moss-heath
Galium saxatile sub-community

On summits at somewhat lower altitudes in southern Scotland and the Pennines, especially where there is some grazing, this vegetation type can be difficult to separate from *Racomitrium*-rich stands of Nardo-Galion grasslands, but the shift there to dominance of fine-leaved grasses and the eventual eclipse of *C. bigelowii*, *P. alpinum*, *Salix herbacea* and lichens should help effect a separation.

F. ovina/vivipara, *D. flexuosa* and *V. myrtillus* can be frequent but not usually in any abundance and *P. erecta* and *C. pilulifera* infrequent; *R. lanuginosum* is typically the most plentiful plant although the cover can be quite open   36

36   *C. bigelowii*, *D. flexuosa* and *V. myrtillus* frequently much reduced in cover although *F. ovina/vivipara* is often plentiful, together with mat or cushion

herbs like *Salix herbacea*, *Alchemilla alpina*, *Silene acaulis*, *Thymus praecox*, *Armeria maritima*, *Polygonum viviparum*, *Luzula spicata*, *Sibbaldia procumbens* and *Minuartia sedoides*; *Rhytidiadelphus loreus*, *Hylocomium splendens* and *Pleurozium schreberi* can be locally abundant where there is intermittent irrigation

U10 *Carex bigelowii-Racomitrium lanuginosum* moss-heath
Silene acaulis sub-community

R. lanuginosum usually the most abundant plant with sometimes plentiful *C. bigelowii* or *F. ovina* but rather patchy *D. flexuosa* and *V. myrtillus*; mat and cushion herbs never prominent but lichen flora can be varied and extensive with frequent *Cetraria islandica*, *Cladonia uncialis*, *C. arbuscula* and *C. gracilis*

U10 *Carex bigelowii-Racomitrium lanuginosum* moss-heath
Typical sub-community

37   Bryophyte dominated carpets with one or more of *Kiaeria starkei*, *Polytrichum sexangulare*, *Oligotrichum hercynicum*, *Conostomum tetragonum*, *Racomitrium heterostichum*, *R. fasciculare*, *Pleurocladula albescens* and *Marsupella brevissima* in high-altitude snow-beds
                                                                                                  38

Above species can be present but only as an infrequent or minor element in vegetation dominated by luxuriant carpets of small herbs, tall herbs, ferns or sub-shrubs   41

38   *K. starkei*, *O. hercynicum* and *P. sexangulare* constant and *R. fasciculare* common with scattered *Deschampsia cespitosa* and *Saxifraga stellaris*; *Salix herbacea* only occasional

U11 *Polytrichum sexangulare-Kiaeria starkei* snow-bed   39

Snow-bed vegetation with a striking local abundance of *Pleurocladula albescens* is best incorporated in this community.

K. starkei and *O. hercynicum* can be quite common but *R. heterostichum* is constant and often abundant with frequent *Conostomum tetragonum*; *Salix herbacea* constant and often plentiful but *D. cespitosa* only occasional and *Saxifraga stellaris* very scarce

U12 *Salix herbacea–Racomitrium heterostichum* snow-bed   40

Snow-bed vegetation with a striking local abundance of *Marsupella brevissima* among *R. fasci-*

culare and *O. hercynicum* is best included in this community.

---

39   *Polytrichum alpinum, C. tetragonum, R. heterostichum, R. lanuginosum, R. canescens* and *Barbilophozia floerkii* very common with scattered plants of *Carex bigelowii, Huperzia selago, Nardus* and *Alchemilla alpina*

> **U11** *Polytrichum sexangulare-Kiaeria starkei* snow-bed
> Typical sub-community

Above species all scarce but *Lophozia sudetica, Pohlia ludwigii* or *Kiaeria falcata* can be prominent

> **U11** *Polytrichum sexangulare-Kiaeria starkei* snow-bed
> Species-poor sub-community

40   Mixtures of *S. herbacea, R. heterostichum* and *K. starkei* usually dominant with frequent *P. alpinum* and *Pohlia nutans* and scattered plants of *C. bigelowii, Luzula spicata, Silene acaulis* and *Juncus trifidus*

> **U12** *Salix herbacea-Racomitrium heterostichum* snow-bed
> *Silene acaulis-Luzula spicata* sub-community

S. herbacea can be abundant among patches of *R. lanuginosum, R. heterostichum* or *R. fasciculare* but much of the ground is usually covered by a crust of leafy hepatics such as *Gymnomitrion concinnatum, Nardia scalaris* and *Diplophyllum albicans*

> **U12** *Salix herbacea-Racomitrium heterostichum* snow-bed
> *Gymnomitrion concinnatum* sub-community

R. fasciculare and *O. hercynicum* can occur with sparse sprigs of *S. herbacea* but *Marsupella brevissima* is very abundant in large, pure patches usually without much *G. concinnatum, N. scalaris* or *D. albicans*

> **U12** *Salix herbacea-Racomitrium heterostichum* snow-bed
> *Marsupella brevissima* sub-community

41   *Luzula sylvatica* constant and often abundant in heathy or tall-herb vegetation with at least some *Deschampsia cespitosa, D. flexuosa, Vaccinium myrtillus, F. ovina/vivipara, Geum rivale, Angelica sylvestris, Rhodiola rosea, Filipendula ulmaria, Alchemilla glabra*   42

L. sylvatica can be quite frequent but not in any abundance or among assemblages of this kind   48

42   *V. myrtillus* very frequent and usually quite abun-

dant with constant *D. flexuosa* and *G. saxatile*, and frequent *Oxalis acetosella* and *Blechnum spicant; F. ulmaria, A. glabra, Angelica, G. rivale* and *Rhodiola* no more than very occasional

> **U16** *Luzula sylvatica-Vaccinium myrtillus* tall-herb community   43

L. sylvatica often less prominent with *V. myrtillus* sometimes scarce and rarely abundant, *D. flexuosa* and *G. saxatile* only locally common; *Angelica, Rhodiola, G. rivale, Alchemilla glabra* and *F. ulmaria* very common

> **U17** *Luzula sylvatica-Geum rivale* tall-herb community   45

43   *L. sylvatica* overwhelmingly dominant with few associates and even *V. myrtillus* and *G. saxatile* of lower frequency and cover

> **U16** *Luzula sylvatica-Vaccinium myrtillus* tall-herb community
> Species-poor sub-community

B. spicant and *O. acetosella* very frequent with *Rhytidiadelphus loreus* common, *Hypnum cupressiforme, Dicranum scoparium, Hylocomium splendens* and *Diplophyllum albicans* occasional to frequent in richer and more diverse vegetation   44

44   Often tall and luxuriant vegetation with frequent and sometimes abundant *Deschampsia cespitosa* and occasional *Anemone nemorosa, Viola riviniana* and *Huperzia selago*; various ferns, most commonly *Dryopteris dilatata* but also *Gymnocarpium dryopteris, Thelypteris phegopteris* and *Athyrium filix-femina*, may be locally prominent; bryophytes often numerous and quite extensive with frequent *Plagiothecium undulatum, Mnium hornum, Dicranum majus, Thuidium tamariscinum* and occasional patches of *Sphagnum subnitens, S. capillifolium* and *S. girgensohnii*

> **U16** *Luzula sylvatica-Vaccinium myrtillus* tall-herb community
> *Dryopteris dilatata-Dicranum majus* sub-community

On less heavily grazed stretches of shady and sheltered slopes with long snow-lie in the north-west Scottish Highlands, it can be difficult to distinguish this vegetation from H21 *Calluna-Vaccinium-Sphagnum* heath, but *L. sylvatica, D. cespitosa* and *O. acetosella* are much less common there, while *Calluna* and *Erica cinerea* become more frequent and abundant.

Often shorter swards with the listed dicotyledons, ferns

and bryophytes only occasional at most, but fine-leaved grasses such as *Anthoxanthum*, *F. ovina/vivipara*, *Agrostis capillaris* and *Nardus* frequent, along with *Potentilla erecta*, and *Calluna* and *V. vitis-idaea* occasional; *Hylocomium splendens*, *Rhytidiadelphus squarrosus*, *Polytrichum commune* and *Pseudoscleropodium purum* very common

> **U16** *Luzula sylvatica-Vaccinium myrtillus* tall-herb community
> *Anthoxanthum odoratum-Festuca ovina* sub-community

This vegetation is often found as patches within stands of Nardo-Galion grasslands and sub-montane heaths, when its boundaries may be very ill-defined.

45   *Saxifraga aizoides* common and locally abundant with occasional *Pinguicula vulgaris* and *Tussilago farfara* and, among the bryophytes, frequent *Philonotis fontana* and *Bryum pseudotriquetrum* and occasional *Campylium stellatum*, *Blindia acuta*, *Scapania undulata* and *Aneura pinguis*; *V. myrtillus*, *Anthoxanthum* and *Alchemilla alpina* very scarce

> **U17** *Luzula sylvatica-Geum rivale* tall-herb community
> *Alchemilla glabra-Bryum pseudotriquetrum* sub-community

On dripping base-rich banks and around limy springs, particularly at lower altitudes where an Arctic-Alpine element in the flora is less well represented, it can be difficult to separate this vegetation type from the *Saxifraga-Alchemilla* community. There, however, *S. aizoides* is frequently accompanied by *S. oppositifolia*, *Thalictrum alpinum*, *Carex pulicaris*, *Thymus praecox* and *Selaginella selaginoides* which are at most occasional in *Luzula-Geum* vegetation, while *L. sylvatica* itself and many taller herbs are rather uncommon.

V. myrtillus, *Anthoxanthum* and *Alchemilla alpina* very frequent with *Mnium hornum*, *Thuidium tamariscinum* and *Rhytidiadelphus triquetrus*; *S. aizoides*, *P. fontana* and *B. pseudotriquetrum* very scarce   46

46   Often luxuriant vegetation with *Geranium sylvaticum*, *Trollius europaeus* and *Heracleum sphondylium* especially common and distinctive among the tall herbs, along with frequent *Solidago virgaurea*, *Thalictrum alpinum*, *Saussurea alpina*, *Oxyria digyna*, montane Hiera-

cia and *Polystichum lonchitis* and scattered bushes of *V. vitis-idaea* and *Calluna*; *Rubus saxatilis*, *Silene acaulis* and *Saxifraga oppositifolia* common among smaller plants; *Dicranum scoparium*, *Plagiochila asplenoides* and *Tortella tortuosa* preferentially frequent among the bryophytes

> **U17** *Luzula sylvatica-Geum rivale* tall-herb community
> *Geranium sylvaticum* sub-community

Combinations of the above species lacking   47

47   *Anthoxanthum*, *Agrostis capillaris*, *A. canina* and *D. flexuosa* constant along with *F. ovina* and *D. cespitosa* in often rank grassy swards with frequent *Ranunculus acris*, *Galium saxatile* and *Oxalis acetosella* and occasional *Cerastium fontanum* and *Euphrasia officinalis*; *Rhytidiadelphus loreus*, *R. squarrosus*, *Dicranum majus*, *Polytrichum alpinum*, *Plagiothecium undulatum* and *Lophocolea bidentata s.l.* all common among the bryophytes

> **U17** *Luzula sylvatica-Geum rivale* tall-herb community
> *Agrostis capillaris-Rhytidiadelphus loreus* sub-community

Where *D. cespitosa* is abundant in this kind of vegetation on rather more base-poor ground where there is some grazing, it can be difficult to distinguish it from the *Anthoxanthum-Alchemilla* sub-community of *Deschampsia-Galium* grassland. In those swards, however, *L. sylvatica* and taller dicotyledons are reduced to very occasional occurrences while *Nardus*, *Viola palustris* and *Sibbaldia procumbens* become common.

L. sylvatica generally abundant with *V. myrtillus* much reduced in frequency but *Calluna* common and sometimes plentiful along with frequent *Blechnum*, *Hypericum pulchrum*, *Primula vulgaris*, *Valeriana officinalis* and occasional, locally prominent, *Athyrium filix-femina*, *Thelypteris limbosperma*, *Dryopteris filix-mas* and *D. borreri*; above-listed grasses, smaller dicotyledons and bryophytes all occasional at most

> **U17** *Luzula sylvatica-Geum rivale* tall-herb community
> *Primula vulgaris-Hypericum pulchrum* sub-community

48   Luxuriant carpets dominated by smaller herbs with constant *Saxifraga aizoides*, *S. oppositifolia*, *Selaginella*, *Pinguicula vulgaris*, *Thalictrum alpinum*, *Polygonum viviparum*, *Alchemilla glabra* and *A. alpina*,

together with *Festuca ovina/vivipara*, *F. rubra*, *Deschampsia cespitosa* and *Carex pulicaris*; *Ctenidium molluscum* and *Bryum pseudotriquetrum* common among an often varied and extensive bryophyte cover

U15 *Saxifraga aizoides-Alchemilla glabra* banks

Vegetation dominated by ferns with the above-listed dicotyledons generally infrequent   49

49   *Pteridium aquilinum* constant and generally very abundant with only *G. saxatile*, *P. erecta*, *F. ovina*, *A. capillaris* and *D. flexuosa* at all common and then usually very sparse

U20 *Pteridium   aquilinum-Galium   saxatile* community   50

Cryptogramma crispa constant and patchily abundant with frequent *D. flexuosa*, *Nardus*, *G. saxatile*, *Oxalis acetosella*, *Racomitrium lanuginosum*, *Diplophyllum albicans* and *Barbilophozia floerkii*; *Pteridium* scarce and never abundant   52

Thelypteris limbosperma and *Blechnum spicant* constant, the former in abundance, with little or no *Pteridium* or *C. crispa*; *Dryopteris dilatata*, *Athyrium filix-femina* and, at higher altitudes, *A. distentifolium* occasional; *Oxalis*, *G. saxatile* and *P. erecta* frequent in a grassy ground

U19 *Thelypteris limbosperma-Blechnum spicant* community

Mixtures of these species remain prominent in transitions to some tall-herb vegetation, particularly of the *Luzula-Geum* community, and to ungrazed stands of certain woodlands, particularly W11 *Quercus-Betula-Oxalis* and W18 *Juniperus-Oxalis* woodlands.

---

50   *F. ovina*, *A. capillaris* and *Anthoxanthum* constant with *Holcus lanatus* common and *Viola riviniana*, *Carex pilulifera*, *Luzula campestris*, *Campanula rotundifolia*, *Rumex acetosa* and *Veronica chamaedrys* occasional to frequent in grassy bracken vegetation with little or no ericoid element.

U20 *Pteridium   aquilinum-Galium   saxatile* community
Anthoxanthum odoratum sub-community

F. ovina, *A. capillaris* and *Anthoxanthum* can be frequent but *D. flexuosa* or, in south-west Britain, *Agrostis curtisii* common and locally abundant after burning, and *V. myrtillus* constant and patchily prominent with scattered *Calluna*; *Dicranum scoparium*, *Pleurozium schreberi* and *Hypnum cupressiforme* frequent

U20 *Pteridium   aquilinum-Galium   saxatile* community
Vaccinium   myrtillus-Dicranum   scoparium sub-community

Pteridium usually overwhelmingly abundant with all the listed associates infrequent and puny

U20 *Pteridium   aquilinum-Galium   saxatile* community
Species-poor sub-community

51   Often open or patchy cover on sparsely-vegetated screes with tussocks of *F. ovina* and clumps of *Campylopus paradoxus*, *Polytrichum formosum* and *Dicranum scoparium* frequent; *Athyrium distentifolium*, *Alchemilla alpina*, *Saxifraga stellaris* and *Polytrichum alpinum* all scarce

U21 *Cryptogramma crispa-Deschampsia flexuosa* community

A. distentifolium constant and often dominant with scattered tussocks of *Deschampsia cespitosa* and frequent *A. alpina*, *S. stellaris*, *Viola palustris* and *Omalotheca supina*; *Rhytidiadelphus loreus* constant and sometimes abundant with frequent *Polytrichum alpinum*, *Hypnum callichroum* and *Kiaeria starkei*

U18 *Cryptogramma crispa-Athyrium distentifolium* snow-bed

Although this vegetation type is most characteristic of snow-beds at high altitudes, it can grade to the *Cryptogramma-Deschampsia* community on lower ground.

COMMUNITY DESCRIPTIONS

U1
Festuca ovina-Agrostis capillaris-Rumex acetosella grassland

Synonymy

Graminetum arenosum Tansley 1911 *p.p.*; Grass heath association Tansley 1911, Farrow 1915; *Festuco-Agrostidetum* Watt 1936 *p.p.*; Grass-heath Tansley 1939 *p.p.*; Breckland Grasslands D-G Watt 1940; *Aira praecox-Teesdalia nudicaulis* vegetation Jarvis 1974.

Constant species

Agrostis capillaris, Festuca ovina, Rumex acetosella.

Rare species

Astragalus danicus, Crassula tillaea, Dianthus deltoides, Lychnis viscaria, Scleranthus perennis, Silene conica, S. otites, Thymus serpyllum ssp. *serpyllum, Veronica spicata.*

Physiognomy

The *Festuca ovina-Agrostis capillaris-Rumex acetosella* grassland is a very diverse but highly distinctive vegetation type, with an open sward of small tussocky grasses, among which there can be an abundance of dicotyledons, many of them diminutive ephemerals, and sometimes an extensive cover of lichens and/or mosses. Of the grasses, *Festuca ovina* (only very locally replaced by *F. rubra*) and *Agrostis capillaris* are the most frequently encountered and abundant overall, though their total cover and proportions in individual stands are very variable. *A. capillaris* is the more sporadic and it certainly seems less universally prominent now than in the days of long-continued and heavy grazing by rabbits, an influence of very great importance when the most renowned stands of the community were first described from Breckland (Farrow 1915, Watt 1940, 1960*a*). *A. stolonifera* and *A. canina* ssp. *montana* have also been recorded less often than seems to have been the case in some areas prior to myxomatosis: then they could be common and sometimes abundant, whereas now they are usually no more than occasional.

Other perennial grasses are generally of limited importance in the community. *Koeleria macrantha* is the most widely distributed overall, though it is never more than moderately frequent and is always of low cover, its relative scarcity here contrasting with its common occurrence in the *Festuca-Hieracium-Thymus* grassland, the community which, in this scheme, includes the more calcicolous of the swards grouped together among the Breckland 'grass-heaths' of the classic accounts (Tansley 1911, Farrow 1915, Watt 1940). In other stands included here, where the *Festuca-Agrostis-Rumex* grassland extends a little way on to less parched soils, *Anthoxanthum odoratum* can be found, together with occasional *Holcus lanatus*, though neither of these plays the important role that they assume among Nardo-Galion swards in north-western Britain. *Deschampsia flexuosa* is likewise of restricted occurrence: it has become locally abundant even among some of the drier Breckland stands following myxomatosis (Ratcliffe 1977), but it only attains any frequency here where the rainfall is higher, as in the western Weald and towards the upland fringes, where vegetation transitional to the *Deschampsia* grassland can be found. Beyond the Weald, too, where the *Festuca-Agrostis-Rumex* grassland becomes much more local in the moist and equable climate of south-west England, *Agrostis curtisii* is sometimes seen in close association with stands of the community, but its absence from this vegetation itself provides a clear floristic boundary with more open patches of the *Agrostis* grassland.

The amount of ground between the grass tussocks, which rarely attain much more than 10 cm in height, is sometimes very extensive, particularly where parching, erosion or continued grazing hold their growth severely in check or cause die-back; and, even in swards that have grown a little more rank, where *F. ovina* is now the usual dominant, there is a pattern of death and decay among the tussocks that exposes a patchwork of bare and litter-lined areas (Watt 1971*a*). Over this ground, a wide variety of associates can gain a hold, coming and going according to their own growth pattern and phenology. Among these, perennial vascular plants are usually in a minority, although a small number are distinctive, most notably *Rumex acetosella* agg., sometimes recorded as

R. tenuifolius (e.g. Watt 1960*a*). This is the only constant of the community apart from *F. ovina* and *A. capillaris* and, although it can sometimes be found among the *Festuca-Hieracium-Thymus* swards, it is much more diagnostic of this community and its small tufts of shoots, produced adventitiously from long horizontal roots, can be very numerous. Then, there is quite often some *Hieracium pilosella*, not so consistent an associate here as in the *Festuca-Hieracium-Thymus* grassland, but still sometimes prominent, on occasion even dominating among the grasses (e.g. Bishop *et al.* 1978) and adding a welcome touch of colour in mid-summer, by which time the vegetation often looks very shrivelled and brown. Stonecrops can also provide a bright splash at this time, though they are rather patchy in their occurrence. *Sedum acre*, for example, is more sporadic than in the *Festuca-Hieracium-Thymus* grassland, although it does occur in abundance on some of the more recently reverted arable land that can be included here, as around the Stanford Practical Training Area (Ratcliffe 1977), and the naturalised *S. album* appears to be spreading along track-side stands of the community in various parts of East Anglia. And, where the *Festuca-Agrostis-Rumex* grassland extends into dry, rocky situations at scattered localities through south-west Britain, the Oceanic West European *S. anglicum* can be found, occasionally with the Oceanic Southern *Umbilicus rupestris*, the two of them giving a rather different floristic stamp to the vegetation than is characteristic through the heart of its range. The other prominent chamaephytes of the *Festuca-Hieracium-Thymus* grassland, *Thymus praecox* and *T. pulegioides* are very scarce in this community, although around Breckland the *Festuca-Agrostis-Rumex* grassland provides the typical British locus for their rare relative, *T. serpyllum* ssp. *serpyllum*. Although recently refound on the Chalk in Cambridgeshire (Perring & Farrell 1977), this is generally not a plant of limestone soils with us (Pigott 1955).

Among hemicryptophytes, *Plantago lanceolata* is the most common overall, and its rosettes can be quite numerous, particularly along trackways and in closely-grazed stands, where *P. media* and *P. coronopus* are also occasionally found, this community providing an important inland locus for the latter species through south-east England. *Taraxacum officinale* agg., often identifiable as *T. laevigatum*, also occurs occasionally in most kinds of *Festuca-Agrostis-Rumex* grassland, with *Hypochoeris radicata* diagnostic of one particular sub-community. Then, *Lotus corniculatus*, *Galium verum* and *G. saxatile* are quite frequent through some of the swards included here, although it should be noted that the last seems to be very much less common in this community now than when the vegetation was heavily rabbit-grazed.

Coarse weedy species sometimes make a prominent

show, too, rendered conspicuous by virtue of their unpalatability. *Senecio jacobaea* is the most frequent of these, but there can also be patches of *Epilobium angusti-folium* and, around rabbit burrows, *Urtica dioica*. Small prostrate or decumbent herbs also occur occasionally as patches among the turf: *Cerastium fontanum*, *Potentilla reptans*, *P. argentea*, *Glechoma hederacea* and *Veronica chamaedrys*, with the more short-lived *Spergularia rubra*, *Stellaria media*, *Filaginella uliginosa* and *Sagina apetala*. *Bilderdykia convolvulus* can also be locally abundant, scrambling over the ground.

More easily missed when sampling is undertaken in summer are diminutive ephemerals, many of which behave as winter annuals, beginning growth in autumn when their rosettes can be found on the open ground among the grasses that have been freshened up in the late rains, but shrivelling after flowering in the following spring. Some annual grasses, notably *Aira praecox* and *Poa annua*, with *Bromus hordeaceus* ssp. *hordeaceus* much less common, make a contribution among this group, with *Teesdalia nudicaulis*, *Aphanes arvensis*, *Erodium cicutarium*, *Erophila verna*, *Myosotis ramosissima* and *Veronica arvensis* the most frequent among the dicotyledons, often occurring together and, with more local enrichment, in one particular sub-community. The rare Continental annual *Silene conica* occurs in this vegetation in East Anglia too, and both there and in the scattered stands around Poole Harbour, the Oceanic Southern *Crassula tillaea* is occasionally seen, characteristically marking out trampled areas where compaction impedes the draining away of rainwater (Watt 1971*b*, Crompton & Sheail 1975). Other Continental rarities occurring in the *Festuca-Agrostis-Rumex* grassland are the perennials *Scleranthus perennis* and *Veronica spicata*, for both of which this kind of open vegetation provides an important locus in Breckland (Pigott & Walters 1954), and *Silene otites*, a grazing-sensitive plant excluded from many East Anglian stands in the days of heavy rabbit-infestation, but increasing somewhat since myxomatosis (Watt 1971*b*). As with the *Festuca-Hieracium-Thymus* grassland, this community also quite commonly has some of the Continental Northern *Astragalus danicus* in East Anglia and, on the rocks of Craig Breidden in Powys, this vegetation and adjacent open heath have *Lychnis viscaria*, another Continental Northern plant but one much rarer and more disjunct in its distribution (Jarvis & Pigott 1973, Jarvis 1974).

A different kind of variety is provided by the scattered occurrence in the *Festuca-Agrostis-Rumex* grassland of sub-shrubs usually kept in check by grazing but now often spreading to form heathy mosaics with more open turf. *Calluna vulgaris* itself can be quite frequent in the community in small amounts, with *Ulex minor* appearing in the Weald and *U. gallii* beyond Poole Harbour. *U.*

europaeus also often spreads where stands have been disturbed and somewhat less impoverished soils may have some *Rubus fruticosus* agg. too. Infestation of surrounding soils by *Pteridium aquilinum* is also very common and scattered fronds are sometimes found in the community. Then, where this kind of grassland has developed over loose sand, erosion of the turf may precipitate a blow-out with *Carex arenaria* spreading on to the exposed material and invading the surrounding grassland, although this kind of pattern is now seen at only a very few inland localities.

The other important element in many of these swards is the cryptogams. In young stands, or in open areas exposed in established grasslands, it is small acrocarpous mosses like *Polytrichum piliferum* with, more occasionally, *P. juniperinum* and *Ceratodon purpureus*, that are important, often becoming locally abundant as colonisers and persisting patchily as the grasses get a hold. Later, larger species such as *Dicranum scoparium* and *Brachythecium albicans* can spread extensively, being better able to cope with the extensive deposition of wind-blown material, and there can also be occasional *Hypnum cupressiforme s.l.*, *Pseudoscleropodium purum* and *Rhytidiadelphus squarrosus*.

Then, in some stands, lichens are very prominent. Species such as *Cornicularia aculeata* and *Cladonia arbuscula* become established early on in colonisation and can remain very frequent and extensive in more closed swards, often with a variety of other species like *C. tenuis*, *C. impexa*, *C. foliacea* and *C. uncialis*. *Peltigera canina* also occurs occasionally.

Sub-communities

***Cornicularia aculeata-Cladonia arbuscula* sub-community.** *F. ovina* tends to be very much the most common of the grasses here, with *A. capillaris* and *K. macrantha* making just an occasional contribution. And, though some stands have quite an extensive cover of their tussocks, the usual picture is of an open sward in which other vascular plants play but a very small part. *R. acetosella* is frequent but otherwise there is just a very occasional plant of *Hieracium pilosella*, *Senecio jacobaea* and *Galium saxatile*.

Mosses and lichens, on the other hand, are often varied and together they quite commonly make up the bulk of the cover. Recently-colonised ground often has patches of *Polytrichum piliferum*, *P. juniperinum* and *Ceratodon purpureus* with *Cornicularia aculeata* and *Cladonia arbuscula*, but *Dicranum scoparium* later becomes frequent with *Cladonia impexa*, *C. tenuis*, *C. foliacea* and *C. uncialis* very common, *C. furcata*, *C. squamosa*, *C. gracilis*, *C. fimbriata* and *C. pyxidata* occasional. *Hypnum cupressiforme s.l.*, *Pohlia nutans*,

Ptilidium ciliare and the introduced moss *Campylopus introflexus* can also be seen in some stands.

Typical sub-community. Small acrocarps and lichens retain occasional representation in this sub-community but the cover of vascular plants is generally more extensive and somewhat more diverse than above. Both *F. ovina* and *A. capillaris* are very common, and each can be quite abundant, with *Koeleria macrantha* occasional, *Anthoxanthum odoratum*, *Holcus lanatus* and *Deschampsia flexuosa* scarce but sometimes showing local prominence. *Aira praecox* and *Poa annua* can also be found quite often and annual dicotyledons such as *Teesdalia nudicaulis*, *Erodium cicutarium*, *Aphanes arvensis*, *Myosotis ramosissima* and *Ornithopus perpusillus* also occur, but with nothing like the coincident frequency characteristic of the next sub-community. More typically, it is perennials such as *Rumex acetosela*, more occasionally *Plantago lanceolata*, *Hieracium pilosella*, *Taraxacum officinale* agg., *Hypochoeris radicata*, *Senecio jacobaea*, *Cerastium fontanum* and *Achillea millefolium*, that are the commonest associated plants, and even these are usually of low cover. Many stands have a rather coarse weedy look, while in others a little *Calluna*, *Pteridium* or *Carex arenaria* can give a distinctive appearance. In the data available, it is in this poorer kind of sward that *Silene conica*, *S. otites* and *Scleranthus perennis* have been recorded and some pathway stands have a local abundance of *Crassula tillaea*.

Apart from the acrocarpous mosses noted above, *Brachythecium albicans* is common here and sometimes quite extensive.

***Erodium cicutarium-Teesdalia nudicaulis* sub-community.** This is the richest and most striking kind of *Festuca-Agrostis-Rumex* grassland in which there is generally a rather open cover of *F. ovina*, with somewhat less frequent *A. capillaris* and occasional *Koeleria macrantha*, *Agrostis stolonifera* and *Holcus lanatus*. *Rumex acetosella*, *Senecio jacobaea*, *Hieracium pilosella*, *Cerastium fontanum* and *Plantago lanceolata* are all very common, with *P. coronopus*, *P. media*, *Cerastium arvense*, *Galium verum* and *Achillea millefolium* occasional, but the sward is rarely rank and indeed often very short with just scattered individuals of these plants.

More distinctive here is the ephemeral flora among which *Aira praecox*, *Erodium cicutarium*, *Teesdalia nudicaulis*, *Aphanes arvensis*, *Myosotis ramosissima*, *Erophila verna*, *Veronica arvensis*, *Trifolium dubium*, *Ornithopus perpusillus*, *Geranium molle* and *Logfia minima* can all be found frequently and often in considerable abundance over the more open ground. Less commonly *Viola tricolor* ssp. *curtisii*, *Veronica agrestis*, *V. polita*, *Filaginella uliginosa* and *Arenaria serpyllifolia* occur, though by summer it often becomes difficult to identify any of these plants as their remains shrivel up. By this time, the vegetation can look rather lifeless, although in June the flowers of *Astragalus danicus* add a touch of purple to many stands, with *Thymus serpyllum* ssp. *serpyllum* following on in its localities here in July and August, and *Sedum acre* and *S. album* providing patches of yellow and white through the summer.

As in the Typical sub-community, the cryptogam flora is usually somewhat restricted in variety and cover here although some stands preserve the pioneer acrocarp flora and others have a reasonably rich mixture of lichens. *Brachythecium albicans* is again the commonest moss, with occasional *Hypnum cupressiforme s.l.*, *Pseudoscleropodium purum* and *Dicranum scoparium*.

Anthoxanthum odoratum-Lotus corniculatus sub-community. The cover of grasses in this sub-community is rather more varied, more extensive and a little ranker than in other kinds of *Festuca-Agrostis-Rumex* grassland, though it is never luxuriant and usually still somewhat open. In addition to *F. ovina* and *A. capillaris* which are both constant, *Anthoxanthum* is very common, *Holcus lanatus* less frequent though still preferential and *Koeleria macrantha* occasional. And among these are some rather distinctive dicotyledons, *Lotus corniculatus*, *Galium verum*, *Achillea millefolium*, *Campanula rotundifolia* and *Plantago lanceolata*, which can give the vegetation a rather mesophytic look. Also interesting is the occasional presence of oak seedlings in the sward.

However, *Rumex acetosella*, *Hieracium pilosella* and *Senecio jacobaea* all remain very common and there are sparse records in open areas of the turf for many of the ephemerals of the *Erodium-Teesdalia* sub-community. Apart from occasional *Brachythecium albicans*, *Dicranum scoparium*, *Hypnum cupressiforme s.l.* and *Pseudoscleropodium purum*, mosses are few, although their cover can be quite extensive. Lichens occur only very rarely and hardly ever show any abundance.

Galium saxatile-Potentilla erecta sub-community. In this sub-community, *F. ovina* and *A. capillaris* are often joined by small amounts of *Deschampsia flexuosa* and occasional *Anthoxanthum*, while *K. macrantha* is quite absent. Among the dicotyledons, *Galium saxatile* and *Potentilla erecta* are both very common here with occasional *Hieracium pilosella*, *Senecio jacobaea*, *Lotus corniculatus* and *Cerastium fontanum*. Although *R. acetosella* remains very frequent, the characteristic small ephemeral herbs of the community are very sparse indeed among the quite extensive cover of perennials. Cryptogams, too, play little part in the flora of this kind of *Festuca-Agrostis-Rumex* grassland. Small plants of

Calluna are sometimes to be seen and it is here that *Ulex* spp. occur most often as occasional scattered bushes.

Hypochoeris radicata sub-community: *Aira praecox-Teesdalia nudicaulis* vegetation Jarvis 1974. *F. ovina* is partly replaced in this sub-community by *F. rubra* and, with *A. capillaris*, these account for the bulk of the perennial grass cover. Generally, though, this is very open and, in the gaps between the tussocks, annual grasses such as *Aira praecox*, *Poa annua*, *Bromus hordeaceus* ssp. *hordeaceus* and *B. mollis* can be quite numerous. Among the dicotyledons, *R. acetosella* is less common than usual and *H. pilosella* and *Senecio jacobaea* are very scarce but rosette hemicryptophytes are very frequent with *Hypochoeris radicata* strongly preferential, *Plantago lanceolata* common and *P. coronopus* and *Leontodon taraxacoides* occasional. Ephemerals are not very numerous but *Teesdalia* occurs in some stands, *Centaurium erythraea* is weakly preferential and there are sparse records for a variety of other winter annuals. Other peculiarities include the occasional occurrence of *Sedum anglicum* and *Umbilicus rupestris* and, on Craig Breidden in Shropshire, this kind of vegetation has the rarities *Lychnis viscaria* and *Veronica spicata*, as well as frequent *Hieracium peleteranum* and occasional *Helianthum nummularium* and *Thymus praecox*.

Polytrichum piliferum, *P. juniperinum*, *Ceratodon purpureus* and *Hypnum cupressiforme s.l.* all occur quite commonly and with locally high cover, but lichens are very scarce.

Habitat

The *Festuca-Agrostis-Rumex* grassland is characteristic of base-poor, oligotrophic and summer-parched soils in the warm and dry lowlands of southern Britain, with grazing and disturbance often very important contributory factors in maintaining the typical aspect of the vegetation. It is a community of open habitats, still most strikingly seen among the swards of Breckland, where the continental climate and a distinctive history of land use give it a special character, but locally congenial conditions extend its range into scattered localities far to the oceanic south-west and around the upland fringes. However, the decline of heath-grazing by stock and the demise of rabbits has led to the loss of many stands, such that this vegetation now often survives as fragments among sub-shrub vegetation, on pathways, along the edges of arable fields and around rock outcrops, with some artificial habitats providing new opportunities for establishment.

The *Festuca-Agrostis-Rumex* grassland is the most widespread calcifuge sward over southern Britain, being found throughout those parts of the country with less than 1000 mm rainfall annually (*Climatological Atlas*

1952) and under 140 wet days yr^{-1} (Ratcliffe 1968) and with a mean annual maximum temperature above 26 °C (Conolly & Dahl 1970), climatic features which help give this vegetation a rather different character from its counterpart on permeable acidic soils in the cool and wet sub-montane zone of the north and west of the country, the *Festuca-Agrostis-Galium* grassland. The broad geographical division between the two vegetation types, and the floristic and physiognomic contrasts they show, reflect the influence which the climate has on the soil moisture regime and surface humidity. The range of the *Festuca-Agrostis-Rumex* grassland coincides closely with those parts of Britain where the warm and dry climate results in a marked potential water deficit (*Climatological Atlas* 1952), the impact of which is especially severe on the more sharply draining profiles of the region, such as are colonised by this community. In comparison with the *Festuca-Agrostis-Galium* grassland, then, *Festuca rubra*, *Anthoxanthum odoratum* and *Potentilla erecta*, which are constant through many north-western swards, have a restricted role here, and such plants as *Viola riviniana*, *Carex pilulifera*, *Hylocomium splendens* and *Pleurozium schreberi*, which are among their characteristic associates there, are hardly ever found in the community.

Water shortage tends to be especially severe in spring and early summer. Over much of the region, the bulk of such rain as there is falls in the second half of the year and, more particularly, between late summer and early autumn (Gregory 1957, Chandler & Gregory 1976) so that, as air temperatures rise, sometimes quite late, but fairly quickly, to relatively high levels, and humidity drops with the ensuing sunny and fairly cloudless skies of summer, the possibility of parching can become severe (Smith 1976, Chandler & Gregory 1976). Even for those perennials able to tolerate the generally dry conditions, therefore, shortage of water during much of the warmer part of the year can markedly restrict growth, acting either directly or by its influence on the availability of nutrients. This is one of the reasons why the grass cover in this vegetation is characteristically open and tussocky, in marked contrast to the often plush swards of the *Festuca-Agrostis-Galium* grassland, and why, by summer, the herbage often looks shrivelled and brown.

The poor competitive ability of the grasses allows for a bigger contribution here than among the north-western swards from low-growing light-demanding chamaephytes like *Hieracium pilosella* (Bishop *et al.* 1978) and the *Sedum* spp., these also well adapted to a xerophytic existence, and from *Rumex acetosella*, which behaves rather like a geophyte in its ability to spring up from underground organs in well-lit gaps. And the community is also one of those vegetation types which offers a locus for rarities such as *Thymus serpyllum* ssp. *serpyllum* (Pigott 1955), *Scleranthus perennis* and *Vero-*

nica spicata (Pigott & Walters 1954) which generally rely on the maintenance of open conditions for their survival.

More widely, the patchwork of bare spaces, periodically renewed by the death of the perennials in old age or because of exceptional drought (Watt 1971*b*, Bishop *et al.* 1978, Bishop & Davy 1984), offers ample opportunity for colonisation by more ephemeral species. Some of these, like *Senecio jacobaea*, are coarse weedy plants, biennial or sometimes longer-lived. But the most distinctive group comprises the diminutive annuals, many of which are specifically adapted to capitalise on the relatively short period very late in the growing season, when the autumn rains moisten the soil surface, but before there is a critical fall in air temperature (Ratcliffe 1961, Newman 1964). For these plants, germination is often of the simultaneous type, and the establishment of a root system quite rapid, with the plants over-wintering as leaf rosettes. In some cases, flower initiation is favoured by the low temperatures of winter and spring, and after-ripening facilitated by the hot summer sun. Among these plants, relatively few, *Teesdalia* and the rare *Silene otites* being the notable exceptions, are strictly Continental in their European range. Some such as *Aira praecox*, *Aphanes arvensis*, *Erophila verna*, *Veronica arvensis* and *Geranium molle* occur widely through the British lowlands; others are essentially southern lowland plants, extending beyond the range of this community largely in drier coastal habitats, for example, *Erodium cicutarium*, *Myosotis ramosissima*, *Ornithopus perpusillus*, *Logfia minima* and *Cerastium arvense*. But, for all of these, the *Festuca-Agrostis-Rumex* grassland provides a major locus within its range, so that the vegetation often looks more like a Thero-Airion community than a Nardo-Galion grassland.

Within its characteristic climatic zone, the distribution of the *Festuca-Agrostis-Rumex* grassland is strictly limited by the occurrence of acidic, free-draining soils, which are of but local occurrence through much of the southern lowlands, and the extremely impoverished nature of these soils accentuates the influence of water shortage on the general floristics and physiognomy of the sward (Watt 1936, 1940, 1971*a*). Pre-eminent among parent materials giving rise to such soils through this part of Britain are various sands and fine gravels, and arenaceous sedimentaries. In the heart of its range, in Breckland, the profiles are of complex make-up and provenance, but the *Festuca-Agrostis-Rumex* grassland typically occupies the more base-poor of the range of soils derived from mixtures of boulder clay and sand variously disturbed by periglacial solifluction and overlain by aeolian deposits. These are now classified as Worlington argillic brown sands and, more frequently, Redlodge humo-ferric podzols, though such soils are often found in quite complex mosaics, sometimes with

Agrostis-Rumex grassland but, where the ground becomes enriched, such places are likely to be directly invaded by *Ulex europaeus* to form *Ulex-Rubus* scrub, or by brambles to produce *Rubus-Holcus* or *Pteridium-Rubus* underscrub, or by rank grasses and tall herbs, all of which shade out the smaller plants of the grassland.

Towards the wetter part of its range, it is often among such vegetation types as these that the *Festuca-Agrostis-Rumex* grassland persists in rocky field corners and in the disturbed ground around settlements, forming part of the untidy patchwork of semi-natural vegetation typical of landscapes which have seen marginal improvement for grazing. Through the drier lowlands, reclamation has been more extensive and drastic, though it often came late because of the inherent difficulties which the soils pose for good cropping, being lime- and often copper-deficient, droughty and liable to wind erosion. But they are readily cultivated and, particularly in East Anglia, have often gone to arable, mostly for sugar-beet and barley (Hodge *et al.* 1984). Afforestation has also been an attractive alternative where agricultural yields are low and, in the first 30 years of its existence, the Forestry Commission had acquired 40% of Breckland where by far the largest and most interesting stands of the community survived. There, then, the *Festuca-Agrostis-Rumex* grassland often remains as fragments on the wider sandy ridges between plantations of pine, or in the corners of ploughed fields which have escaped planting and treatment with chemical fertilisers and herbicides. Stands taken in to airfields and military training grounds sometimes preserve more extensive and semi-natural zonations and mosaics by virtue of the prohibition on such improving activities, and the community can develop over a surprising range of man-made surfaces there. Within urban or industrial areas, too, in the warm dry lowlands, the *Festuca-Agrostis-Rumex* grassland can occur over sharply-draining derelict ground among a wide variety of more eutrophic grasslands, tall-herb vegetation and scrub.

Where the most extensive stands of the community remain, in parts of Breckland, there is the additional striking influence of variations in base-richness among the sharply-draining, impoverished soils depending on the proximity to the surface of the underlying Chalk and the amount of calcareous material among the superficials (Figure 31). Such differences exert the major control over the renowned suite of grass-heaths, first fully described by Watt (1936, 1940), and grouped in this scheme within the *Festuca-Hieracium-Thymus* grassland (Watt's A and B swards) and the *Festuca-Agrostis-Rumex* grassland (D–G). The zonations, which can be seen still in whole or part, on the Lakenheath–Elveden, Icklingham and Stanford–Wretham heaths (Ratcliffe 1977), run from the most calcifuge swards on Redlodge humo-ferric podzols, through transitions over Worl-

ington and Methwold sands, to the most calcicole on Newmarket brown rendzinas (Corbett 1973, Hodge *et al.* 1984). Where *Calluna-Festuca* heath occupies the more acidic profiles, its boundaries are generally very obvious but zonations among the grasslands can be very gradual or form complex mosaics, often depending directly on heterogeneity in the superficials; where grazing has been relaxed, the boundaries often become further blurred by expansion throughout of *F. ovina*. Even in the close-cropped swards, some of the perennials and many of the ephemerals and cryptogams run throughout but the preference of *Thymus praecox*, *T. pulegioides*, *Leontodon hispidus*, *Cirsium acaule*, *Sanguisorba minor*, *Avenula pratensis* and, less obviously, *Koeleria macrantha*, is for the *Festuca-Hieracium-Thymus* grassland; and that of *Rumex acetosella*, *Teesdalia* and *Galium saxatile* is for the *Festuca-Agrostis-Rumex* grassland. In some Breckland sites, too, *Carex arenaria* can become prominent among these mosaics, marking a shift to *Carex* or *Carex-Cladonia* dune communities where sand has been eroded and re-deposited (Watt 1937, Noble 1982), with some possibility of redevelop-

Figure 31. Breckland grass-heaths in relation to soils and treatments.

CG7a *Festuca-Hieracium-Thymus* grassland, *Koeleria* sub-community
CG7b *Festuca-Hieracium-Thymus* grassland, *Cladonia* sub-community
CG7c *Festuca-Hieracium-Thymus* grassland, *Ditrichum-Diploschistes* sub-community
U1a *Festuca-Agrostis-Rumex* grassland, *Cornicularia-Cladonia* sub-community
U1b *Festuca-Agrostis-Rumex* grassland, Typical sub-community
U1c *Festuca-Agrostis-Rumex* grassland, *Erodium-Teesdalia* sub-community

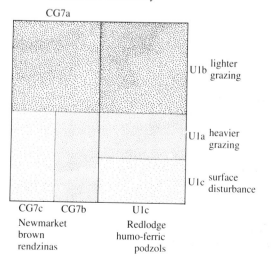

CG7a

U1b lighter grazing

U1a heavier grazing

U1c surface disturbance

CG7c   CG7b   U1c
Newmarket brown rendzinas   Redlodge humo-ferric podzols

ment of the *Festuca-Agrostis-Rumex* grassland over stabilised surfaces (Watt 1938). Similar zonations can be seen at scattered localities around the East Anglian coast, too.

Distribution

The *Festuca-Agrostis-Rumex* grassland occurs widely over suitable substrates throughout the warm and dry lowlands of England and Wales, becoming increasingly local towards the wetter west and north of the country. The most extensive and distinctive stands, of the *Erodium-Teesdalia* and *Cornicularia-Cladonia* sub-communities, are found in Breckland, although it is possible that much Typical *Festuca-Agrostis-Rumex* grassland could approach their richness had not the habitat of the community become fragmented and overgrown through much of its range. The *Anthoxanthum-Lotus*, *Potentilla-Galium* and *Hypochoeris* sub-communities each have distinctive distributions in relation to regional differences in climate and soils.

Affinities

This vegetation received scarcely any attention prior to the classic accounts by Watt (1936, 1940; see also Tansley 1939) of the Breckland stands, where the emphasis was on the floristic variation through the whole range of grass-heaths. Certainly, the continuity in composition and physiognomy through the entire suite of vegetation types described by Watt is very clear, but there is no doubt that, in a national context, they are best seen as closely juxtaposed examples of two rather distinct communities, with the major disjunction occurring in his grassland C. In this scheme, the more calcicolous swards are readily included within the *Festuca-Hieracium-Thymus* grassland, our nearest approach in Britain to the more basiphile steppe-grasslands of the Festucion valesiacae and the Koelerio-Phleion phleiodis (Ellenberg 1978, Oberdorfer 1978). The more calcifuge, on the

other hand, Watt's grasslands D–G, are best seen as the most distinctive examples of this rather widely distributed and, for the most part, fairly nondescript *Festuca-Agrostis-Rumex* grassland, a vegetation type which can then also clearly include other kinds of acidophilous swards of sandy soils which received passing mention in Tansley (1911).

The community represents an obvious geographical counterpart to the *Festuca-Agrostis-Galium* grassland, the major calcifuge sward of the wet and cool submontane zone of north-west Britain, occupying a similar ecological position in relation to woodland and heath vegetation. But, whereas that is clearly a kind of Nardo-Galion community, the *Festuca-Agrostis-Rumex* grassland is perhaps better located among the vegetation types of sandy soils included in the Sedo-Scleranthetea, of which *Myosotis ramosissima*, *Potentilla argentea*, *Rumex tenuifolius*, *Scleranthus perennis*, *Sedum anglicum*, *S. acre*, *Taraxacum laevigatum*, *Veronica verna*, *Brachythecium albicans*, *Ceratodon purpureus* and *Polytrichum piliferum* are regarded as character species (Ellenberg 1978). More particularly, it can find a place in the Thero-Airion alliance, of which *Aira praecox*, *Logfia minima*, *Ornithopus perpusillus* and *Teesdalia nudicauli* are diagnostic, with Continental equivalents such as the *Festuco-Thymetum serpylli* R. Tx. (1928) 1937 described from The Netherlands and north-west Germany (Tüxen 1937, Westhoff & den Held 1969), the *Airetum praecoci* (Schwickerath 1944) Krausch 1967 and *Airo caryophylleae-Festucetum ovinae* R. Tx. 1955 from various part of Germany (Oberdorfer 1978) and Poland (Matuszkiewicz 1981). Among the range of such vegetation type described from Britain, the *Festuca-Agrostis-Rumex* grassland and its maritime counterpart the *Armeria-Cerastium* community, can be seen as transitiona between the grassier calcifuge swards of the Nardo-Galion and the stonecrop-dominated vegetation of fractured rocky outcrops.

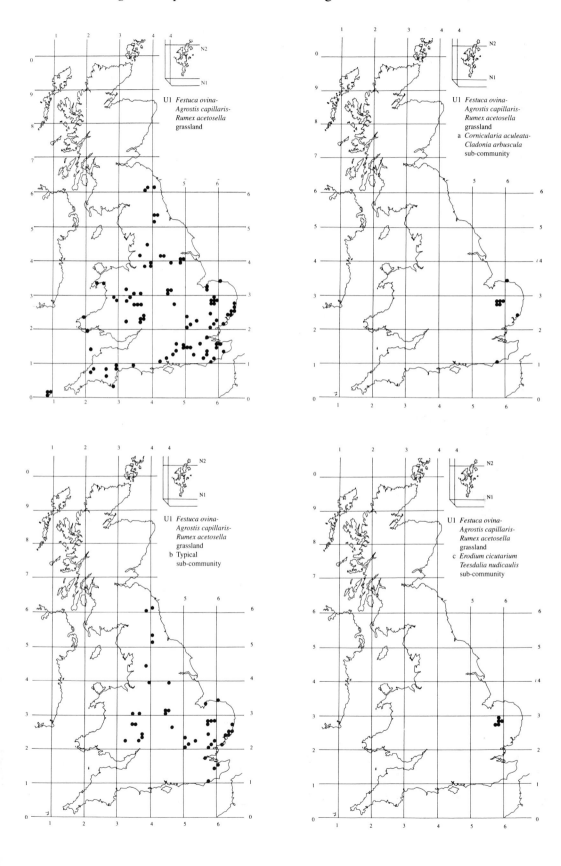

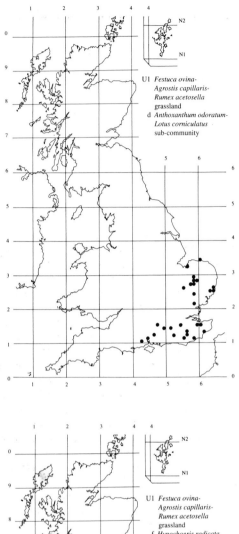

U1 *Festuca ovina-*
Agrostis capillaris-
Rumex acetosella
grassland
d *Anthoxanthum odoratum-*
Lotus corniculatus
sub-community

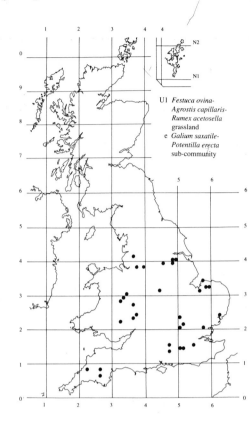

U1 *Festuca ovina-*
Agrostis capillaris-
Rumex acetosella
grassland
e *Galium saxatile-*
Potentilla erecta
sub-community

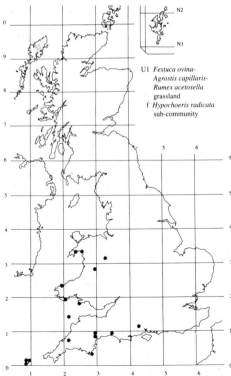

U1 *Festuca ovina-*
Agrostis capillaris-
Rumex acetosella
grassland
f *Hypochoeris radicata*
sub-community

changes can be seen in this basic pattern. The *Festuca-Agrostis* sub-community of the *Deschampsia* grassland is replaced by the *Vaccinium* sub-community, and there are similar shifts in both the associated heath and woodland types, with bilberry-rich sub-communities becoming the norm, thus giving a somewhat montane character to the entire sequence of vegetation types. Usually these are disposed over the scarps and valley sides that mark the first substantial rise to the Pennine uplands, the mosaic of communities being again largely dependent on the history of treatments, with a widespread continuance of grazing helping to maintain extensive tracts of the *Deschampsia* grassland against the spread of sub-shrubs and trees. Patches of grassland and heath can recur at quite high altitudes on the grit edges, though wherever there is a lessening of slope with the shift on to the great shelving dips, poor drainage and the accumulation of ombrogenous peat curtail the extent of these communities. A common pattern is for the *Deschampsia* grassland to give way to *Nardus-Galium* grassland on the stagnopodzol intergrades, a zonation which can be interrupted by the occurrence of flushed areas with poor-fen vegetation like grassy *Carex-Sphagnum* mire, and then for there to be a switch to *Calluna-Eriophorum* bog. *D. flexuosa* can run throughout such sequences and, indeed, where blanket peats have been drained, the *Deschampsia* grassland itself can become established on the drier fringes and ramparts, with a local spread of *V. myrtillus* where

grazing is withdrawn or of the unpalatable *E. nigrum* ssp. *nigrum*. The *Deschampsia* grassland can also figure over the drying surfaces of raised mires with the *Eriophorum* bog, though with the shift into more oceanic areas it is *Molinia* which usually assumes dominance with the loss of waterlogging on the bog surface.

Distribution
The *Deschampsia* grassland has a fairly widespread but local distribution through the moderately oceanic lowlands, becoming more common towards the upland fringes of northern England.

Affinities
Although *D. flexuosa* has long been acknowledged as a major component of heaths throughout much of the lowland zone, this grassland has figured little in descriptions, except as part of a compendious 'grass heath'. And its phytosociological status is unclear. It shows floristic affinities, in the more open stands of the *Festuca-Agrostis* sub-community, with the *Festuca-Agrostis-Rumex* grassland, though clearly it cannot itself be considered as a Thero-Airion community. A better solution, while acknowledging that it can have a variety of remnant associated floras depending on its origin, is to locate it among the Nardo-Galion swards which stresses its close floristic and ecological relationships with a range of sub-shrub communities.

Floristic table U2

	a	b	2
Deschampsia flexuosa	V (4–9)	V (4–9)	V (4–9)
Calluna vulgaris	IV (1–4)	IV (1–5)	IV (1–5)
Galium saxatile	III (1–5)	II (1–3)	III (1–5)
Potentilla erecta	III (1–3)	II (1–4)	III (1–4)
Festuca ovina	III (2–6)	I (4)	II (2–6)
Pteridium aquilinum	III (1–8)		II (1–8)
Agrostis capillaris	III (1–5)		II (1–5)
Polytrichum piliferum	III (1–5)		II (1–5)
Rumex acetosella	III (1–6)		II (1–6)
Dicranum scoparium	II (1–5)	I (1–3)	II (1–5)
Rumex acetosa	II (4–5)	I (1)	I (1–5)
Betula pendula sapling	II (1–4)		I (1–4)
Epilobium angustifolium	II (1–3)		I (1–3)
Rubus fruticosus agg.	II (1–3)		I (1–3)
Campylopus pyriformis	I (1–3)		I (1–3)
Orthodontium lineare	I (1–3)		I (1–3)
Holcus mollis	I (1–3)		I (1–3)
Quercus hybrid sapling	I (1–3)		I (1–3)

Floristic table U2 (*cont.*)

	a	b	2
Betula pendula seedling	I (1–2)		I (1–2)
Quercus sp. seedling	I (1–3)		I (1–3)
Poa annua	I (3–4)		I (3–4)
Cladonia gracilis	I (3–5)		I (3–5)
Ulex europaeus	I (2–4)		I (2–4)
Vaccinium myrtillus		V (1–4)	II (1–4)
Pleurozium schreberi	I (4)	IV (1–5)	II (1–5)
Empetrum nigrum nigrum		III (1–5)	II (1–5)
Eriophorum vaginatum		III (1–5)	I (1–5)
Molinia caerulea	I (2–4)	II (1–4)	I (1–4)
Juncus squarrosus	I (5)	II (1–3)	I (1–5)
Juncus effusus	I (1)	II (1–6)	I (1–6)
Polytrichum commune		II (1–4)	I (1–4)
Agrostis canina		II (1–4)	I (1–4)
Anthoxanthum odoratum		II (1–4)	I (1–4)
Carex nigra		II (1–3)	I (1–3)
Festuca rubra		I (4–5)	I (4–5)
Nardus stricta		I (1–3)	I (1–3)
Barbilophozia floerkii		I (1–4)	I (1–4)
Luzula multiflora		I (1–3)	I (1–3)
Hypnum cupressiforme s.l.	II (1–4)	II (1–6)	II (1–6)
Plagiothecium undulatum	I (1)	I (1–3)	I (1–3)
Rhytidiadelphus squarrosus	I (1)	I (1–4)	I (1–4)
Pohlia nutans	I (1–2)	I (1–2)	I (1–2)
Number of samples	19	11	30
Number of species/sample	9 (3–16)	11 (6–15)	9 (3–16)
Herb/shrub height (cm)	21 (4–60)	20 (15–30)	21 (4–60)
Herb/shrub cover (%)	87 (50–100)	86 (30–100)	87 (50–100)
Ground layer height (mm)	14 (10–30)	42 (10–100)	21 (10–100)
Ground layer cover (%)	19 (0–80)	7 (0–15)	17 (0–80)
Altitude (m)	110 (20–168)	449 (305–539)	254 (20–539)
Slope (°)	4 (0–20)	7 (0–20)	5 (0–20)

a   *Festuca ovina-Agrostis capillaris* sub-community
b   *Vaccinium myrtillus* sub-community
2   *Deschampsia flexuosa* grassland (total)

Figure 32. Floristic transition from calcifugous to mesotrophic grassland with agricultural improvement on more free-draining brown soils around the upland fringes.

	U4a		U4b		MG6b	
Agrostis capillaris	V	(1–10)	V	(1–8)	V	(4–8)
Anthoxanthum odoratum	V	(1–10)	V	(1–8)	V	(1–7)
Festuca rubra	III	(1–8)	IV	(1–8)	IV	(2–8)
Festuca ovina	V	(1–19)	III	(1–8)	I	(2–4)
Potentilla erecta	V	(1–6)	III	(1–4)	I	(1–3)
Galium saxatile	IV	(1–6)	III	(1–4)		
Pleurozium schreberi	III	(1–6)	I	(1–2)		
Deschampsia flexuosa	II	(1–6)	I	(1–6)		
Nardus stricta	II	(1–6)	I	(1–4)		
Vaccinium myrtillus	II	(1–6)				
Holcus lanatus	II	(1–6)	V	(1–6)	V	(2–6)
Achillea millefolium	III	(1–6)	IV	(1–6)	II	(2–5)
Trifolium repens	II	(1–6)	IV	(1–6)	V	(1–8)
Cerastium fontanum	II	(1–4)	IV	(1–4)	V	(1–5)
Poa pratensis	II	(1–6)	III	(1–7)	III	(1–5)
Cynosurus cristatus	I	(1–3)	II	(1–3)	V	(2–7)
Lolium perenne			II	(1–5)	V	(2–7)
Ranunculus acris	I	(1–4)	II	(1–4)	IV	(1–5)
Dactylis glomerata	I	(1–3)	II	(1–4)	III	(1–4)
Taraxacum officinale agg.	I	(1–3)	II	(1–4)	II	(1–3)
Bellis perennis			II	(1–6)	II	(1–4)
Trifolium pratense			II	(2–5)	II	(1–7)
Cirsium arvense	I	(1–3)	I	(1–4)	II	(1–5)
Cirsium arvense	I	(1–3)	I	(1–4)	II	(1–5)
Leontodon autumnalis					II	(1–4)
Poa trivialis					II	(2–5)
Bromus hordeaceus hordeaceus					II	(1–6)
Rhytidiadelphus squarrosus	IV	(1–10)	III	(1–8)	II	(1–6)
Luzula campestris	IV	(1–4)	III	(1–3)	II	(1–5)
Plantago lanceolata	II	(1–4)	III	(1–4)	III	(1–5)
Rumex acetosa	II	(1–4)	III	(1–4)	III	(1–4)
Number of samples	172		35		43	
Number of species/sample	22	(7–62)	20	(11–39)	14	(4–26)

U4a *Festuca–Agrostis–Galium* grassland, Typical sub-community
U4b *Festuca–Agrostis–Galium* grassland, *Holcus–Trifolium* sub-community
MG6b *Lolio–Cynosuretum*, *Anthoxanthum* sub-community

certain that, in the more accessible parts of the uplands, as through the south-west, in much of Wales, the Lake District and the Southern Uplands, substantial areas of some kind of pasture were established as early as the medieval period, with an extension over the next few centuries to remoter regions like the Highlands (Roberts 1959, Steven & Carlisle 1959, King & Nicholson 1964, Pearsall & Pennington 1973, Harvey & St Leger-Gordon 1974).

The exact character of these grasslands, the more immediate forebears of our present rough grazings, can only be guessed at, but it is evident from the records that they were often treated in a different fashion from today and this may be safely assumed to have affected their composition. Frequently, in these earlier days, the upland grazings were used as summer pasture in a system of transhumance and, among the stock, cattle were at least as important as sheep, sometimes more so (Franklin 1952, Roberts 1959). From the mid-1700s, with the introduction of improved sheep breeds, the balance among the grazing animals began to shift in the other direction, often substantially so, and, with the coincidental spread of enclosure of the lower slopes, the foundations of the modern kinds of sheep-rearing that prevail today through most of the uplands were established. Often, now, the sheep are left on the unenclosed grazings for much of the year, as in the 'heaf' or 'heft' system that has been the basis of pastoralism through the Southern Uplands and Pennines for more than two centuries, with the flocks largely self-sustaining in home ranges on the hills, and with some wintering of vulnerable stock, like first-year ewe lambs, on the enclosed pasture below (Trow-Smith 1957, King & Nicholson 1964, Pearsall & Pennington 1973).

Of the different kinds of grasslands, heaths and mires that make up the varied patchworks of rough grazing through the uplands, the *Festuca-Agrostis-Galium* swards are among the vegetation most favoured by sheep. In those parts of the country where the community is proportionately important, as in the Southern Uplands, considerably higher stocking densities can be sustained on the open hills: there, Hunter (1961) recorded values of one ewe per 0.6 ha, as opposed to one ewe per more than 4 ha in those regions where poorer-quality vegetation predominated. At its best in this area, it can experience grazing intensities twice those felt by rushy vegetation like the *Juncus-Galium* pasture and three times those on grassy *Molinia-Potentilla* mire (Hunter 1954, 1962, see also Boulet 1939). Compared with these communities, the *Festuca-Agrostis-Galium* grassland presents generally palatable mixtures of plants, and its composition and physiognomy now reflect the sustained impact of the close and choosy cropping characteristically associated with sheep, particularly with the breeding ewes which have largely

replaced the less selective wethers popular in the days when mutton, rather than young lamb, was favoured for the table (Roberts 1959, Spedding 1971, Grant *et al.* 1985). And, where the vegetation types occur in mosaics, there seems to be marked seasonality in the way in which present-day flocks graze these different elements, with the *Festuca-Agrostis-Galium* grassland being cropped more in summer than in winter, the less palatable swards turned to more often as herbage becomes scarce (Hunter 1954, 1962).

The overall effect of this kind of predation is to keep the vegetation short and varied, with the distinctive mixed dominance of a number of more or less fine-leaved grasses, together with grazing-resistant hemicryptophytes, and to set back repeatedly any spread of sub-shrubs or invasion by shrubs and trees. Behind such a general influence, however, there is undoubtedly a great diversity of subtle and shifting interactions between sward and stock which give every stand of this really rather well defined vegetation its own peculiarities of history and structure. For one thing, there is within the community itself considerable variety in the grazing value of the herbage that is dependent partly on edaphic differences. The best quality swards fall mostly within the *Holcus-Trifolium* sub-community, some of which are good enough to qualify as 'third-grade rye-grass pastures' and, whether more productive because of lying on naturally more mesotrophic soils or by virtue of improvement by manuring and top-sowing, these are often of great importance in supplementing the available grazing over generally poor tracts of pasture on the open hills (e.g. Hughes 1958) or during the lean months of winter where stock are brought down on to enclosed stands. Close behind among the more widely distributed kinds of *Festuca-Agrostis-Galium* grassland are the swards of the Typical sub-community and probably, also, some of the *Vaccinium-Deschampsia* type, where these provide some relief among mosaics of heath, a common occurrence (Hunter 1962). To the opposite extreme, lies the coarser herbage in the *Vaccinium-Deschampsia* sub-community, which can extend on to soils highly favourable for the expansion of *Nardus*: such swards can experience grazing intensities less than half those on the better quality grassland (Hunter 1962).

And probably very important, too, for the way in which the kinds of vegetation included here are preferred, one against the other, by the grazing stock, and selected over other components of rough grazings, are the proportions and patterns in which all the various grasslands, heaths and mires are disposed before the flocks on a particular stretch of the uplands. This means that the floristic and physiognomic definition of the more favoured elements, like the different types of *Festuca-Agrostis-Galium* grassland, is partly dependent upon the character of the whole mosaic of vegetation

overwhelmingly so, and there is often quite abundant *F. ovina* and *A. capillaris*, with some *Anthoxanthum* and occasional *D. flexuosa* and, preferentially frequent here, *Danthonia decumbens*. *V. myrtillus* is common but it is generally *Calluna* that gives a distinctive character to this kind of *Nardus-Galium* grassland and, in some stands, it is quite abundant, though typically grazed down to short sprigs. Much more locally, *Erica cinerea* or *E. tetralix* can be found at moderately high cover. *G. saxatile* and *P. erecta* remain very frequent and there is occasional *C. pilulifera* and *C. binervis*, but other herbaceous associates are few. *Hypnum cupressiforme s.l.*, *Dicranum scoparium* and *Pleurozium* are the commonest bryophytes, although *Pseudoscleropodium purum* is preferential at low frequency.

***Racomitrium lanuginosum* sub-community:** *Nardus-Trichophorum* nodum McVean & Ratcliffe 1962 *p.p.*; *Nardus* is usually dominant here and all other grasses, particularly *Agrostis capillaris* and *Anthoxanthum*, are reduced in frequency and cover: generally there is just a little *F. ovina* with occasional *D. flexuosa* and *Agrostis canina* between the tussocks. *J. squarrosus* is fairly common, but more distinctive in this vegetation is the frequency of *Scirpus cespitosus*, and this is often quite abundant, sometimes co-dominant with the *Nardus*. Both *V. myrtillus* and *Calluna* occur commonly and each can have moderately high cover; occasionally, too, there is some *Empetrum nigrum*, *Erica tetralix* or *E. cinerea*, and local abundance of these can add to the heathy appearance of the vegetation, though characteristically the sward is short. *Galium saxatile* tends to be rather patchy in its representation but *Potentilla erecta* remains frequent and there is occasional *Huperzia selago* and *Narthecium ossifragum* along with scattered plants of *C. pilulifera*, *C. binervis* and *C. nigra*. The other striking feature of this sub-community is the common occurrence of *Racomitrium lanuginosum*, sometimes in considerable quantity, and though there can be a little *Rhytidiadelphus loreus* along with *Hypnum cupressiforme s.l.* pleurocarps are generally sparse. Occasional *Diplophyllum albicans*, *Scapania gracilis* and *Pleurozia purpurea* can give a somewhat oceanic look to the vegetation and, unusually among *Nardus-Galium* grasslands, there are quite often some lichens here, most frequently *Cladonia uncialis* with occasional *Cetraria islandica*.

Habitat

The *Nardus-Galium* grassland is typical of moist, peaty mineral soils, usually base-poor and infertile, over the higher hill slopes of the cool, wet north and west of Britain. It is a secondary vegetation type which contributes extensively to our poorer-quality upland grazings and, though climate and soils exert a strong influence on the general floristic features of the community and its internal variation, its spread owes much to particular kinds of pastoral treatment.

Even more so than the *Festuca-Agrostis-Galium* grassland, this is a vegetation type of the rainy and cool uplands. Swards with some *Nardus* are not unknown in the warmer and drier lowlands of Britain – the species makes a modest contribution there to a variety of grasslands and heaths – but the abundance of the plant in this kind of vegetation is strongly concentrated within the 1200 mm isohyet (*Climatological Atlas* 1952), where there are more than 160 wet days yr^{-1} (Ratcliffe 1968), and where the mean annual maximum isotherm is 26 °C or less (Conolly & Dahl 1970), a zone which takes in much of the South-West Peninsula, Wales, northern England and Scotland. Through these regions, then, conditions are generally cloudy and humid, with very small potential water deficits, less than 25 mm across most of the range of the community (Page 1982), and though there is a strong tendency to leaching, even over parent materials that are not wholly lime-deficient, soils are kept moist for much of the year. Very much as in the *Festuca-Agrostis-Galium* grassland, therefore, these swards take much of their general floristic character from the occurrence of plants like *Anthoxanthum*, *Agrostis canina*, *Deschampsia flexuosa*, *Danthonia*, *Carex binervis* and *C. pilulifera*, along with the more universally distributed *Festuca ovina*, *Agrostis capillaris*, *Galium saxatile* and *Potentilla erecta*; from the abundance of pleurocarpous mosses like *Rhytidiadelphus squarrosus*, *Hylocomium splendens* and *Pleurozium*; and from the vigorous growth of these associates, where the dominant allows, helping to exclude the ephemerals and light-demanding chamaephytes so characteristic of the open Thero-Airion swards of the drought-prone acid soils in the lowlands.

The floristic and environmental overlap between the *Nardus-Galium* and *Festuca-Agrostis-Galium* grasslands is thus very considerable: both are essentially submontane Nardo-Galion communities and indeed it is likely that the former occupies some ground which could carry the latter and has been derived there from it. However, although both can extend at their upper extreme to 800 m or so, the altitudinal distributions of the two vegetation types are rather different, the *Festuca-Agrostis-Galium* grassland being typically found between 150 and 500 m, the *Nardus-Galium* grassland being concentrated from 300 to 700 m, with a mean altitude more than 200 m above that of the other community. There is little obvious direct effect of the cooler and more humid conditions prevailing over these higher slopes on the flora of the *Nardus-Galium* grassland apart from the encouragement of the generally montane *V. myrtillus*: Arctic-Alpines like *Carex bigelowii*, *V. uliginosum* and *Empetrum nigrum* ssp. *herm-*

aphroditum still play an insignificant role here and only beyond 700 m or so, where mean annual maxima often fall below 22 °C (Conolly & Dahl 1970), do they become important with *Nardus* in the *Nardus-Carex* community.

The more evident effects of climate on the *Nardus-Galium* grassland are indirect ones felt through pedogenesis, because the association of this vegetation with higher ground is essentially a reflection of its preference for moister acid soils, frequently podzolised and often gleyed, with substantial accumulations of surface mor. The *Festuca-Agrostis-Galium* grassland does extend some little way on to such profiles in the *Vaccinium-Deschampsia* sub-community, which is floristically transitional to the *Nardus-Galium* grassland, but it is by and large a vegetation type of more free-draining brown podzolics or podzols proper, with moder or but a thin humose topsoil. Beneath the *Nardus-Galium* grassland, by contrast, the commonest profiles are of the stagno-podzolic or stagnohumic gley type (Avery 1980), in which there is often more than 10 cm, sometimes up to 30 cm, of peaty topsoil, with varying degrees of leaching and gleying below, according to the particular character of the parent material and the drainage conditions of the ground (Tansley 1939, McVean & Ratcliffe 1962, King 1962, King & Nicholson 1964, Birse & Robertson 1976, Hill & Evans 1978, Birse 1980, 1984). Such soils are very widely distributed throughout the uplands, occurring extensively as intergrades on the gentle transitions between steeper, shedding slopes, which provide a typical location for the *Festuca-Agrostis-Galium* grassland, and the summit plateaus above, with their deep ombrogenous peats from which the *Nardus-Galium* grassland is characteristically excluded except where the material has been eroded and redistributed over a mineral base (Smith 1918, Pearsall 1968). But they are also commonly found over gentle receiving slopes throughout the altitudinal range and in transitions to flushes where the ground is back-gleyed.

The more poorly drained among these soils are frequently derived from impervious bedrocks such as the Ordovician and Silurian shales that are so extensive beneath the community through the Southern Uplands and West Wales, and the Carboniferous shales that provide an important substrate in the Pennines. The softness of such argillaceous rocks also means that they tend to weather to more subdued landscapes on to which run-off is channelled from the upstanding hill slopes around and over which drainage is fairly sluggish. The extensive deposition of heavy-textured drift, such as glacial till, over gentler slopes through the sub-montane zone exacerbates such stagnation and, by insulating the vegetation from the underlying bedrocks, such superficials can carry the *Nardus-Galium* grassland on to ground which would be otherwise too sharply drained.

But the community also occurs widely on a variety of more coarse-textured and initially pervious parent materials, both bedrocks and drift, where a measure of drainage impedence has resulted from the formation of a thin iron pan with long-continued leaching. In other cases, profiles can be kept suitably moist by irrigation and, among tracts of dry soils, the *Nardus-Galium* grassland is quite often an indicator of modest flushing: it can even extend on to free-draining alluvium along stream sides (McVean & Ratcliffe 1962, Welch 1967). Except very locally, the ground waters are lime-poor and, although the community can be found over calcareous rocks, the strong tendency to leaching means that, in most cases, the superficial pH is between 3.5 and 5 (McVean & Ratcliffe 1962, King 1962, King & Nicholson 1964). The generally heavy rainfall at these higher altitudes is also important in encouraging the accumulation of the thick layers of mor, even over quite steep and more freely draining ground, and in guarding against any strong tendency to droughting.

It is this combination of edaphic conditions that is so conducive to the vigour of *Nardus* in this vegetation: first and foremost, it is a plant that favours a humose topsoil moist for much of the year but in an oxidising state in summer, and a mineral base, generally strongly leached, preferably highly impoverished and often gleyed (Pearsall 1950, Chadwick 1960). In such circumstances it is able to outcompete its potential rivals, particularly where, as is often the case here, treatments have put these other plants at an additional disadvantage. Among the associates which are more typical of this community than of the *Festuca-Agrostis-Galium* grassland, *V. myrtillus*, *D. flexuosa* and *Pleurozium*, *J. squarrosus*, *L. multiflora* and *P. commune* are also favoured by the soil conditions that prevail here, the first three where the profiles tend to be more free-draining, the last three where they are distinctly peaty and moister. Variations in the proportions of these species account for much of the difference between the two widely-distributed kinds of *Nardus-Galium* grassland, the Typical and the *Agrostis-Polytrichum* sub-communities and a general edaphic contrast may underlie these distinctions: certainly the latter sub-community is more common at higher altitudes – its mean is some 75 m above that of the former type – and in places where redistributed peat provides a moist and highly humose substrate, as around some of the summits in the Southern Uplands.

There is strong edaphic continuity between the conditions characteristic of these two types of *Nardus-Galium* grassland and the *Calluna-Danthonia* and *Racomitrium* sub-communities, where the soils are likewise generally base-poor and impoverished. The *Calluna-Danthonia* swards essentially continue the floristic trends of Typical *Nardus-Galium* grassland into some of the drier and warmer upland fringes where the community occurs. Its mean altitude is more than 100 m less than that of most stands of the community and it is the usual form of *Nardus-Galium* grassland to be found on podzolic soils on Exmoor, through southern Wales, the Peak District and the Pennine fringes, where rainfall is usually less than 1600 mm yr^{-1} (*Climatological Atlas* 1952) with often fewer than 180 wet days yr^{-1} (Ratcliffe 1968), and where mean annual maxima are generally over 25 °C (Conolly & Dahl 1970). Here, peat-loving plants tend to make their minimal contribution and the vegetation grades into sub-montane heathy grassland.

Towards the opposite extreme, the *Racomitrium* sub-community takes the *Nardus-Galium* grassland into some of the wettest and coolest situations where generally suitable soils are to be found. Through the range of this kind of vegetation, the annual rainfall is usually well over 1600 mm (*Climatological Atlas* 1952) with more than 200 wet days yr^{-1} (Ratcliffe 1968) and mean annual maxima of less than 23 °C (Conolly & Dahl 1970). In some places, a sunless northerly or easterly aspect provides a local enhancement of these cool, moist conditions with some prolongation of snow-lie. With the shelter that this offers, the *Racomitrium* sub-community can extend to well over 600 m in the central Highlands, grading to certain kinds of chionophilous *Nardus-Carex* vegetation (McVean & Ratcliffe 1962). Other stands occur at lower altitudes than this, over slopes that are quite exposed, but where the general climatic conditions are more equable, such that the cover resembles a grassy moss-heath: extensive tracts of such vegetation occur on North Harris (Ratcliffe 1977) and Shetland. The cool oceanic climate is reflected in the *Racomitrium* sub-community in the frequency of *R. lanuginosum*, *Cladonia uncialis* and the occasional Atlantic bryophytes, with *Scirpus cespitosus*, *Molinia* and *Empetrum nigrum* providing continuity with the run-down wet heaths so extensive through those parts of Britain where thin ombrogenous peats have been eroded and their vegetation burned and grazed.

The very different swards of the *Carex-Viola* sub-community extend the *Nardus-Galium* grassland on to soils that are considerably more base-rich than usual, where the ground is irrigated by run-off from calcareous rocks. Its distribution thus tends to be local, though it can be widespread where such substrates make an important contribution to the landscape, as with the more lime-rich of the Dalradian metasediments between Breadalbane and Clova, and especially over the flanks of Ben Lawers (McVean & Ratcliffe 1962). More isolated stands have been recorded from similar rocks on Shetland (Birse 1980), over Tertiary basalts on Skye (Birks 1973) and where there is flushing from limy partings among Silurian shales in the Southern Uplands (Ferreira 1978) and from Carboniferous Limestone in the

Pennines (Welch 1967). The soils under the *Carex-Viola* sub-community tend to have little raw humus and they can sometimes be distinctly silty where fine material has been washed downslope or deposited alongside flooding streams. The tendency to gleying varies but it is often pronounced and irrigation keeps the pH nearer 6 than 5, the typical upper limit for other kinds of *Nardus-Carex* grassland. Under such conditions, species such as *J. squarrosus* and *L. multiflora* can persist but *Nardus* loses some of its competitive edge and the more strongly calcifugous among the associates fade in importance to be replaced by the characteristic diversity of mildly basiphile and mesophytic plants.

Throughout its range, over this variety of soil types and substrates, the *Nardus-Galium* grassland is typically an element in the open hill grazings that occupy the lower and middle slopes beyond the limits of enclosure and, among the patchworks of vegetation types available to stock there, it is among the least valuable for sustaining the kinds of animals favoured these days (Stapledon 1937, Fenton 1953). For one thing, much more so than with the *Festuca-Agrostis-Galium* grassland, the community prefers soils that are not only for the most part base-poor, but also highly oligotrophic, so the amounts of lime and major nutrients in the herbage are relatively low (e.g. Pearsall 1968). More obviously, there is the characteristic prevalence of *Nardus* in the swards and, though this grass is of similar digestibility to other fine-leaved species (Thomas & Fairburn 1956, Hodgson & Grant 1981), it has a much higher proportion of fibrous tissue in its wiry foliage (Burr & Turner 1933, Pearsall 1968), so it represents a very unrewarding bite. Indeed, most stock, particularly sheep, are reluctant to graze it when offered a choice and, though the community may receive a little more attention than usual in the winter months when herbage is in generally short supply (Hunter 1954, 1962), it can suffer grazing intensities less than half those experienced by the *Festuca-Agrostis-Galium* grassland (Boulet 1939, Hunter 1962).

Cattle, which were the more important animals throughout the uplands until the eighteenth century, are not quite so choosy as sheep and have been reported to turn to *Nardus* sooner, though they do not graze it very closely (Nicholson *et al.* 1970, Grant *et al.* 1985), while ponies, not often pastured in the uplands now but still locally important, can nip out the growing centres of the tussocks and discard the tough leaf bases (Havinden & Wilkinson 1970). Significantly, where there has remained some diversity among the stock, as on Dartmoor where both cattle and ponies are still pastured along with sheep, the *Nardus-Galium* grassland is noticeably uncommon among the open grazings, despite the presence of eminently suitable soils (Ward *et al.* 1972*a*). In Scotland, too, Fenton (1936, 1937) and

Wilson (1936) noted that the contribution of *Nardus-Galium* grassland declined where Galloway cattle were put out to graze. Then, there is the striking description from Roberts (1959) of the avid way in which all-winter wethers will attack *Nardus* (and *J. squarrosus*) just as late-winter leaf initiation is occurring, biting out the new growth and leaving the old tussock surrounds to be whipped away by the wind.

Such evidence, together with historical accounts of what has happened to hill grazings over the last few centuries, has led to the convincing suggestion that the widespread entrenchment of the *Nardus-Galium* grassland among *Festuca-Agrostis-Galium* and other better-quality swards has been strongly favoured by the switch from cattle to sheep as the predominant upland grazing stock and, more recently, from wethers to the more choosy breeding ewes (Fenton 1937, Roberts 1959). But the differences in the timing of the more intense bouts of grazing between the older and newer pastoral systems may be just as important as the variations in selectivity between the animals (King & Nicholson 1964, Grant *et al.* 1985) and it is also very clear that the success of *Nardus* in spreading through particular tracts of pasture is strongly influenced by interaction between treatments and soil conditions. Generally speaking, the species does best where uncontrolled but selective grazing has been applied over long periods to swards on moist, peaty and infertile ground. Where it is less able to compete for edaphic reasons, the response of *Nardus* to a favourable pastoral regime will be muted; conversely, on soils of intermediate suitability it is most readily affected by treatment changes (Ratcliffe 1959*a*). Then, as a general background to such responses, there may be a tendency for the *Nardus-Galium* grassland to mark the progressive deterioration of upland grazings with centuries of exploitation since clearance (Ratcliffe 1959*a*, King & Nicholson 1964).

Zonation and succession

The *Nardus-Galium* grassland occurs widely through the uplands of northern and western Britain in zonations and mosaics with a variety of other grasslands, heaths and mires, where the major influences on the vegetation patterns are soil differences and treatments. Regional climatic variations across the range of the community affect the particular components of these patchworks and local climatic differences mediate transitions to windswept moss-heath and snow-beds. In most situations, however, the *Nardus-Galium* grassland is an anthropogenic vegetation type, derived by the burning and grazing of cleared land in the forest zone. Although relaxation of pasturing might allow a ready reversion to heath in many places, the run-down of ground long occupied by the community may hinder any succession to forest at lower altitudes.

Some of the clearest edaphically-related zonations involving the *Nardus-Galium* grassland can be seen where the moist, peaty soils that it favours occur as intergrades between more sharply draining podzols and rankers on steeper, shedding slopes and thicker peats kept moist by heavy rainfall or locally impeded drainage over flatter ground. Quite commonly, such patterns find clear expression in an altitudinal banding with a zone of *Nardus-Galium* grassland occupying the gentler ground on transitions between the steeper hillsides below and the summit plateaus above, but more complex patterns can be seen over stepped topography or more broken slopes. The zonation over the drier podzols is very often to the *Festuca-Agrostis-Galium* grassland and, over graded transitions on higher hills, the *Vaccinium-Deschampsia* sub-community of that vegetation, which has occasional *Nardus*, can pass almost imperceptibly into Species-poor *Nardus-Galium* grassland. In other cases, sharper topographical changes from slope to plateau, or scarp to dip, can show a more abrupt zonation of these vegetation types: these sometimes mark geological shifts from resistant pervious bedrocks to softer, impervious ones, as over grit/shale alternations, while elsewhere the deposition of heavy drift sharpens up drainage differences by enhancing impedence over gentler slopes. Then, the sudden change in the abundance of *Nardus* may provide a much better indication of the boundaries between the communities, the belts and patches of the *Nardus-Galium* grassland showing up especially clearly in winter, although there is still often considerable qualitative continuity among the associates in the swards. Patterns of these kinds are very widespread through the British uplands, and well illustrated in the account of the Moffat Hills (Smith 1918) and, in more fragmentary fashion, in maps of the Carneddau (Ratcliffe 1959*a*) and Cader Idris (Edgell 1969). They have been clearly described too from the southern Pennines, although here the swards of the steeper, better-drained ground tend to be of the *Deschampsia flexuosa* type (Adamson 1918), while in the warm oceanic south-west of England such *Nardus-Galium* grassland as does occur is often found in association with the *Agrostis curtisii* grassland (e.g. Ward *et al.* 1972*a*).

In the other direction in edaphic transitions of this kind, the *Nardus-Galium* grassland often passes to some kind of bog or related vegetation on ombrogenous peat that has accumulated over summit plateaus or high-level terraces. Ultimately, it is blanket mire that usually terminates such sequences, with the *Calluna-Eriophorum* bog occurring over the flatter summits of the Grampians, the higher ground in the Southern Uplands and down the Pennines, the more oceanic *Scirpus-Eriophorum* bog being found in association with the community in more westerly parts of Scotland, in

U10
Carex bigelowii-Racomitrium lanuginosum moss-heath

Synonymy
Rhacomitrium heath   Smith 1900*b*, Smith 1911*b*, Price-Evans 1932, Tansley 1939, Ratcliffe 1959*a*, Edgell 1969; Moss-lichen associes   Smith 1911*b*, Watson 1925, Price-Evans 1932, Tansley 1939, *p.p.*; *Rhacomitrium-Carex bigelowii* nodum Poore 1955*c*, Poore & McVean 1957, Huntley 1979; *Dicranum fuscescens-Carex bigelowii* sociation Poore 1955*c*; *Cariceto-Rhacomitretum lanuginosi* McVean & Ratcliffe 1962, Birks 1973; *Polygoneto-Rhacomitretum lanuginosi* McVean & Ratcliffe 1962; *Juncus trifidus-Festuca ovina* nodum   McVean & Ratcliffe 1962; *Festuca ovina-Luzula spicata* nodum   Birks 1973; *Agrostis montana-Rhacomitrium lanuginosum* community Birse & Robertson 1976; *Rhacomitrium lanuginosum-Dicranum fuscescens* nodum Huntley 1979; *Festuco-Rhacomitretum lanuginosi* Birse 1980; *Carex bigelowii-Festuca vivipara* Association (Birse & Robertson 1976) Birse 1980 *p.p.*

Constant species
Carex bigelowii, Deschampsia flexuosa, Festuca ovina/vivipara, Vaccinium myrtillus, Racomitrium lanuginosum, Cladonia uncialis.

Rare species
Artemisia norvegica, Diapensia lapponica, Koenigia islandica, Loiseleuria procumbens, Luzula arcuata, Minuartia sedoides, Sibbaldia procumbens, Aulacomnium turgidum, Hypnum hamulosum, Kiaeria starkei, Nephroma arctica.

Physiognomy
The *Carex bigelowii-Racomitrium lanuginosum* community takes in both continuous carpets of mossy heath and much more open vegetation in which *Racomitrium lanuginosum* remains an important distinguishing feature. In the closed swards included here, this moss is often truly dominant, forming an extensive, sometimes total, cover of densely-packed shoots, frequently curled over all in one direction by relentless winds, but growing together as a vigorous mat up to 5 cm or so thick, which can be peeled off the rocky substrate beneath. From this kind of vegetation, which can stretch for many hectares over broad plateaus, there is a complete gradation through broken rocky ground with more patchy carpets, to almost barren stone-littered surfaces on which small clumps of *R. lanuginosum* are virtually the only cover.

Some other mosses play a more infrequent, but locally prominent, role in the community, though this variation is not of itself sufficient to characterise different kinds of *Carex-Racomitrium* heath (cf. Poore 1955*c*, Huntley 1979). Most obvious among these is *Dicranum fuscescens* which is only occasional throughout but sometimes patchily abundant within masses of *R. lanuginosum*, often where there are slight depressions, perhaps just a few centimetres deep, which catch and hold a little snow in the winter. Some of these spots are clearly transitional to late snow-beds, but often the effect is just to produce a mosaic within the moss carpet of what is otherwise fairly uniform vegetation. *Polytrichum alpinum* can behave in the same fashion, though it is rarely as extensive as *D. fuscescens*, and, more locally, *Rhytidiadelphus loreus* and other bulky pleurocarps, or the rare *Kiaeria starkei*, can pick out sheltered places. Then, scattered through the carpet, there can be occasional shoots of *Dicranum scoparium, Hypnum cupressiforme, Polytrichum piliferum, P. alpestre, Campylopus paradoxus* and *Andreaea alpina*. Some other *Racomitrium* spp. may occur infrequently too: *R. heterostichum, R. fasciculare* and *R. canescens* have all been recorded here, the last once noted in abundance by McVean & Ratcliffe (1962) over an area where fresh sand had been blown among rocks, but having a very restricted role in general here compared with, say, Icelandic moss-heaths (McVean 1955). One particular sub-community also provides a locus for the rare montane mosses *Aulacomnium turgidum* and *Hypnum hamulosum*. Frequent hepatics are much less numerous than mosses but *Diplophyllum albicans* and *Anastrepta orcadensis* occur occasionally and assiduous searching, especially of damper places, sometimes turns up uncommon taxa such as *Anthelia juratzkana* and *Gymnomitrion corallioides* (Watson 1925, Birks 1973).

Lichens are not usually of high cover in the carpet, although a number are found frequently throughout and in some stands there is marked local enrichment of this element in the flora. Most common are *Cladonia uncialis* and *Cetraria islandica*, with *Cladonia arbuscula*, *C. gracilis*, *Cornicularia aculeata* and *Sphaerophorus globosus* more occasional and uneven in their representation. Among a variety of infrequent lichen associates is the very rare foliose species *Nephroma arcticum* (McVean & Ratcliffe 1962) and a large number of saxicolous taxa, particularly of the genera *Lecidea*, *Lecanora*, *Parmelia* and *Umbilicaria*, including some strict Arctic-Alpines, growing on exposed rock fragments (Watson 1925). From these, James *et al.* (1977) tentatively defined two rare montane associations of the Rhizocarpon alpicolae alliance.

Scattered through this ground, or dotted about in the shelter of rocks or moss clumps in the more open kind of vegetation, vascular plants are sometimes reduced to sparse wind-clipped individuals of a very few species. However, there is generally some *Carex bigelowii* and, though this is nothing like so luxuriant or floriferous as in flushed and less windswept situations, it can be quite abundant in the community, its rhizomes spreading protected in or beneath the moss mat. Then, there are frequent small tussocks of *Festuca ovina*, very often clearly *F. vivipara*, and, particularly towards lower altitudes, *Deschampsia flexuosa* with small sprigs of *Vaccinium myrtillus*. More occasional, or rather more unevenly distributed among the sub-communities, are *Galium saxatile*, *V. vitis-idaea*, *Agrostis canina*, *A. capillaris*, *Alchemilla alpina* and *Salix herbacea*, and, where these become a little more frequent, together with *Carex pilulifera* and *Potentilla erecta*, the community begins to look transitional to the sort of grassy or sub-shrub heath that occurs in windswept places at lower levels. In other stands, towards the more exposed extreme to which the community penetrates, there can be a very striking enrichment, in what is often an open and heterogenous cover, with *Luzula spicata*, *Polygonum viviparum*, *Thymus praecox* and various cushion herbs, notably *Silene acaulis*, *Armeria maritima* and the rare *Minuartia sedoides* and *Sibbaldia procumbens*. It is in this kind of *Carex-Racomitrium* heath, too, that another rarity, *Juncus trifidus*, is most often seen, usually on wind-blasted ablation surfaces, in vegetation which McVean & Ratcliffe (1962) described as a distinct *Juncus-Festuca ovina* nodum, but which can be readily accommodated here.

Three further extremely rare, and only fairly recently discovered, members of the British mountain flora are also found in more open, rocky stands of this moss-heath (Raven & Walters 1956). *Artemisia norvegica*, a plant that is otherwise known only from parts of Norway and the Urals (Hultén 1954), occurs in this vegetation in numerous small colonies spread over three localities in Ross (Blakelock 1953, Perring & Farrell 1977), while *Diapensia lapponica*, an altogether more spectacular plant when in flower and one with a widespread Arctic-Subarctic range, is restricted to a single rocky crest near Fort William, where it is fairly plentiful but damaged by collectors and deer (Perring & Farrell 1977). *Koenigia islandica*, which occurs on Skye and Mull, is not restricted to *Carex-Racomitrium* heath, being found also in wet, stony *Carex-Koenigia* flushes, but some of its drier stations north of The Storr on Skye belong here (Birks 1973).

Sub-communities

***Galium saxatile* sub-community:** *Cariceto-Rhacomitretum typicum* Birks 1973; *Festuceto-Rhacomitretum*, Typical & *Nardia scalaris* subassociations Birse 1980 *p.p.*; *Carex bigelowii-Festuca vivipara* Association, *Rhacomitrium* degeneration phases Birse 1980. *R. lanuginosum* is still usually the most abundant plant in these more closed swards, but the cover and variety of the vascular associates are greater than in much Typical *Carex-Racomitrium* heath, and mixtures of *C. bigelowii*, *F. ovina/vivipara*, *D. flexuosa* and *V. myrtillus* quite often attain sub-dominance. Among the community occasionals, *V. vitis-idaea*, *Agrostis canina* and *A. capillaris* are all quite frequent, and each can show modest abundance, adding to the character of a grassy sub-shrub heath. Moreover, *Galium saxatile* is strongly preferential here, with *Carex pilulifera* and *Potentilla erecta* also good diagnostic plants but at lower frequencies. Some stands have *Alchemilla alpina* at moderately high cover, very occasionally in association with such herbs as *Campanula rotundifolia*, *Viola riviniana* and *Succisa pratensis*, while in others there is some *Salix herbacea*, which can give an altogether more chionophilous look.

Apart from the dominant *R. lanuginosum* and fairly frequent scattered tufts of *Dicranum scoparium* and *Polytrichum alpinum*, mosses are not very numerous, with just scarce records for *P. piliferum*, *P. alpestre*, *P. longisetum*, *Racomitrium heterostichum*, *Rhytidiadelphus loreus* and *Campylopus paradoxus*. And hepatics are generally limited to sparse individuals of *Diplophyllum albicans*. The commonest lichens are *Cladonia uncialis* and *Cetraria islandica*, with occasional *Cladonia arbuscula* and, preferential at low frequencies, *C. coccifera*, *C. impexa*, *C. cervicornis* and *C. crispata*, but even taken together these rarely have appreciable cover in the carpet.

Typical sub-community: *Rhacomitrium* heath Smith 1900*b*, Smith 1911*b*, Price-Evans 1932, Tansley 1939, Ratcliffe 1959*a*, Edgell 1969; *Rhacomitrium-Carex bigelowii* nodum Poore 1955*c*, Poore & McVean 1957, Huntley 1979; *Dicranum fuscescens-Carex bigelowii* ociation Poore 1955*c*; *Cariceto-Rhacomitretum*, typical facies McVean & Ratcliffe 1962; *Rhacomitrium lanuginosum-Dicranum fuscescens* nodum Huntley 1979; *Festuceto-Rhacomitretum, Cladonia arbuscula* sub-association Birse 1980. *R. lanuginosum* is often strongly dominant here, although the sub-community also includes many very open stands in which no species has more than sparse cover. The other constants of the heath all remain frequent, *C. bigelowii* in particular making some moderate contribution to the swards in some places, *F. ovina/vivipara* in others, but *D. flexuosa* is rather patchy in its occurrence and *V. myrtillus* is frequently very sparse and sometimes altogether absent. Other vascular associates are at most occasional, but here can be a little *V. vitis-idaea, Empetrum nigrum* ssp. *hermaphroditum, Huperzia selago* and *Galium saxatile,* and *J. trifidus* is found in some stands.

Along with *R. lanuginosum,* there are common records for *D. fuscescens* and, over undulating ground, the two can form a mosaic in the carpet. *Polytrichum alpinum* and *Dicranum scoparium* occur occasionally and there is sometimes locally abundant *Kiaeria starkei.* More striking, compared with the *Galium* sub-community, is the variety of the lichen flora, with *Cladonia uncialis* and *Cetraria islandica* both more frequent here, and often joined by *Cladonia arbuscula* and *C. gracilis,* less commonly by *C. squamosa* and *Cornicularia aculeata.*

Silene acaulis sub-community: Moss-lichen associes Smith 1911*b*, Watson 1925, Price-Evans 1932, Tansley 1939, *p.p.*; *Rhacomitrium-Carex bigelowii* nodum, *Polygonum viviparum-Salix herbacea* facies Poore & McVean 1957; *Cariceto-Rhacomitretum,* cushion-herb facies McVean & Ratcliffe 1962, Birks 1973; *Cariceto-Rhacomitretum, Juncus* facies McVean & Ratcliffe 1962; *Polygoneto-Rhacomitretum lanuginosi* McVean & Ratcliffe 1962; *Juncus trifidus-Festuca ovina* nodum McVean & Ratcliffe 1962; *Festuca ovina-Luzula spicata* nodum Birks 1973. In this, the most distinctive and heterogenous kind of *Carex-Racomitrium* heath, the cover of *R. lanuginosum* can be extensive, but much vegetation included here is very open, when the moss carpet is fragmented, sometimes to virtual non-existence, and even in the more closed swards there can be local diversity in the mat. In particular, where there is intermittent irrigation, say from snow-melt, species such as *Rhytidiadelphus loreus,*

Pleurozium schreberi and *Hylocomium splendens* increase in prominence, sometimes sharing dominance with *R. lanuginosum* over small patches. *Dicranum fuscescens, D. scoparium* and, more especially, *Polytrichum alpinum* are also common, though not generally with any appreciable cover. Then, the rare *Aulacomnium turgidum* and *Hypnum hamulosum* are preferential to this kind of vegetation and, more occasionally, there is some *H. callichroum* or *Drepanocladus uncinatus.* Among hepatics, *Ptilidium ciliare* is much more frequent than usual. Lichens are not usually so obvious a feature as in the Typical sub-community, but there is frequently a little *Cladonia uncialis* and *Cetraria islandica,* and *Sphaerophorus globosus* is strongly preferential.

More striking than these elements, though, are the vascular associates which can form anything from a dense scatter of plants through the moss carpet to a very sparse cover over what looks at first sight like a wilderness of rocks and gravel. Among the constants, *D. flexuosa* and *V. myrtillus* are both somewhat reduced in frequency and hardly ever of any abundance, and even *C. bigelowii* is usually limited to scattered, stunted tufts. *F. ovina/vivipara* remains very common, however, and is fairly often a co-dominant. Most of the other abundant herbs, though, belong to a group of preferentials, many of which grow as small mats or cushions, giving the vegetation a very distinctive appearance. Commonest among these are *Salix herbacea, Alchemilla alpina* and, more strictly confined to this kind of *Carex-Racomitrium* heath, *Silene acaulis, Thymus praecox, Armeria maritima, Polygonum viviparum, Luzula spicata, Minuartia sedoides, Sibbaldia procumbens* and *Omalotheca supina.* Rarely are all of these represented over a single small area, and indeed there tends to be a continuous transition between stands of the Typical sub-community in which a very few of these can occur, usually one or other of *S. acaulis, A. maritima* or *M. sedoides,* and richer swards, although for simplicity the cushion-herb facies of McVean & Ratcliffe's (1962) *Cariceto-Rhacomitretum* is entirely subsumed here. Other stands which can be accommodated within this kind of *Carex-Racomitrium* heath have small tussocks of *Deschampsia cespitosa* and occasional basiphilous plants such as *Thalictrum alpinum, Selaginella selaginoides* and *Pinguicula vulgaris,* species which give some floristic continuity with the *Festuca-Alchemilla-Silene* dwarf-herb community. Then, the very open vegetation with a sparse cover of cushion herbs and tussocks of *J. trifidus,* which McVean & Ratcliffe (1962) characterised as the *Juncus-Festuca* nodum, can also be taken into this sub-community. Apart from these features, there is usually nothing that is distinctive enough to suggest separating these stands from the *Carex-Racomitrium* heath, although the lack of competition from higher plants, that is typical of the extremely exposed environment, sometimes encourages cryptogams that are otherwise most characteristic of late snow-beds: the foliose lichen *Solorina crocea* is one of these and there can also be occasional records for *Conostomum tetragonum* and *Gymnomitrion concinnatum*.

Habitat

The *Carex-Racomitrium* moss-heath is characteristic of windswept, cloud-ridden plateaus at moderate to very high altitudes through the cold, humid mountains of north-west Britain. It is strongly concentrated in the Scottish Highlands, where very large stands can be found over ridges and summits that are mostly blown clear of snow, but it also occurs more locally on moderately exposed cols and spurs, and can extend into situations where wind erosion and bitter temperatures maintain some of the most inhospitable upland scenery in the country. In general, harsh climatic conditions make this a climax community, although stands at lower altitudes have sometimes been affected by grazing.

The combination of exposure to cold with high humidity is of great importance for the development of this kind of vegetation. It is largely a community of the low- to middle-alpine zones in our mountains, extending down to below 500 m along the western edge of its range, where it is represented on the Inner Isles from Skye down to Arran, but confined to progressively higher altitudes moving through the north-west Highlands into the Grampians: in Ross, for example, its base lies at around 750 m, whereas in the central Highlands it is not found much below 900 m but extends in fragmentary fashion to over 1200 m (McVean & Ratcliffe 1962, Birse 1980). Throughout these regions, at these levels, the summers are brief and cool, with mean annual maxima usually below 21 °C (Conolly & Dahl 1970). Outlying stations further south, where the community is of local importance over summits in the Southern Uplands, the Lake District, the north Pennines and, in more attenuated form, in north Wales, are just a little warmer, with maxima of 23–24 °C, but these still present some of the bleakest tracts of high ground outside the Highlands. Winter minima, on the other hand, and thus the annual range of temperatures, vary considerably across these parts of Britain, with the most bitter and more continental conditions being experienced over the higher summit plateaus of the east-central Highlands, the climate further west being noticeably more equable, particularly towards the lowest altitudinal limits of the community, over the spurs of mountains along the Atlantic seaboard of Scotland (*Climatological Atlas* 1952).

Towards the former extreme, the *Carex-Racomitrium* heath only survives where there is some degree of shelter from the very harshest exposure, its hold becoming increasingly tenuous the less oceanic the general climatic

conditions: over the eastern slopes of the Cairngorms, for example, the lower limit of the community can be up to 250 m below that on the more exposed western spurs (McVean & Ratcliffe 1962). In general, however, this is a vegetation type of open, relatively unsheltered conditions, being most extensive over tracts of flat or gently-sloping ground, on what Smith (1900), in the first, classic account of the community, called 'alpine plateaus'. In such places, away from hollows and lee slopes, there is little relief from the strong, unrelenting winds that blow at these altitudes, so the ground is for the most part kept free of any but a patchy cover of snow through the winter. Over the range of the *Carex-Racomitrium* heath the amount of snow can be substantial with over 100 days observed snow- or sleet-fall in some places (Manley 1940), but this is caught and held only very lightly and locally here, being mostly swept off into more sheltered situations. There can sometimes be a light covering of *verglas* over the moss carpet, with icy pennants frozen on to upstanding sedge and grass leaves (Poore 1955c), but generally speaking, the vegetation and soils are fully exposed to the influence of fluctuating temperatures and to the drying effect of the wind on ground already deprived of much of its winter moisture by the redistribution of precipitation in snow drifted elsewhere.

Potentially, then, the flora of such situations is likely to have a strong Arctic-Alpine character, consisting of plants adapted to the very short growing season and tolerant of bouts of bitter cold and desiccation alternating with drenching mist, and of the environmental instability resulting from freeze-thaw and solifluction. Such plants must also be able to survive in soils that are often of a very fragmentary character and highly impoverished, occurring locally in deep pockets, but often shallow and stony, usually sharply-draining and sometimes strongly podzolised and with at most a thin humic crust (Smith 1911b, Poore 1955c, Poore & McVean 1957, McVean & Ratcliffe 1962). What inhibits the expression of such a floristic aspect throughout the *Carex-Racomitrium* heath is the formidable competitive power of *R. lanuginosum* in all but the more exposed and disturbed situations here. Of all the many kinds of upland vegetation to which this moss contributes, it is in this community, characteristic of usually snow-free, but less continental, conditions, and with little or no grazing, that its growth is most vigorous and effective in ousting montane species susceptible to being crowded out.

Some plants tolerant of the generally montane climate, able to extend their rhizomes beneath the moss carpet, as do *C. bigelowii* and *V. myrtillus*, or growing as small tussocks which push aside the densely-growing moss shoots, as with *D. flexuosa* and *F. ovina/vivipara*, can withstand the competition to some extent and may actually benefit from the shelter that the mat provides

community, but *D. cespitosa*, *C. bigelowii* and *Alchemilla alpina* can run on with some frequency into the late snow-bed and, where they remain patchily abundant, blur the boundary between the vegetation types. *D. cespitosa*, *A. alpina*, *G. saxatile*, *D. flexuosa*, *R. loreus* and *P. alpinum* also all remain very common in another strongly chionophilous community frequent in this part of the Scottish Highlands, the *Cryptogramma-Athyrium* vegetation, but mixtures of *C. crispa* and *A. distentifolium* typically dominate there and stands almost always mark out stretches of block scree, patches of tumbled boulders or rocky ledges.

More difficult to delineate precisely, or to interpret environmentally, are transitions to the *Alchemilla-Sibbaldia* community which are commonly seen in these mountains over the irrigated ground around late snow-beds. Mixtures of *A. alpina* and *S. procumbens* are the usual dominants, sometimes with *Minuartia sedoides*, *Thymus praecox* and *Omalotheca supina* in modest abundance, but grasses can be quite plentiful, among them *D. cespitosa*, *A. capillaris*, *Anthoxanthum* and *Nardus*, and there may be quite prominent patches of hypnaceous mosses. Such stretches of ground are often fed by the distinctive kinds of snow-bed springs, the *Pohlietum glacialis* or *Sphagno-Anthelietum*: scattered shoots of *D. cespitosa* are frequent in such vegetation but the striking dominance of the respective bryophytes is usually sufficient to mark them out clearly.

Finally, in this kind of pattern, the *Deschampsia-Galium* vegetation, in the form of the *Rhytidiadelphus* sub-community, frequently extends up to its highest altitudinal limits as small stands marking out sheltered hollows in *Carex-Racomitrium* heath, the characteristic moss-dominated sward of exposed summits in the more oceanic parts of the Highlands. Separation from stretches of the Typical sub-community of this vegetation, the most widely distributed form, is usually fairly easy, the change from abundance of *R. lanuginosum* to hypnaceous mosses being well defined. But increasingly towards the north-west, the *Carex-Racomitrium* heath is represented by the *Silene* sub-community where, under conditions of severe exposure, the carpet of *R. lanuginosum* often becomes fragmented, and where intermittent irrigation from snow-melt is marked by patches of *R. loreus*, *P. schreberi* and *H. splendens* and diminutive tussocks of *D. cespitosa*. The high frequency of mats and cushions of plants such as *Salix herbacea*, *Silene acaulis*, *Thymus praecox* and *Armeria maritima* generally helps define the *Carex-Racomitrium* heath but, over ablation fields, the cover is often broken and heterogenous and intimate mosaics of this vegetation with mossy *Deschampsia-Galium* snow-beds can be seen.

In shifting northwards, the balance among the elements in these sequences changes somewhat. The *Polytrichum-Kiaeria* snow-bed, for example, can be seen with

the *Deschampsia-Galium* community as far north as Beinn Dearg and Ben Wyvis, but even on Beinn Eighe and the Letterewe Hills, and more obviously on Ben Hope, Foinaven and Ben More Assynt in Sutherland, the *Rhytidiadelphus* sub-community of the *Deschampsia-Galium* vegetation is often among the most chionophilous swards. It is in this part of the Highlands, too, that the similarities and contrasts between the community and the tall-herb *Luzula-Geum* vegetation, mediated perhaps by climatic and edaphic conditions, perhaps also by grazing, are to be seen in the swards disposed over slopes and ledges of the corries. *D. cespitosa*, *Alchemilla alpina*, *Agrostis capillaris*, *Anthoxanthum*, *D. flexuosa*, *G. saxatile*, *R. acetosa* and hypnaceous mosses all remain very frequent in Typical *Luzula-Geum* vegetation and there are transitional stands in which *D. cespitosa* occurs with particular abundance. But, generally, it is a subordinate plant in the tall-herb stands, and there is typically a good representation of species such as *Luzula sylvatica*, *Geum rivale*, *Rhodiola rosea*, *Angelica sylvestris*, *Filipendula ulmaria* and *Succisa pratensis*, and a ground carpet in which *Rhytidiadelphus triquetrus*, *Thuidium tamariscinum*, *Plagiomnium undulatum* and *Rhizomnium punctatum* are very common. Comparable transitions to this can also be seen where the substrates and seeping waters tend to be more base-poor, when the tall-herb vegetation is generally of the *Luzula-Vaccinium* type. Again, here, *D. cespitosa* may remain frequent and locally of modest abundance, but *V. myrtillus*, *Dryopteris dilatata*, *Blechnum spicant* and *Gymnocarpium dryopteris* are usually very frequent and, apart from the hypnaceous mosses, the floristic overlap is less than with the *Luzula-Geum* community.

East of Creag Meagaidh and Ben Heasgarnich in the central Highlands, the *Rhytidiadelphus* sub-community of the *Deschampsia-Galium* grassland becomes very fragmentary and is replaced geographically in moderately late snow-beds by the *Carex-Polytrichum* sedge-heath. Both kinds of vegetation are to be seen together on Creag Meagaidh and Ben Alder and in some localities, on Ben Lawers for example, there are quite extensive tracts of heath of a somewhat intermediate character (McVean & Ratcliffe 1962). Also in the central Highlands, on Ben Alder and Beinn Laoigh, are some snow-bound slopes where the *Deschampsia-Galium* community can be found with *Salix-Racomitrium* snow-beds as well as those of the *Polytrichum-Kiaeria* type, but there is usually little difficulty in distinguishing the patches of this strongly chionophilous vegetation with its cover of dwarf willow and diagnostic bryophytes. Further east, the *Anthoxanthum-Alchemilla* sub-community continues to find a place beyond the range of the *Rhytidiadelphus* sub-community among sequences of moderately chionophilous heaths and grasslands, as around Lochnagar and Caenlochan (Huntley 1979), but it is a minor element in such patterns and sometimes looks very much like a replacement for tall-herb and fern vegetation that has been derived by grazing.

Distribution

The *Deschampsia-Galium* grassland, particularly the *Rhytidiadelphus* type, is strongly concentrated in the western Highlands. The *Anthoxanthum-Alchemilla* sub-community occurs somewhat more widely, extending into parts of the eastern Grampians, and in more fragmentary form into southern Scotland and north Wales.

Affinities

As understood here, the *Deschampsia-Galium* community includes the more species-poor kinds of *D. cespitosa* grassland and snow-bed first described in detail by McVean & Ratcliffe (1962). Even in their account, the very close relationship of their *Deschampsietum* and *Deschampsieto-Rhytidiadelphetum* was recognised and, in the light of further sampling (including Huntley 1979), it makes good sense to combine the less basiphilous stands among this vegetation, and to transfer the remainder to other communities where a more species-rich and calcicolous character is the norm.

The wider perspective that is now available also enables the *Deschampsia-Galium* community to be set in the context of other swards in which *D. cespitosa* plays an important part, and of the full range of dwarf- and tall-herb vegetation occurring over moderately snow-bound and irrigated ground at high altitudes. As far as the first perspective is concerned, the *Deschampsia-Galium* grassland can be seen as continuing a floristic trend that is already visible among sub-montane Nardo-Galion swards, where *D. cespitosa* attains local prominence when climate and soil combine to create congenial conditions. On this view, then, the community comprises an upland and calcifuge counterpart to the mesophytic *Holcus-Deschampsia* grassland, the major vegetation type with this grass on moist circumneutral soils through the British lowlands. This latter is clearly anthropogenic, usually replacing Alno-Padion or damp Carpinion forest, but it shows some edaphic continuity with the more montane swards and, if these too are sometimes biotically derived, the vegetation types are perhaps best seen as rather different products of the same ability of *D. cespitosa* to exploit situations which become suitable for it. It would be good to know how far the taxonomic difference between ssp. *cespitosa* and ssp. *alpina* corresponds to the definition of the various vegetation types between the climatic extremes across which they spread in this country.

In floristic terms, the affinities of the *Holcus-Deschampsia* grassland are with the lowland, mesophytic swards of the Arrhenatheretalia, among which this kind of vegetation has sometimes been thought worthy of separating off into a distinct Deschampsion alliance. The *Deschampsia-Galium* community, on the other hand, though its montane character is muted, clearly belongs among the kinds of vegetation which have been grouped in the Deschampsieto-Myrtilletalia, an order typical of moderately snow-bound or seasonally-irrigated ground with oligotrophic soils not liable to solifluction (Dahl 1956). A Deschampsieto-Anthoxanthion has been erected within this to contain swards floristically distinct from the chionophilous bilberry heaths of the Phyllodoco-Vaccinion and the moderately late snow-bed heaths with *Nardus* and *C. bigelowii* gathered into the Nardeto-Caricion. In Scandinavia, vegetation similar to the *Anthoxanthum-Alchemilla* sub-community has been included in the *Deschampsietum caespitosa alpicolum* of Nordhagen (1928, 1943) described from Sylene and Sikilsdalen, but counterparts of the *Rhytidiadelphus* sub-community are less obvious. Icelandic heaths sometimes have dominant *Hylocomium splendens* (McVean 1955) but *Rhytidiadelphus* spp. appear to be absent there and the nearest equivalent in Scandinavia looks to be the hydrophilous meadow vegetation of the *Ranunculetum acris acidophilum* of Gjaerevøll (1956).

An interesting unresolved question is how such swards in Britain relate to the dwarf-herb vegetation of irrigated ground such as is included in the *Alchemilla Sibbaldia* and *Festuca-Alchemilla-Silene* communities on the one hand, and the tall-herb *Luzula-Geum* and *Luzula-Vaccinium* communities on the other. The former, which can take in the *triquetrosum* of McVean & Ratcliffe's (1962) *Deschampsieto-Rhytidiadelphetum* show considerable floristic overlap with the *Deschampsia-Galium* grassland, but are probably distinguished environmentally by a preference for more base-rich ground subject to some solifluction, and belong among the Ranunculo-Anthoxanthion of the Salicetalia herbaceae (Gjaerevøll 1956). The latter, which accommodate the more species-rich of McVean & Ratcliffe's (1962) *Deschampsietum*, are also floristically close, but these are among our Cicerbition alpinae communities, typical of damp ledges inaccessible to grazing animals and transitional to sub-alpine willow scrub where there is a measure of continued protection and stability. How far the *Deschampsia-Galium* grasslands represent impoverished derivatives of such rich vegetation produced by grazing was an issue raised by McVean & Ratcliffe (1962) but still unanswered.

Floristic table U13

	a	b	13
Deschampsia cespitosa	V (4–10)	V (1–4)	V (1–10)
Galium saxatile	V (1–4)	V (1–6)	V (1–6)
Agrostis capillaris	V (1–4)	V (1–4)	V (1–4)
Rhytidiadelphus loreus	IV (1–8)	V (1–10)	V (1–10)
Polytrichum alpinum	IV (1–4)	IV (1–10)	IV (1–10)
Hylocomium splendens	IV (1–6)	IV (1–6)	IV (1–6)
Rumex acetosa	IV (1–4)	III (1–4)	III (1–4)
Alchemilla alpina	IV (1–6)	III (1–4)	III (1–6)
Rhytidiadelphus squarrosus	IV (1–9)	II (1–6)	III (1–9)
Anthoxanthum odoratum	IV (1–4)	II (1–6)	III (1–6)
Sibbaldia procumbens	III (1–4)	I (1–3)	II (1–4)
Ranunculus acris	III (1–4)		II (1–4)
Festuca rubra	II (1–4)	I (1–4)	II (1–4)
Luzula sylvatica	II (1–4)	I (1)	II (1–4)
Diphasium alpinum	II (1–6)	I (1)	II (1–6)
Hypnum callichroum	II (1–3)	I (1)	II (1–3)
Sphagnum capillifolium	II (1–3)	I (1)	I (1–3)
Peltigera canina	II (1–3)	I (1–3)	I (1–3)
Saxifraga stellaris	II (1–3)	I (1)	I (1–3)
Agrostis canina	II (1–4)	I (1–3)	I (1–4)
Luzula multiflora	II (1–3)	I (1)	I (1–3)
Achillea millefolium	II (1–3)	I (1)	I (1–3)
Thymus praecox	II (1–4)	I (1–4)	I (1–4)
Polytrichum commune	II (1–6)	I (1)	I (1–6)
Plagiothecium undulatum	II (1–3)	I (1)	I (1–3)
Blechnum spicant	II (1–3)		I (1–3)
Dryopteris dilatata	II (1–4)		I (1–4)
Alchemilla glabra	II (1–4)		I (1–4)
Barbilophozia lycopodiodes	II (1–3)		I (1–3)
Alchemilla filicaulis filicaulis	II (1–3)		I (1–3)
Campanula rotundifolia	II (1–3)		I (1–3)
Viola riviniana	II (1–4)		I (1–4)
Saxifraga hypnoides	II (1–3)		I (1–3)
Drepanocladus uncinatus	II (1–3)		I (1–3)
Rhytidiadelphus triquetrus	II (1–5)		I (1–5)
Cerastium alpinum	I (1–3)		I (1–3)
Saussurea alpina	I (1–3)		I (1–3)
Carex bigelowii	III (1–3)	IV (1–6)	III (1–6)
Vaccinium myrtillus	III (1–4)	IV (1–4)	III (1–4)
Ptilidium ciliare	II (1–3)	IV (1–3)	III (1–3)
Racomitrium lanuginosum	II (1–4)	III (1–4)	II (1–4)
Barbilophozia floerkii	I (1–4)	II (1–3)	I (1–4)
Anastrepta orcadensis	I (1–3)	II (1–3)	I (1–3)
Cladonia gracilis	I (1)	II (1–3)	I (1–3)
Cetraria islandica		II (1–3)	I (1–3)

Floristic table U13 (*cont.*)

	a	b	13
Minuartia sedoides		I (1–6)	I (1–6)
Sphagnum russowii		I (1–3)	I (1–3)
Festuca ovina/vivipara	III (1–4)	III (1–5)	III (1–5)
Viola palustris	III (1–4)	III (1–3)	III (1–4)
Nardus stricta	III (1–4)	III (1–5)	III (1–5)
Deschampsia flexuosa	III (1–4)	III (1–4)	III (1–4)
Pleurozium schreberi	III (1–3)	III (1–3)	III (1–3)
Cerastium fontanum	II (1–3)	II (1–3)	II (1–3)
Potentilla erecta	II (1–4)	II (1–4)	II (1–4)
Oxalis acetosella	II (1–3)	II (1–3)	II (1–3)
Euphrasia frigida	II (1–3)	II (1–3)	II (1–3)
Pohlia nutans	II (1–6)	II (1–4)	II (1–6)
Huperzia selago	I (1–3)	I (1–3)	I (1–3)
Nardia scalaris	I (1–3)	I (1–3)	I (1–3)
Salix herbacea	I (1)	I (1–3)	I (1–3)
Carex pilulifera	I (1–3)	I (1–3)	I (1–3)
Hypnum cupressiforme	I (1–4)	I (1–4)	I (1–4)
Armeria maritima	I (1–3)	I (1–3)	I (1–3)
Vaccinium vitis-idaea	I (1–3)	I (1–4)	I (1–4)
Polygonum viviparum	I (1–3)	I (1–3)	I (1–3)
Silene acaulis	I (1–3)	I (1)	I (1–3)
Kiaeria starkei	I (1)	I (1)	I (1)
Selaginella selaginoides	I (1–3)	I (1)	I (1–3)
Omalotheca supina	I (1)	I (1)	I (1)
Number of samples	29	18	47
Number of species/sample	26 (8–53)	20 (12–30)	24 (8–53)
Vegetation height (cm)	20 (2–75)	6 (5–8)	15 (2–75)
Vegetation cover (%)	95 (30–100)	94 (80–100)	95 (30–100)
Altitude (m)	723 (294–1220)	871 (692–1100)	780 (294–1220
Slope (°)	21 (0–50)	12 (0–30)	17 (0–50)

a   *Anthoxanthum odoratum-Alchemilla alpina* sub-community

b   *Rhytidiadelphus loreus* sub-community

13   *Deschampsia cespitosa-Galium saxatile* grassland (total)

ominated stands of the *Cryptogramma-Athyrium* snow-bed on stretches of block scree, are very characteristic of the upper snow-bound slopes of the Fannich and Affric–Cannich hills, Beinn Dearg, and to a lesser extent of the Monar Forest, though there the more chionophilous elements tend to be fragmentarily developed. Northwards from this part of Scotland, moss-dominated stands of the *Deschampsia-Galium* community tend to occupy the latest snow-beds and around Letterewe and Ben More Assynt *Alchemilla-Sibbaldia* vegetation can be found on irrigated ground around these, grading to the centre with an increase in hypnaceous mosses and a loss of less chionophilous herbs. Beyond here, however, the community becomes of very patchy occurrence on the mountains of the far north-west.

Moving south across the Great Glen into the central Highlands broadly similar patterns to those described above can be seen on Bidean Nam Bian, Ben Alder and Creag Meagaidh, though to the east of here, the *Deschampsia-Galium* vegetation tends to be replaced in the sequences by the *Carex-Polytrichum* sedge-heath, the swards of which are usually quite distinct from the *Alchemilla-Sibbaldia* community. The *Alchemilla-Sibbaldia* vegetation itself continues to be well represented even into the drier east-central Highlands, where is an important element in the varied and extensive suites of chionophilous vegetation over the northern slopes of the Cairngorms.

The other striking kind of zonation which is best seen in the central Highlands involves transitions between the *Alchemilla-Sibbaldia* community and other more calcicolous vegetation types of moderately snow-bound slopes and irrigated ground. The *Festuca-Alchemilla-Silene* community in particular is very close in its floristics and physiognomy and is most often found with *Alchemilla-Sibbaldia* vegetation over the limestones and calcareous mica-schists of the Breadalbane Mountains. It is distinctly more calcicolous in its total flora, though it is often the abundance of less demanding species like *Silene acaulis* and *Minuartia sedoides* that provide the most obvious indication of its extent, with *Sibbaldia* taking their place on moving into the *Alchemilla-Sibbaldia* community. And environmentally, it is the longer duration of snow lie that determines the shift to the latter

vegetation type. Quite often the sequence continues into stands of the distinctive montane flush vegetation of the *Caricetum saxatilis* on waterlogged ground in hollows or around permanent *Cratoneuron-Festuca* springs. Continuously irrigated slopes can have rich stands of *Deschampsia-Galium* grassland and where more inaccessible banks or ledges occur the dwarf-herb stands can pass to the luxuriant *Luzula-Geum* or *Dryas-Silene* communities or dripping *Saxifraga-Alchemilla* vegetation. Freedom from grazing often mediates the shift to tall-herb assemblages and the *Alchemilla-Sibbaldia* swards may sometimes experience some cropping and trampling. Ultimately, though, it is climatic and edaphic conditions which maintain the community as a climax.

Distribution
The *Alchemilla-Sibbaldia* community occurs widely through the Scottish Highlands.

Affinities
As defined here, the community is an expanded but essentially similar vegetation type to the *Alchemilla-Sibbaldia* nodum of McVean & Ratcliffe (1962), the first authors to provide a description of this kind of assemblage. With the more comprehensive account of related communities that is available now, the syntaxon retains its integrity as a close relative of the *Festuca-Alchemilla-Silene* dwarf-herb vegetation, though its relationship to the *Deschampsia-Galium* swards is perhaps a little closer than in McVean & Ratcliffe (1962). Nonetheless, it is probably best placed among the more herb-rich communities of the Salicetalia herbaceae which Gjaerevøll (1956) gathered into a Ranunculo-Anthoxanthion alliance to separate them from the moss-dominated late snow-beds. The nearest equivalents described from Scandinavia are the *Alchemilletum alpinae* of Rondane (Dahl 1956) and the *Sibbaldietum procumbentis* which Gjaerevøll (1956) characterised from various parts of Norway. These occur in the kind of irrigated situations commonly occupied by our *Alchemilla-Sibbaldia* community, while similar vegetation in Gjaerevøll's (1956) *Anthoxantho-Deschampsietum flexuosae* marks out the snow-cornice habitat in which the Scottish swards are sometimes found.

Floristic table U14

Alchemilla alpina	V (1–6)		*Thymus praecox*	IV (1–6)
Sibbaldia procumbens	V (1–8)		*Omalotheca supina*	IV (1–4)
Nardus stricta	V (1–4)		*Viola palustris*	IV (1–3)
Deschampsia cespitosa	V (1–6)			
Agrostis capillaris	V (1–6)		*Euphrasia officinalis* agg.	III (1–3)
Galium saxatile	V (1–4)		*Anthoxanthum odoratum*	III (1–3)
Polytrichum alpinum	IV (1–3)		*Festuca ovina/vivipara*	III (1–4)

Floristic table U14 (*cont.*)

Selaginella selaginoides	III (1–3)	*Alchemilla glabra*	I (1–4)
Carex bigelowii	III (1–4)	*Racomitrium heterostichum*	I (1–4)
Racomitrium fasciculare	III (1–4)	*Rhytidiadelphus triquetrus*	I (1–6)
Luzula spicata	III (1–3)	*Diplophyllum albicans*	I (1–3)
Pogonatum urnigerum	III (1–4)	*Ctenidium molluscum*	I (1–4)
Racomitrium canescens	III (1–8)	*Saxifraga oppositifolia*	I (1–3)
Hylocomium splendens	II (1–6)	*Cladonia pyxidata*	I (1–3)
Rhytidiadelphus loreus	II (1–4)	*Sagina saginoides*	I (1–3)
Deschampsia flexuosa	II (1–4)	*Saxifraga stellaris*	I (1–3)
Potentilla erecta	II (1–4)	*Trollius europaeus*	I (1–2)
Racomitrium lanuginosum	II (1–4)	*Diphasium alpinum*	I (1–3)
Rhytidiadelphus squarrosus	II (1–4)	*Festuca rubra*	I (1)
Pleurozium schreberi	II (1–3)	*Carex pilulifera*	I (1–2)
Polygonum viviparum	II (1–4)	*Alchemilla filicaulis vestita*	I (1–3)
Nardia scalaris	II (1–4)	*Cladonia arbuscula*	I (1–3)
Cerastium fontanum	II (1–3)	*Achillea millefolium*	I (1–3)
Huperzia selago	II (1–3)	*Drepanocladus uncinatus*	I (1–5)
Oligotrichum hercynicum	II (1–4)	*Leontodon autumnalis*	I (1–3)
Barbilophozia floerkii	II (1–5)	*Hypnum callichroum*	I (1–3)
Vaccinium myrtillus	II (1–3)	*Armeria maritima*	I (1–3)
Agrostis canina	II (1–6)	*Botrychium lunaria*	I (1–2)
Carex pilulifera	II (1–4)	*Saxifraga hypnoides*	I (1–3)
Dicranum scoparium	II (1–3)	*Racomitrium aquaticum*	I (1–2)
Ranunculus acris	II (1–3)	*Cladonia bellidiflora*	I (1–3)
Thalictrum alpinum	II (1–4)	*Lophozia sudetica*	I (1–4)
Silene acaulis	II (1–8)	*Andreaea alpina*	I (1)
Pohlia nutans	II (1–4)	*Polytrichum piliferum*	I (1–3)
Minuartia sedoides	II (1–7)	*Blechnum spicant*	I (1–3)
Juncus trifidus	II (1–4)		

Number of samples	27
Number of species/sample	30 (15–52)

Vegetation height (cm)	5 (1–10)
Vegetation cover (%)	92 (50–100)

Altitude (m)	888 (640–1116)
Slope (°)	16 (3–60)

Luzula multiflora	I (1–3)
Ptilidium ciliare	I (1–3)
Campanula rotundifolia	I (1–3)
Cetraria islandica	I (1–3)
Salix herbacea	I (1–4)
Rumex acetosa	I (1–3)
Cerastium alpinum	I (1–2)

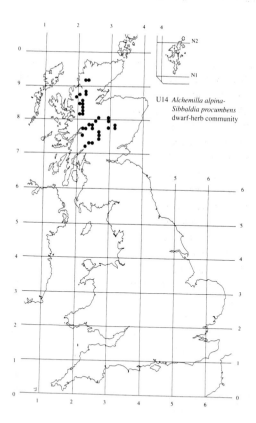

U14 *Alchemilla alpina-
Sibbaldia procumbens*
dwarf-herb community

U15
Saxifraga aizoides-Alchemilla glabra banks

Synonymy

Saxifragetum aizoidis McVean & Ratcliffe 1962, Birks 1973; Mixed Saxifrage facies McVean & Ratcliffe 1962; *Saxifraga aizoides-Festuca-Deschampsia* nodum Huntley 1979; *Saxifraga aizoides-Tussilago farfara* nodum Huntley 1979.

Constant species

Alchemilla alpina, A. glabra, Carex pulicaris, Deschampsia cespitosa, Festuca ovina/vivipara, F. rubra, Pinguicula vulgaris, Polygonum viviparum, Ranunculus acris, Saxifraga aizoides, S. oppositifolia, Selaginella selaginoides, Thalictrum alpinum, Bryum pseudotriquetrum, Ctenidium molluscum.

Rare species

Alchemilla filicaulis ssp. *filicaulis, Carex capillaris, Cerastium alpinum, Cystopteris montana, Epilobium alsinifolium, Juncus biglumis, Oxytropis campestris, Poa alpina, P. glauca, Potentilla crantzii, Polystichum lonchitis, Salix lapponum, Sibbaldia procumbens, Barbilophozia quadriloba, Hylocomium pyrenaicum, Hypnum baumbergeri, H. callichroum.*

Physiognomy

The *Saxifraga aizoides-Alchemilla glabra* community forms highly distinctive banks of vegetation disposed over steep, rocky or earthen slopes. Typically, there is a dripping wet carpet of plants, sometimes just a decimetre or so thick, but growing very luxuriantly, in which *Saxifraga aizoides* is generally the most abundant plant, looking especially striking in summer with its bright yellow flowers. *S. oppositifolia* is constant too and, though it is usually found in small clumps, stands can be seen in which it attains co-dominance with *S. aizoides*: McVean & Ratcliffe (1962) separated these off into a 'mixed Saxifrage facies' but, with further sampling, it is not really possible to justify this. *S. hypnoides* also occurs quite commonly and *S. stellaris* very occasionally, but neither has high cover.

The other abundant element in the vegetation consists of grasses, though neither these nor the sedges of the community ever attain the prominence here that is characteristic of the flushed grasslands or mires in which *S. aizoides* is important. Nonetheless, *Deschampsia cespitosa, Festuca rubra* and *F. ovina/vivipara* are all very frequent, and the first two sometimes make up quite a proportion of the sward. Also common, though not so extensive, are *Anthoxanthum odoratum* and *Agrostis capillaris*, with *Nardus stricta* and *Agrostis canina* occurring occasionally. The community provides a locus, too, for the rare Arctic-Alpine grasses *Poa alpina* and the particularly handsome *P. glauca*, a taxon now subsuming *P. balfouri* (Tutin *et al.* 1980). Sedges are usually less numerous and less prominent, except in transitions to stony flushes, but *Carex pulicaris* is very frequent and there are occasional scattered shoots of *C. flacca, C. lepidocarpa, C. demissa* and *C. panicea*, with some stands having a little of the rare *C. capillaris. Juncus triglumis* occurs at low frequency and the much rarer *J. biglumis* has also been recorded. *Luzula multiflora* and *L. sylvatica* are both occasional, but the latter never has the abundance here that is so typical of much tall-herb ledge vegetation.

Scattered through this carpet is a variety of herbs usually growing in fairly diminutive form, locally a little more bushy or taller. Among these, *Alchemilla glabra* and *A. alpina* are especially common and likely to become prominent, with generally more sparse individuals of *Thalictrum alpinum, Polygonum viviparum, Ranunculus acris, Selaginella selaginoides, Pinguicula vulgaris, Thymus praecox, Oxyria digyna, Euphrasia officinalis, Campanula rotundifolia, Geum rivale, Cerastium fontanum, Viola riviniana* and *Linum catharticum.* More occasionally, there is some *Rhodiola rosea, Potentilla erecta, Oxalis acetosella, Angelica sylvestris, Rumex acetosa*, alpine Hieracia, *Galium saxatile* and *Rhinanthus minor.* Rare plants that sometimes find a place here include *Cerastium alpinum, Epilobium alsinifolium, Alchemilla filicaulis* ssp. *filicaulis, Potentilla crantzii*

	a	b	c	16
Nardus stricta	I (1)	III (1–5)	I (1–4)	II (1–5)
Carex binervis	I (1–3)	III (1–4)	I (1–3)	II (1–4)
Pseudoscleropodium purum	I (1–3)	II (1–3)	I (1)	I (1–3)
Vaccinium vitis-idaea	I (1–3)	II (1–3)		I (1–3)
Calluna vulgaris	I (1–4)	II (1–3)		I (1–4)
Juncus squarrosus		II (1–6)	I (1)	I (1–6)
Festuca rubra		I (1–4)		I (1–4)
Cladonia impexa		I (1–3)		I (1–3)
Cladonia uncialis		I (1–3)		I (1–3)
Rhytidiadelphus loreus	III (1–4)	III (1–4)	I (1–3)	III (1–4)
Blechnum spicant	III (1–4)	III (1–4)	I (1–3)	III (1–4)
Agrostis canina	III (1–4)	III (1–4)	I (1)	III (1–4)
Rumex acetosa	II (1–4)	II (1–4)	II (1–4)	II (1–4)
Lophocolea bidentata s.l.	II (1–3)	II (1–3)	I (1–3)	II (1–3)
Hypnum cupressiforme	II (1–4)	II (1–4)	I (1–4)	II (1–4)
Diplophyllum albicans	II (1–3)	II (1–3)		II (1–3)
Alchemilla alpina	II (1–4)	II (1–4)		II (1–4)
Erica cinerea	I (1–4)	I (1–3)	I (1–3)	I (1–4)
Carex bigelowii	I (1–3)	I (1–3)	I (1–3)	I (1–3)
Dicranum fuscescens	I (5)	I (1–3)		I (1–5)
Thelypteris limbosperma	I (1–4)	I (1–4)		I (1–4)
Digitalis purpurea	I (1–3)	I (1–3)		I (1–3)
Campanula rotundifolia	I (1–3)	I (1–3)		I (1–3)
Racomitrium lanuginosum	I (1–5)	I (1–4)		I (1–5)
Empetrum nigrum	I (1–3)	I (1–4)		I (1–4)
Carex nigra		I (1–3)	I (1)	I (1–3)
Number of samples	28	25	12	55
Number of species/sample	29 (19–43)	24 (14–52)	9 (2–22)	24 (2–52)
Vegetation height (cm)	39 (18–60)	19 (10–25)	24 (15–40)	29 (10–60)
Vegetation cover (%)	90 (50–100)	98 (90–100)	100	94 (50–100)
Altitude (m)	493 (30–915)	483 (120–710)	379 (270–566)	469 (30–915)
Slope (°)	39 (0–85)	32 (1–60)	15 (2–35)	30 (0–85)

a   *Dryopteris dilatata-Dicranum majus* sub-community
b   *Anthoxanthum odoratum-Festuca ovina* sub-community
c   Species-poor sub-community
16 *Luzula sylvatica-Vaccinium myrtillus* tall-herb community

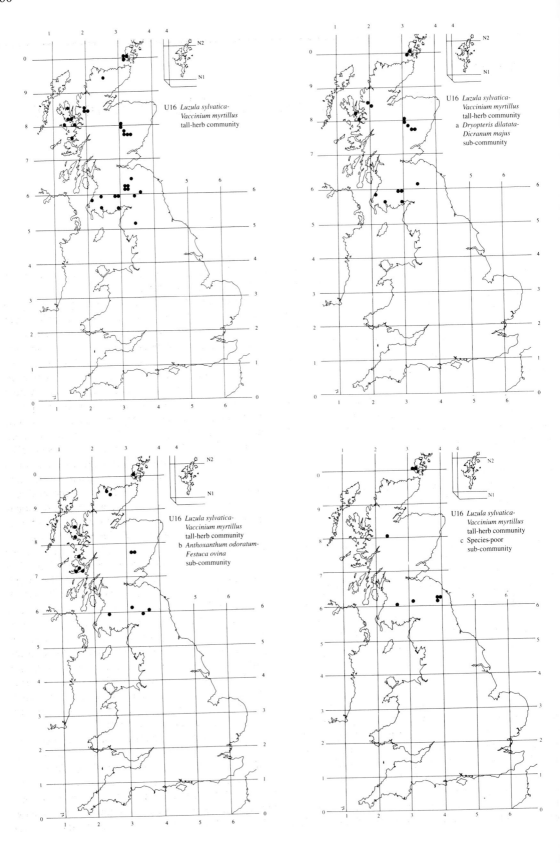

U16 *Luzula sylvatica-*
Vaccinium myrtillus
tall-herb community

U16 *Luzula sylvatica-*
Vaccinium myrtillus
tall-herb community
a *Dryopteris dilatata-*
Dicranum majus
sub-community

U16 *Luzula sylvatica-*
Vaccinium myrtillus
tall-herb community
b *Anthoxanthum odoratum-*
Festuca ovina
sub-community

U16 *Luzula sylvatica-*
Vaccinium myrtillus
tall-herb community
c Species-poor
sub-community

U17
Luzula sylvatica-Geum rivale tall-herb community

Synonymy
Ledge vegetation Smith 1911*a p.p.*, Ratcliffe 1960,
Edgell 1969; *Luzula-Angelica-Rumex* Community
Spence 1960; *Luzula-Dryopteris, Luzula-Deschampsia
fluexuosa-Rhinanthus* and *Luzula-Blechnum-Solidago*
Communities Spence 1960; Tall-herb nodum McVean
& Ratcliffe 1962, Prentice & Prentice 1975; *Deschamp-
sietum caespitosae alpinum* McVean & Ratcliffe 1962;
Sedum rosea-Alchemilla glabra Association Birks
1973; *Luzula sylvatica-Silene dioica* Association Birks
1973; Cliff ledge communities Jermy & Crabbe 1978;
Saxifraga aizoides-Festuca-Deschampsia nodum
Huntley 1979; *Alchemilla glabra-Sedum rosea* nodum
Huntley 1979.

Constant species
*Angelica sylvestris, Deschampsia cespitosa, Geum rivale,
Luzula sylvatica, Rhodiola rosea, Hylocomium splendens.*

Rare species
Alchemilla filicaulis ssp. *filicaulis, Carex atrata, C. rupestris, C. vaginata, Cerastium alpinum, Draba incana,
Epilobium alsinifolium, Meconopsis cambrica, Orthilia
secunda, Poa alpina, P. glauca, Polystichum lonchitis,
Potentilla crantzii, Salix lanata, S. lapponum, S. myrsinites, S. reticulata, Leptodontium recurvifolium, Mastigophora woodsii, Oxystegus hibernicus, Plagiochila carringtonii, Scapania ornithopodioides.*

Physiognomy
The *Luzula sylvatica-Geum rivale* community takes in
varied and often species-rich assemblages of plants,
among which taller and bulkier herbs predominate,
frequently making luxuriant growth and giving the
vegetation the appearance of 'hanging gardens' disposed over the ledges and crags that provide the typical
habitat. Stands are commonly of irregular shape, often
small and fragmentary, frequently with local peculiarities of floristics and structure and characteristically hard
of access, all features which make it difficult to provide a
comprehensive and succinct account of this vegetation,
but which nevertheless themselves contribute to its
highly distinctive appearance.

As in the *Luzula-Vaccinium* community, *Luzula sylvatica* is a constant plant here, often attaining big stature
and flowering profusely, but it is not generally so
abundant, showing consistently high cover in only one
particular kind of *Luzula-Geum* vegetation. Indeed,
there is no fixed pattern of dominance in this community, and any of the other constants, together with a
number of the occasionals and sub-community preferentials, can show such local prominence as to give
individual stands a peculiar stamp. Among the most
frequent plants, *Angelica sylvestris, Geum rivale, Rhodiola rosea, Alchemilla glabra, Filipendula ulmaria* and
Succisa pratensis together provide a reliable distinction
from the *Luzula-Vaccinium* community and each can
have high cover in dense patches, or they may dominate
together in various mixtures, forming the bulk of lush,
jumbled herbage, 3–4 dm tall. *Deschampsia cespitosa*
can be fairly abundant, too: indeed as defined here, the
Luzula-Geum community takes in some of the less
closely grazed and more species-rich vegetation dominated by this grass, which McVean & Ratcliffe (1962)
included in their *Deschampsietum caespitosae alpinum*.
And it is in such stands, also, that other grasses, such as
*Festuca ovina, F. vivipara, Anthoxanthum odoratum,
Agrostis capillaris, A. canina, Deschampsia flexuosa* and
Nardus stricta make their most obvious contribution
though, in general, these more fine-leaved species are not
structurally important here, occurring usually as spar-
sely-scattered individuals or as small tussocks in more
open places around the taller herbs.

Other characteristic plants in this major physiognomic element of the flora include the Northern Montane
Trollius europaeus and, south of the Great Glen, *Geranium sylvaticum* and *Heracleum sphondylium*, each able
to show local dominance. Rather more occasional
throughout and not usually so abundant are *Solidago
virgaurea, Rumex acetosa,* the Continental Northern

Crepis paludosa and *Cirsium helenioides*, and the Arctic-Alpine *Thalictrum alpinum*, *Polygonum viviparum*, *Saussurea alpina* and *Oxyria digyna*. In some stands, strikingly handsome montane Hieracia, notably of the sections Subalpina, Alpina and Cerinthoidea, are a conspicuous feature, often occurring in locally distinctive mixtures in different mountain ranges (e.g. Raven & Walters 1956, Perring 1968, Kenneth & Stirling 1970). The community also provides an important locus for the rare Arctic-Alpine fern *Polystichum lonchitis* though, on ledges here, this plant does not generally exhibit the gregarious habit found locally among boulders in block scree, and indeed it may be crowded out among dense herbage (Page 1982). Other ferns can be more abundant, particularly where the *Luzula-Geum* vegetation extends its range into the more equable western Highlands and to lower altitudes around the oceanic seaboard of Scotland where, along with scattered *Blechnum spicant*, there can be occasional local prominence of *Athyrium filix-femina*, *Dryopteris dilatata*, *D. borreri* and *Thelypteris limbosperma*. Some strongly fern-dominated vegetation of this general kind is probably best included here (e.g. Prentice & Prentice 1975, Jermy & Crabbe 1978), but further sampling is needed to provide a clear picture of these assemblages and, for example, the *Osmunda regalis* stands described from South Uist by Spence (1960). Ledge vegetation with abundant *Silene dioica* or *Epilobium angustifolium*, often found around the nesting sites of golden eagles or peregrines, can also be provisionally included in the *Luzula-Geum* community.

Mixed in among the tall herbs of many stands of this vegetation are some sub-shrubs. *Vaccinium myrtillus* is the commonest of these, though it only exceptionally has the abundance characteristic of much of the *Luzula-Vaccinium* community. *V. vitis-idaea* can also be found, though it is frequent only in one particular kind of *Luzula-Geum* vegetation and even there sparse, and very occasionally there can be some *Empetrum nigrum*. *Calluna vulgaris* is somewhat commoner, though it becomes abundant only where the community extends to lower altitudes, where transitions to heath vegetation are more frequent. It is in such situations, too, that occasional saplings of *Sorbus aucuparia* and *Populus tremula* tend to be found in this community. *Juniperus communis* can sometimes occur, as well, while on higher crags Arctic-Alpine willows like *Salix lapponum*, *S. lanata* and *S. myrsinites* are occasionally recorded, thickening up locally to form a low bushy cover in transitions to *Salix-Luzula* scrub.

Smaller herbs tend to be not so numerous or abundant among the tall and luxuriant herbage of the *Luzula-Geum* community as in the *Dryas-Silene* vegetation that replaces it on more exposed calcareous outcrops, but some species are quite common throughout and, around the less shaded margins of stands, where the cover becomes fragmented into crevice vegetation, or where it

grades off into grazed swards, richer mixtures of associates can be seen. Among the most frequent plants of this element overall are *Alchemilla alpina*, *Viola riviniana*, *Ranunculus acris*, *Campanula rotundifolia*, *Potentilla erecta*, *Galium saxatile*, *Thymus praecox*, *Selaginella selaginoides* and *Anemone nemorosa*. More unevenly distributed among the different sub-communities but sometimes quite common there are *Saxifraga aizoides*, *S. oppositifolia* and *S. hypnoides*, the first in particular able to show local abundance, *Carex pulicaris*, *Rubus saxatilis* and *Oxalis acetosella*. Then, more occasional throughout are *Coeloglossum viride*, *Carex binervis*, *Taraxacum officinale* agg., *Parnassia palustris* and *Huperzia selago*, with small ferns such as *Cystopteris fragilis* and *Asplenium viride* often marking transitions to narrow shelves and crevices. Various kinds of *Luzula-Geum* vegetation also provide a locus for the rare Arctic-Alpines *Poa glauca*, *P. alpina* and *Potentilla crantzii*, with a number of other rarities strongly preferential to the richest sub-community.

Even among the denser herbage, bryophytes are often an important component of this vegetation, frequently forming lush patches over the stools and litter of the tall herbs and occasionally thickening up to become virtual dominants in more extensive carpets. *Hylocomium splendens*, *Thuidium tamariscinum* and *Ctenidium molluscum* are among the commonest and most abundant species, with *Rhytidiadelphus triquetrus* and *R. squarrosus* also very frequent, though not usually of high cover, *R. loreus* somewhat patchy in its occurrence but locally prominent. Quite common, too, though generally as scattered individuals, are *Mnium hornum*, *Rhizomnium punctatum*, *Calliergon cuspidatum* and *Pseudoscleropodium purum*, with occasional *Drepanocladus uncinatus*, *Fissidens adianthoides*, *Breutelia chrysocoma*, *Pellia epiphylla*, *P. endiviifolia*, *Cratoneuron commutatum*, *Racomitrium lanuginosum*, *Plagiothecium denticulatum*, *Plagiomnium rostratum*, *Rhizomnium pseudopunctatum* and *Ditrichum flexicaule*. In addition to these, particular sub-communities can show considerable further enrichment with species such as *Dicranum scoparium*, *D. majus*, *Polytrichum alpinum*, *Plagiomnium undulatum*, *Bryum pseudotriquetrum*, *Philonotis fontana*, *Lophocolea bidentata s.l.* and *Plagiochila asplenoides*, while especially towards the more oceanic north-west of Scotland, the *Luzula-Geum* community can provide a locus for such North Atlantic rarities as *Leptodontium recurvifolium*, *Mastigophora woodsii*, *Oxystegus hibernicus*, *Plagiochila carringtonii* and *Scapania ornithopodioides*. Lichens are scarce but sometimes occasional thalli of *Peltigera canina* are to be found.

Sub-communities

Alchemilla glabra-Bryum pseudotriquetrum sub-community: Ledge vegetation Ratcliffe 1960; *Sedum rosea-*

INDEX OF SPECIES IN GRASSLANDS AND MONTANE COMMUNITIES

The species are listed alphabetically, with the code numbers of the NVC communities in which they occur thereafter. Bold codes indicate that a species is constant throughout the community, italic codes that a species is constant in one or more sub-communities.

Acer pseudoplatanus sapling CG5

Aceras anthropophorum CG2, CG3

Achillea millefolium *MG1*, MG2, MG3, MG4, *MG5*, MG6, MG7, MG9, CG2, CG3, CG4, CG5, CG6, CG7, CG8, CG9, CG10, CG11, CG12, U1, *U4*, U6, U10, U13, U14, U20

Achillea ptarmica MG8, MG9, U6

Acinos arvensis CG1, CG7

Aegopodium podagraria MG1

Agrimonia eupatoria MG1, MG5, MG9, CG2, CG3, CG6, CG7

Agropyron donianum CG13

Agrostis canina CG10, CG11, CG12, CG13, U2, *U5*, U6, U7, U10, U12, U13, U14, U15, U16, *U17*, U19, U20, U21

Agrostis canina canina CG10, U4

Agrostis canina montana CG10, U1, U3, U4

Agrostis capillaris MG1, **MG3**, MG4, **MG5**, *MG6*, MG7, MG8, MG9, MG10, MG11, MG12, CG1, CG2, CG3, CG4, CG6, CG7, CG8, CG9, **CG10**, **CG11**, **CG12**, CG13, CG14, **U1**, U2, U3, **U4**, **U5**, *U6*, U7, U8, U9, U10, U11, **U13**, **U14**, U15, U16, *U17*, U18, U19, *U20*, U21

Agrostis curtisii **U3**, U20

Agrostis stolonifera MG1, MG4, MG5, MG6, MG7, MG8, MG9, **MG10**, **MG11**, **MG12**, **MG13**, CG2, CG3, CG4, CG6, CG7, U1, U19, U20

Aira caryophyllea CG1

Aira praecox CG1, CG10, *U1*

Ajuga chamaepitys CG2

Ajuga reptans MG3

Alchemilla acutiloba MG3

Alchemilla alpina CG10, **CG11**, **CG12**, **CG14**, U5, U7, U9, *U10*, U11, U12, *U13*, **U14**, **U15**, U16, *U17*, **U18**

Alchemilla filicaulis filicaulis MG3, CG10, CG11, CG12, CG14, U13, U15, U17

Alchemilla filicaulis vestita MG3, MG5, CG10, CG11, CG12, CG14, U14, U17

Alchemilla glabra MG2, **MG3**, MG5, MG8, CG10, CG14, U13, U14, **U15**, U16, *U17*

Alchemilla glaucescens CG13

Alchemilla glomerulans MG3

Alchemilla monticola MG3

Alchemilla subcrenata MG3

Alchemilla wichurae MG3, CG10, CG11

Alchemilla xanthochlora MG2, MG3, MG5, MG8, CG10, CG11

Alectoria nigricans U10

Alopecurus geniculatus MG6, MG7, MG10, MG11, **MG13**

Alopecurus pratensis MG1, MG3, **MG4**, MG5, MG6, *MG7*, MG9, MG10, MG13, U4

Amblystegium compactum CG13

Ammophila arenaria MG11

Anacamptis pyramidalis CG1, CG3, CG4, *CG5*, CG6

Anastrepta orcadensis U7, U10, U13, U18

Anastrophyllum donianum U7

Andreaea alpina U10, U14

Andreaea rupestris U12

Anemone nemorosa MG3, CG11, CG12, CG14, U4, U5, U16, U17

Aneura pinguis CG14, U15, U17

Angelica sylvestris **MG2**, MG8, MG9, MG10, CG10, CG14, U15, **U17**

Anoectangium aestivum U15

Antennaria dioica CG9, CG10, CG11, CG12, CG13, CG14, U10

Anthelia juratzkana U11

Anthoxanthum odoratum MG1, MG2, **MG3**, MG4, **MG5**, *MG6*, MG7, **MG8**, MG9, MG10, MG12, MG13, CG2, CG3, CG4, CG5, CG6, CG7, CG9,

Park, Shrewsbury: Nature Conservancy Midland Region.

Shimwell, D.W. (1971a). Festuco-Brometea Br.-Bl. & R.Tx 1943 in the British Isles: the phytogeography and phytosociology of limestone grasslands. I. General introduction; Xerobromion in England. *Vegetatio*, **23**, 1–28.

Shimwell, D.W. (1971b). Festuco-Brometea Br.-Bl. & R.Tx 1943 in the British Isles: the phytogeography and phytosociology of limestone grasslands. Eu-Mesobromion in the British Isles. *Vegetatio*, **23**, 29–60.

Shrub, M. (1973). Chalk grassland conservation: a farmer's point of view. In *Chalk Grassland: Studies on its Conservation and Management in South-east England*, ed. A.C. Jermy & P.A. Stott, pp. 39–41. Maidstone: Kent Trust for Nature Conservation.

Sinker, C.A., Packham, J.R., Trueman, I.C., Oswald, P.H., Perring, F.H. & Prestwood, W.V. (1985). *Ecological Flora of the Shropshire Region*. Shrewsbury: Shropshire Trust for Nature Conservation.

Sissingh, G. & Tideman, P. (1960). De Plantengemeenschappen uit de omgeving van Didam en Zevenaar. *Mededeelingen Landbouw-hoogeschool, Wageningen*, **60**, 1–30.

Sjögren, E. (1971). The influence of sheep grazing on limestone heath vegetation on the Baltic island of Öland. In *The Scientific Management of Animal and Plant Communities for Conservation*, ed. E. Duffey & A.S. Watt, pp. 487–95. Oxford: Blackwell.

Sjörs, H. (1954). Sletterangar: grangärde Finnmark. *Acta Phytogeographica Suecica*, **34**, 1–135.

Smith, A.G. (1970). The influence of mesolithic and neolithic man on British vegetation: a discussion. In *Studies on the Vegetational History of the British Isles*, ed. D. Walker & R.G. West, pp. 81–90. Cambridge: Cambridge University Press.

Smith, A.G., Grigson, C., Hillman, G. & Tooley, M.J. (1981). The Neolithic. In *The Environment in British Prehistory*, ed. E.G. Simmons & M.J. Tooley, pp. 125–209. London: Duckworth.

Smith, A.J.E. (1978). *The Moss Flora of Britain and Ireland*. Cambridge: Cambridge University Press.

Smith, C.J., Elston, J. & Bunting, A.H. (1971). The effects of cutting and fertiliser treatments on the yield and botanical composition of chalk turf. *Journal of the British Grassland Society*, **26**, 213–17.

Smith, C.J. (1980). *Ecology of the English Chalk*. London: Academic Press.

Smith, L.P. (1976). *The Agricultural Climate of England and Wales*. Ministry of Agriculture, Fisheries and Food Technical Bulletin 35. London: HMSO.

Smith, R. (1900). Botanical Survey of Scotland. *Scottish Geographical Magazine*, **16**.

Smith, R.T. (1986). Opportunistic behaviour of bracken (*Pteridium aquilinum* L. Kuhn) in moorland habitats: Origins and Constraints. In *Bracken*, ed. R.T. Smith & J.A. Taylor, pp. 215–24. Carnforth: Parthenon.

Smith, R.T. & Taylor, J.A. (ed.) (1986). *Bracken*. Carnforth: Parthenon.

Smith, U.K. (1979). Biological Flora of the British Isles: *Senecio integrifolius* (L) Clairv. *Journal of Ecology*, **67**, 1109–24.

Smith, W.G. (1905). Botanical Survey of Scotland. *Scottish Geographical Magazine*, **21**.

Smith, W.G. (1911b). Arctic-Alpine Vegetation. In *Types of British Vegetation*, ed. A.G. Tansley, pp. 288–329. Cambridge: Cambridge University Press.

Smith, W.G. (1918). The distribution of *Nardus stricta* in relation to peat. *Journal of Ecology*, **6**, 1–13.

Smith, W.G. & Moss, C.E. (1903). Geographical Distribution of Vegetation in Yorkshire. Part I. Leeds and Halifax District. *Geographical Journal*, **21**, 375–401.

Smith, W.G. & Rankin, W.M. (1903). Geographical distribution of vegetation in Yorkshire. Part II. Harrogate and Skipton District. *Geographical Journal*, **22**, 149–78.

Smithson, F. (1953). The micro-mineralogy of North Wales soils. *Journal of Soil Science*, **4**, 194–210.

Soil Survey (1983). 1:250,000 Soil Map of England and Wales: six sheets and legend. Harpenden: Soil Survey of England and Wales.

Soper, D. (1986). Lessons from fifteen years of bracken control with asulam. In *Bracken*, ed. R.T. Smith & J.A. Taylor, pp. 351–7. Carnforth: Parthenon.

South Gower Coast Report (1981). Vegetation Survey and Monitoring on the hard coast areas of the South Gower Coast NCR site, West Glamorgan. Nature Conservancy Council, Wales Field Unit.

Sparke, C.J. (1982). *Factors affecting the improvement of hill land dominated by bracken (Pteridium aquilinum (L.) Kuhn)*. Glasgow University: PhD thesis.

Sparke, C.J. & Williams, G.H. (1986). Sward changes following bracken clearance. In *Bracken*, ed. R.T. Smith & J.A. Taylor, pp. 225–31. Carnforth: Parthenon.

Spedding, C.R.W. (1971). *Grassland Ecology*. Oxford: Oxford University Press.

Spence, D.H.N. (1960). Studies on the vegetation of Shetland. III. Scrub in Shetland and in South Uist, Outer Hebrides. *Journal of Ecology*, **48**, 73–85.

Stace, C.A. (ed.) (1975). *Hybridization and the Flora of the British Isles*. London: Botanical Society of the British Isles and Academic Press.

Staines, S.J. (1984). *Soils in Cornwall III: Sheets SW 61, 71 and parts of SW 62, 72, 81 and 82 (The Lizard)*. Harpenden: Soil Survey of England and Wales.

Stapledon, R.G. (1925). Permanent Grass. *Farm Crops*, **3**, 74–136.

Stapledon, R.G. (1937). *The Hill Lands of Britain*. London: Faber & Faber.

Steindorsson, S. (1945). Studies on the vegetation of the Central Highlands of Iceland. *The Botany of Iceland*, **3**, 345–547.

Steven, H.M. & Carlisle, A. (1959). *The Native Pinewoods of Scotland*. Edinburgh: Oliver & Boyd.

Stevens, J.H. & Atkinson, K. (1970). Soils and their capability. In *Durham County and City with Teesside*, ed. J.C. Dewdney, pp. 46–57. Durham: British Association.

Stieperaere, H. (1978). Quelques aspects des pelouses tourbeuses du Juncion squarrosi (Oberd. 1957) Pass. 1964 en France. *Colloques Phytosociologiques*, **7**, 359–69.

Stott, P.A. (1970). The study of chalk grassland in northern France. An historical review. *Biological Journal of the Linnean Society*, **2**, 173–207.

Streeter, D.T. (1971). The effects of public pressure on the vegetation of chalk downland at Box Hill, Surrey. In *The Scientific Management of Animal and Plant Communities for Conservation*, ed. E. Duffey & A.S. Watt, pp. 459–68. Oxford: Blackwell.

Tallis, J.H. (1958). Studies in the Biology and Ecology of *Rhacomitrium lanuginosum* Brid. I. Distribution and Ecology. *Journal of Ecology*, **46**, 271–88.

Tansley, A.G. (ed.) (1911). *Types of British Vegetation*. Cambridge: Cambridge University Press.

Tansley, A.G. (1922). Studies on the vegetation of the English Chalk. II. Early stages of redevelopment of woody vegetation on chalk grassland. *Journal of Ecology*, **10**, 168–77.

Tansley, A.G. (1939). *The British Islands and their Vegetation*. Cambridge: Cambridge University Press.

Tansley, A.G. & Adamson, R.S. (1925). Studies of the vegetation of the English Chalk. III. The chalk grasslands of the Hampshire–Sussex border. *Journal of Ecology*, **13**, 177–223.

Tansley, A.G. & Adamson, R.S. (1926). Studies of the vegetation of the English Chalk. IV. A preliminary survey of the chalk grasslands of the Sussex Downs. *Journal of Ecology*, **14**, 1–32.

Tansley, A.G. & Rankin, W.M. (1911). The plant formation of calcareous soils. B. The sub-formation of the Chalk. In *Types of British Vegetation*, ed. A.G. Tansley, pp. 161–86. Cambridge: Cambridge University Press.

Taylor, C. (1975). *Fields in the English Landscape*. London: Dent.

Taylor, J.A. (1978). The British upland environment and its management. *Geography*, **63**, 338–53.

Taylor, J.A. (1985). The relationship between land-use change and variations in bracken encroachment rates in Britain. In *The Biogeographical Impact of Land Use Change*, ed. R.T. Smith, pp. 19–28. Norwich: BSc/Geo Books.

Taylor, J.A. (1986). The Bracken Problem: A Local Hazard and Global Issue. In *Bracken*, ed. R.T. Smith & J.A. Taylor, pp. 21–42. Carnforth: Parthenon.

Thomas, A.S. (1960). Changes in vegetation since the advent of myxomatosis. *Journal of Ecology*, **48**, 287–306.

Thomas, A.S. (1962). Ant-hills and termite-mounds in pastures. *Journal of the British Grassland Society*, **17**, 103–8.

Thomas, A.S. (1963). Further changes in vegetation since the advent of myxomatosis. *Journal of Ecology*, **51**, 151–86.

Thomas, A.S., Rawes, M. & Banner, W.J.L. (1957). The vegetation of the Pewsey Vale escarpment, Wiltshire. *Journal of the British Grassland Society*, **12**, 39–48.

Thomas, B. & Fairburn, C.B. (1956). The white bent, its composition, digestibility and probable nutritive value. *Journal of the British Grassland Society*, **11**, 230–4.

Thomas, M.T. (1936). Investigations on the improvement of hill grazings. II. The introduction and maintenance of nutritious and palatable species and strains. *Bulletin of the Welsh Plant Breeding Station, Series H*, **14**, 4–57.

Thompson, H.V. & Worden, A.M. (1956). *The Rabbit*. London: Collins.

Thompson, T.R.E., Rudeforth, C.C., Hartnup, R., Lea, J.W. & Wright, P.S. (1986). Soil and slope conditions under Bracken in Wales. In *Bracken*, ed. R.T. Smith & J.A. Taylor, pp. 101–7. Carnforth: Parthenon,.

Tidmarsh, C.E.M. (1939). *The ecology of Carex arenaria*.

Cambridge University: PhD thesis.

Tinsley, H.M. & Grigson, C. (1981). The Bronze Age. In *The Environment in British Prehistory*, ed. I.G. Simmons & M.J. Tooley, pp. 210–49. London: Duckworth.

Tittensor, R.M. & Steele, R.C. (1971). Plant communities of the Loch Lomond oakwoods. *Journal of Ecology*, **59**, 561–82.

Trapnell, C.G. (1933). Vegetation types in Godthaab Fjord. *Journal of Ecology*, **21**, 294–334.

Trow-Smith, R. (1957). *A History of British Livestock Husbandry*. London: Routledge.

Tubbs, C.R. (1968). *The New Forest: An Ecological History*. Newton Abbot: David and Charles.

Tubbs, C.R. (1986). *The New Forest*. London: Collins.

Turner, J. (1970). Post-Neolithic disturbance of British vegetation. In *Studies in the Vegetational History of the British Isles*, ed. D. Walker & R. G. West, pp. 98–116. Cambridge: Cambridge University Press.

Turner, J. (1978). History of vegetation and flora. In *Upper Teesdale*, ed. A.R. Clapham, pp. 88–101. London: Collins.

Turner, J. (1981). The Iron Age. In *The Environment in British Prehistory*, ed. I.G. Simmons & M.J. Tooley, pp. 250–81. London: Duckworth.

Turner, J. & Hodgson, J. (1979). Studies in the vegetational history of the Northern Pennines. I. Variations in the composition of the early Flandrian forests. *Journal of Ecology*, **67**, 629–46.

Tutin, T.G. (1980). *Umbellifers of the British Isles*. London: Botanical Society of the British Isles.

Tutin, T.G., Heywood, V.H., Burges, N.A., Valentine, D.H., Walters, S.M. & Webb, D.A. (1964). *Flora Europaea, Volume 1*. Cambridge: Cambridge University Press.

Tutin, T.G., Heywood, V.H., Burges, N.A., Moore, D.M., Valentine, D.H., Walters, S.M. & Webb, D.A. (1968). *Flora Europaea, Volume 2*. Cambridge: Cambridge University Press.

Tutin, T.G., Heywood, V.H., Burges, N.A., Moore, D.M., Valentine, D.H., Walters, S.M. & Webb, D.A. (1972). *Flora Europaea, Volume 3*. Cambridge: Cambridge University Press.

Tutin, T.G., Heywood, V.H., Burges, N.A., Moore, D.M., Valentine, D.H., Walters, S.M. & Webb, D.A. (1976). *Flora Europaea, Volume 4*. Cambridge: Cambridge University Press.

Tutin, T.G., Heywood, V.H., Burges, N.A., Moore, D.M., Valentine, D.H., Walters, S.M. & Webb, D.A. (1980). *Flora Europaea, Volume 5*. Cambridge: Cambridge University Press.

Tüxen, R. (1937). Die Pflanzengesellschaften Nordwestdeutschlands. *Mitteilungen der Floristisch-soziologischen Arbeitsgemeinschaft*, **3**, 1–170.

Tüxen, R. (1955). Das System der nordwestdeutschen Pflanzengesellschaften. *Mitteilungen der Floristisch-soziologischen Arbeitsgemeinschaft N.F.*, **5**, 155–76.

Tüxen, R. & Preising, E. (1951). Erfahrungsgrundlagen für die pflanzensoziologische Kartierung des westdeutschen Grünlandes. *Angewandte Pflanzensoziologie*, **4**, 1–28.

Van der Meulen, F. & Wiegers, J. (1972). *A Phytosociological Research of some Chalk grasslands in Southern England*. Utrecht: Instituut voor Systematische Plantkunde.

van Schaik, C.P. & Hogeweg, P.A. (1977). A numerical–

syntaxonomical study of the Calthion palustris Tx. 1937 in the Netherlands. *Vegetatio*, **35**, 65–80.

Volk, H. (1937). Über einige Trockenrasengesellschaften des Würzburger Wellenkalkgebietes. *Beihefte zum Botanischen Zentralblatt*, **57**, 577 ff.

von Horn, A. (1935). *Die Rasenschmiele (Aira (Deschampsia) caespitosa). Eine Untersuchung ihrer Lebensbedingungen im Hinblick auf die Bekämpfungsmöglichkeit*. Instute für Kulturtechnik der Friedrich-Wilhems-Universität Berlin: Inaugural Dissertation.

Walter, H. & Lieth, H. (1967). *Klimadiagramm Weltatlas*.

Walters, S.M. (1949). *Alchemilla vulgaris* L. agg. in Britain. *Watsonia*, **1**, 6–18.

Walters, S.M. (1952). *Alchemilla subcrenata* Buser in Britain. *Watsonia*, **2**, 277–8.

Ward, S.D. (1971a). The phytosociology of *Calluna-Arctostaphylos* heaths in Scotland and Scandinavia. II. The north-east Scottish heaths. *Journal of Ecology*, **59**, 679–96.

Ward, S.D. (1971b). The phytosociology of *Calluna-Arctostaphylos* heaths in Scotland and Scandinavia. III. A critical examination of the Arctostaphyleto-Callunetum. *Journal of Ecology*, **59**, 697–712.

Ward, S.D., Evans, D.F. & Millar, R.O. (1972b). *A Vegetation Survey of the Moorfoot Hills Grade 1 Site*. Bangor: Nature Conservancy Montane Grasslands Habitat Team.

Ward, S.D., Jones, A.D. & Manton, M. (1972a). The vegetation of Dartmoor. *Field Study*, **3**, 505–34.

Warwick Percy, C. (1970). Agriculture. In *Durham County and City with Teesside*, ed. J.C. Dewdney, pp. 284–93. Durham: British Association.

Watson, E.V. (1960). A quantitative study of the bryophytes of chalk grassland. *Journal of Ecology*, **48**, 397–414.

Watson, W. (1925). The bryophytes and lichens of Arctic-Alpine vegetation. *Journal of Ecology*, **13**, 1–26.

Watt, A.S. (1936). Studies in the ecology of Breckland. I. Climate, soils and vegetation. *Journal of Ecology*, **24**, 117–38.

Watt, A.S. (1937). Studies in the ecology of Breckland. II. On the origin and development of blow-outs. *Journal of Ecology*, **25**, 91–112.

Watt, A.S. (1938). Studies in the ecology of Breckland. III. Development of the *Festuco-Agrostidetum*. *Journal of Ecology*, **26**, 1–37.

Watt, A.S. (1940). Studies in the ecology of Breckland. IV. The grass heath. *Journal of Ecology*, **28**, 42–70.

Watt, A.S. (1945). Contributions to the ecology of bracken (*Pteridium aquilinum*). III. Frond types and the make-up of the population. *New Phytologist*, **44**, 156–78.

Watt, A.S. (1947a). Contributions to the ecology of bracken (*Pteridium aquilinum*). IV. The structure of the community. *New Phytologist*, **46**, 97–121.

Watt, A.S. (1947b). Pattern and process in the plant community. *Journal of Ecology*, **35**, 1–22.

Watt, A.S. (1955). Bracken versus heather, a study in plant sociology. *Journal of Ecology*, **43**, 490–506.

Watt, A.S. (1956). Contributions to the ecology of bracken (*Pteridium aquilinum*). VII. Bracken and litter. 1. The origin of rings. *New Phytologist*, **55**, 369–81.

Watt, A.S. (1957). The effect of excluding rabbits from Grassland B (Mesobrometum) in Breckland. *Journal of Ecology*, **45**, 861–78.

Watt, A.S. (1960a). The effect of excluding rabbits from acidophilous grassland in Breckland. *Journal of Ecology*, **48**, 601–4.

Watt, A.S. (1962). The effect of excluding rabbits from Grassland A (Xerobrometum) in Breckland, 1936–60. *Journal of Ecology*, **50**, 181–98.

Watt, A.S. (1969). Contributions to the ecology of bracken (*Pteridium aquilinum*). VII. Bracken and Litter. 2. Crown form. *New Phytologist*, **68**, 841–59.

Watt, A.S. (1971a). Factors controlling the floristic composition of some plant communities in Breckland. In *The Scientific Management of Animal and Plant Communities for Conservation*, ed. E. Duffey & A.S. Watt, pp. 137–52. Oxford: Blackwell.

Watt, A.S. (1971b). Rare species in Breckland: their management for survival, *Journal of Applied Ecology*, **8**, 593–609.

Watt, A.S. (1974). Senescence and rejuvenation in ungrazed chalk grassland (Grassland B) in Breckland: the significance of litter and moles. *Journal of Applied Ecology*, **11**, 1157–71.

Watt, A.S. (1976). The ecological status of bracken. In *The Biology of Bracken*, ed. F.H. Perring & B.G. Gardiner, pp. 217–40. London: Linnean Society.

Watt, A.S. (1981a). A comparison of grazed and ungrazed grassland A in East Anglian Breckland. *Journal of Ecology*, **69**, 499–508.

Watt, A.S. (1981b). Further observations on effects of excluding rabbits from grassland A in East Anglian Breckland: the pattern of change and factors affecting it (1936–73). *Journal of Ecology*, **69**, 509–36.

Watt, A.S. & Jones, E.W. (1948). The ecology of the Cairngorms. I. The environment and the altitudinal zonation of the vegetation. *Journal of Ecology*, **36**, 283–304.

Webb, N. (1986). *Heathlands*. London: Collins.

Welch, D. (1965). A change in the upper altitudinal limit of *Coleophora alticolella* Zell. (Lep.). *Journal of Animal Ecology*, **34**, 725–9.

Welch, D. (1966a). The reproductive capacity of *Juncus squarrosus*. *New Phytologist*, **65**, 77–86.

Welch, D. (1966b). Biological Flora of the British Isles: *Juncus squarrosus* L. *Journal of Ecology*, **54**, 535–48.

Welch, D. (1967). Communities containing *Juncus squarrosus* in Upper Teesdale, *Vegetatio*, **14**, 229–40.

Welch, D. & Rawes, M. (1964). The early affects of excluding sheep from high-level grasslands in the north Pennines. *Journal of Applied Ecology*, **1**, 281–300.

Welch, D. & Rawes, M. (1969). Moisture regime of soils on metamorphosed limestone in Upper Teesdale. *Transactions of the Natural History Society of Northumberland*, **17**, 57–67.

Wells, T.C.E. (1967a). Changes in a population of *Spiranthes spiralis* (L.) Chevall. at Knocking Hoe National Nature Reserve, Bedfordshire, 1962–65. *Journal of Ecology*, **55**, 83–99.

Wells, T.C.E. (1967b). Changes in the botanical composition of a sown pasture on the Chalk in Kent, 1956–64. *Journal of the British Grassland Society*, **22**, 277–81.

Wells, T.C.E. (1968). Land-use changes affecting *Pulsatilla vulgaris* in England. *Biological Conservation*, **1**, 37–43.

Wells, T.C.E. (1969). Botanical aspects of conservation managment of chalk grasslands. *Biological Conservation*, **2**, 36–44.

Wells, T.C.E. (1971). A comparison of the effects of sheep grazing and mechanical cutting on the structure and botanical composition of chalk grassland. In *The Scientific Management of Animal and Plant Communities for Conservation*, ed. E. Duffey & A.S. Watt, pp. 497–515. Oxford: Blackwell.

Wells, T.C.E. (1973). Botanical aspects of chalk grassland management. In *Chalk Grassland. Studies on its Conservation and Management in South-east England*, ed. A.C. Jermy & P.A. Stott, pp. 10–15. Maidstone: Kent Trust for Nature Conservation.

Wells, T.C.E. (1975). The floristic composition of chalk grassland in Wiltshire. In *Supplement to the Flora of Wiltshire*, ed. L.F. Stearn, pp. 99–125. Devizes: Wiltshire Archaeological and Natural History Society.

Wells, T.C.E. (1976). Biological Flora of the British Isles: *Hypochoeris maculata* L. *Journal of Ecology*, **64**, 757–74.

Wells, T.C.E. (1981). Population ecology of terrestrial orchids. In *The Biological Aspects of Rare Plant Conservation*, ed. H. Synge, pp. 281–95. London: Wiley.

Wells, T.C.E. & Barling, D.M. (1971). Biological Flora of the British Isles: *Pulsatilla vulgaris* Mill. *Journal of Ecology*, **59**, 275–92.

Wells, T.C.E. & Morris, M.G. (1970). *Conservation Research and Management of Calcareous Grassland*. Guide to tour No. 5 for the British Ecological Society International Symposium on 'The Scientific Management of Animal and Plant Communities for Conservation'.

Wells, T.C.E., Sheail, J., Ball, D.F. & Ward, L.K. (1976). Ecological studies on the Porton Ranges: relationships between vegetation, soils and land-use history. *Journal of Ecology*, **64**, 589–626.

Werger, M.J.A. (1973). On the use of Association Analysis and Principal Components Analysis in interpreting a Braun-Blanquet phytosociological table of a Dutch grassland. *Vegetatio*, **28**, 129–44.

Westhoff, V. & den Held, A.J. (1969). *Plantengemeenschappen in Nederland*. Zutphen: Thieme.

West Yorkshire Biological Data Bank (1983). *Vascular Plant Communities of Magnesian Limestone Grassland in West Yorkshire*. Keighley: West Yorkshire Biological Data Bank.

Wheeler, B.D. (1975). *Phytosociological studies on Rich Fen Systems in England and Wales*. University of Durham: PhD thesis.

White, G. (1788). *The Natural History of Selbourne*. London: Bensley.

Whittow, J.B. (1979). *Geology and Scenery in Scotland*. Harmondsworth: Penguin Books.

Wigginton, M.J. & Graham, G.G. (1981). *Guide to the Identification of some Difficult Plant Groups*. Banbury: Nature Conservancy Council, England Field Unit.

Willems, J.H. (1978). Observations on North-West European limestone grassland communities: phytosociological and ecological notes on Chalk grasslands of southern England. *Vegetatio*, **37**, 141–50.

Williams, G.H. & Foley, A. (1976). Seasonal variations in the carbohydrate content of bracken. *Botanical Journal of the Linnean Society*, **73**, 87–94.

Williams, J.T. & Varley, Y.W. (1967). Phytosociological studies of some British grasslands. *Vegetatio*, **15**, 169–89.

Williams, T.E. & Davies, A.G. (1946). A grasslands survey of the Monmouthshire 'moors'. *Journal of the British Grasslands Society*, **1**.

Wilson, A.S.B. (1936). The improvement of rough hill pastures by cattle grazing. *Scottish Journal of Agriculture*, **19**.

Woodward, W.B. (1970). Conservation. In *Durham County and City with Teesside*, ed. J.C. Dewdney, pp. 169–78. Durham: British Association.

Wooldridge, S.W. & Goldring, F. (1953). *The Weald*. London: Collins.

Worth, R.H. (1933). The Vegetation of Dartmoor. *Report of the Transactions of the Plymouth Institute*, **17**, 285–96.

Yates, E.M. (1972). The management of heathlands for amenity purposes in south-east England. *Geographica Polonica*, **24**, 227–40.